T0312745

Smart Use of State Public Health Data for Health Disparity Assessment

Smart Use of State Public Health Data for Health Disparity Assessment

Ge Lin and Ming Qu

CRC Press
Taylor & Francis Group
Boca Raton London New York

CRC Press is an imprint of the
Taylor & Francis Group, an **informa** business

A PRODUCTIVITY PRESS BOOK

CRC Press
Taylor & Francis Group
6000 Broken Sound Parkway NW, Suite 300
Boca Raton, FL 33487-2742

© 2016 by Taylor & Francis Group, LLC
CRC Press is an imprint of Taylor & Francis Group, an Informa business

No claim to original U.S. Government works

Printed on acid-free paper
Version Date: 20160120

International Standard Book Number-13: 978-1-4822-0531-2 (Hardback)

This book contains information obtained from authentic and highly regarded sources. Reasonable efforts have been made to publish reliable data and information, but the author and publisher cannot assume responsibility for the validity of all materials or the consequences of their use. The authors and publishers have attempted to trace the copyright holders of all material reproduced in this publication and apologize to copyright holders if permission to publish in this form has not been obtained. If any copyright material has not been acknowledged please write and let us know so we may rectify in any future reprint.

Except as permitted under U.S. Copyright Law, no part of this book may be reprinted, reproduced, transmitted, or utilized in any form by any electronic, mechanical, or other means, now known or hereafter invented, including photocopying, microfilming, and recording, or in any information storage or retrieval system, without written permission from the publishers.

For permission to photocopy or use material electronically from this work, please access www.copyright.com (http://www.copyright.com/) or contact the Copyright Clearance Center, Inc. (CCC), 222 Rosewood Drive, Danvers, MA 01923, 978-750-8400. CCC is a not-for-profit organization that provides licenses and registration for a variety of users. For organizations that have been granted a photocopy license by the CCC, a separate system of payment has been arranged.

Trademark Notice: Product or corporate names may be trademarks or registered trademarks, and are used only for identification and explanation without intent to infringe.

Library of Congress Cataloging-in-Publication Data

Names: Lin, Ge, 1959- , author. | Qu, Ming, 1964- , author.
Title: Smart use of state public health data for health disparity assessment
/ Ge Lin and Ming Qu.
Description: Boca Raton : Taylor & Francis, 2016. | Includes bibliographical
references and index.
Identifiers: LCCN 2016000611 | ISBN 9781482205312 (hard cover : alk. paper)
Subjects: | MESH: Research Design | Health Status Disparities | Datasets as
Topic | Epidemiological Monitoring | Socioeconomic Factors | United States
Classification: LCC RA408.5 | NLM WA 20.5 | DDC 362.1072/3--dc23
LC record available at http://lccn.loc.gov/2016000611

Visit the Taylor & Francis Web site at
http://www.taylorandfrancis.com

and the CRC Press Web site at
http://www.crcpress.com

Contents

SECTION II   Health Disparity Surveillance Based on Hospital and Emergency Department Data

SECTION III   Data Integrations and Their Applications in Health Disparity Assessments

Preface

There is a saying that "if you build it, they will come." This book attempts to test that motto. Most state public health data are administratively collected for specific mandates and funding agency requirements. Each dataset has some strengths and weaknesses. The smart use of public health data is to enhance administrative data through data linkage, so that strengths from each dataset can be leveraged and the weaknesses can be compensated. We build this data infrastructure so that public health programs and research communities can use it effectively at a local level.

There is another saying: "all politics is local." When applied to public health, we can say that all public health is local. National public health campaigns, such as enacting smoke-free legislation, are organized locally. Important public health concerns differ from place to place. Some places face air pollution, some face physical inactivity, and some face an insufficient healthy food supply. When public health policies and programs are designed and implemented at the local level, they are most effective if public health issues can be articulated using local data. We now realize the importance of individualized medicine, but public health has always been localized. In public health research, however, there is a tendency to ignore local data. Researchers tend to seek national-level data and refine research issues that are generalizable and transferable, which directly contradicts localized public health. In this book, we attempt to show that it can be rewarding to use localized public health data, in this case from Nebraska, that may or may not be generalized to other states. In addition, all data used in this book have been geocoded to the census tract level. Hence, if we want to go more local, we can go down to the neighborhood level.

Finally, Rome wasn't built in a day. Building public health data infrastructure takes time, and one has to be patient. In this book, we propose a bottom-up approach to building an integrated public health data warehouse. The book has three sections. Section I has seven chapters devoted to knowledge and skill preparations for recognizing disparity issues and integrating and analyzing local public health data. Section II provides a systematic surveillance effort by linking census tract poverty to other health disparity dimensions. Section III provides in-depth studies related to Sections I and II.

Although I believe that the greatest use of my life is to spend it devoted to something that will outlast it, I am sure that I will have some regrets when this book is in print. As a nonnative English speaker, and someone who wants to do all analyses by myself, I am sure there are some mistakes that I have not been able to catch. I take full responsibility for them, and I hope to have a chance to make corrections. A nice thing about doing it all by myself is that I have saved all the SAS codes for data analyses. With proper data use authorization, all the data in the book can be considered secondary data that can be readily used for training and other data analyses.

Acknowledgments

This book was written when I codirected the Nebraska Public Health Joint Data Center from 2011 to 2015. During this period, I represented the College of Public Health at the University of Nebraska Medical Center, and Dr. Ming Qu represented the Division of Public Health, Department of Health and Human Services. In the beginning, I took advice from the Division of Public Health leadership to rotate among as many programs as possible and spend some quality time with them. During the rotation, I learned a lot from many individuals in the various public programs. After a number of years, I was able to use the multiprogram perspectives and languages to communicate with each program separately and together. Initially an outsider, I was eventually viewed and treated as an insider within the Division of Public Health.

During the process of establishing the Joint Data Center, I received unfettered support from Sue Medinger (Community and Rural Health Planning Unit), Sue Semerena (Environmental Health Unit), Paula Eurek (Lifespan Health Services Unit), Chris Newlon (Public Health Emergency Preparedness and Emergency Response), and Ming Qu (Epidemiology and Informatics Unit). The support from the public health programs was unlimited, but I benefitted the most from that of Kathy Ward and Melissa Leypoldt (Every Woman Matters program), Jamie Hahn (cardiovascular disease program), Ying Zhang and Ashley Newmyer (Crash Outcome Data Evaluation System), Michelle Hood (health statistics), Rachel Cooper (trauma registry), Jill Krause (Parkinson's disease registry), Mark Miller (vital statistics), Brenda Coufal (Pregnancy Risk Assessment Monitoring System), Jeff Armitage (Office of Community Health and Performance Management), and Julie Miller (newborn screening and genetics).

On a personal level, Norm Nelson lent his statistical experience for manipulating various public health datasets. He also shared many SAS codes that he accumulated over 30 years. Dr. Tom Safranek, the state epidemiologist, provided guidance and tips for conducting surveillance. He was instrumental in justifying large-scale geocoding for many public health datasets under the framework of social determinants. David Palm, who chaired the advisory board of the Joint Data Center, provided much feedback on the center's projects that bridged academic interests with public health program interests.

Many program data analysts and surveillance specialists provided data and sometimes summary statistics for the book. I most frequently contacted Guangming Han, Mengqian Li, Robin Williams, Bryan Rettig, Jianping Daniels, Jihyun Ma, and David DeVries, but I am sure I failed to mention many people who also provided data support. Dorothy Smiley read many chapters and provided some comments. Corey Levitan provided editorial support for Chapter 8.

Finally, many students, data analysts, and research associates supported Joint Data Center activities that directly benefited the book writing. Among them were Jianhua Qin, Karis Bowen, Xiaqing Jiang, Jiajun Wen, Neng Wan, Soumitra Bhuyan, and Sam Opoku. Some of them provided data linkage support, some geographic information system support, and some preliminary data analyses and reports.

Most data analysts and students from both the Division of Public Health and the Joint Data Center have contributed to publications; some of them have been included in the Section III of this book. However, I should make it clear that there was no financial support from any individual or organization to the book writing process.

Authors

Ge Lin is a professor of epidemiology in the School of Community Health Sciences, University of Nevada, Las Vegas. He was trained in spatial demography and geographic information systems. He is known for his works in spatial modeling, spatial statistics for count data, and spatial disparities in health. His most recent research focuses on the science of public health data. He uses the infrastructure approach to develop integrated data marts, data analysis utilities, and training modules for public health data specialists. He has been supported by the National Institutes of Health, National Institute of Justice, National Science Foundation, and Nebraska Department of Health and Human Services.

Ming Qu is the administrator of the Epidemiology and Informatics Unit, Nebraska Department of Health and Human Services, a unit that provides statistical information, epidemiology, and geographic information system services that support public health actions and policies to improve the health and safety of Nebraskans. Prior to this position, he was an injury epidemiologist and Crash Outcome Data Evaluation System administrator for the Department of Health and Human Services, where he was instrumental in the development of the Nebraska Injury Surveillance System. Dr. Ming administers the Nebraska Crash Outcome Data Evaluation System. He previously was a research assistant for the Nebraska Prevention Center for Alcohol and Drug Abuse, University of Nebraska, Lincoln. He is the author of numerous papers and book chapters.

1 Enhanced Public Health Program Collaboration through Data Integration

1.1 INTRODUCTION

Public health assessment and surveillance require programs to seamlessly collaborate and work together, not only on emergency issues, but also on daily issues such as chronic disease prevention and health risk reduction. Historically, public health functions were served by a few entities or individuals and, in many cases, by the same individuals, or a team led by the same individuals. Two thousand years ago, Hippocrates studied the impacts of our surrounding environment, such as climate, soil, and water, on health. He used his knowledge, such as avoiding dampness and swamps in residential locations, to guide the settlement of Greek colonies in other parts of the Mediterranean. Quarantine, which was first introduced in the Middle Ages in Venice, also requires collective actions or collaborations among different individuals or programs. In medieval times, food inspection, disinfection, waste disposal isolation, and quarantine were regulated by medieval city councils.

In the United States, collective actions and cross-referencing multiple datasets for disease surveillance have a long history (Rosen, 1993). Shattuck's 1850 report recommended that health prevention programs in Massachusetts use mortality data, population census data, and location-specific incidence, occupation, and socioeconomic status (SES) data. Based on John Snow's study in London, the New York Board of Health in 1866 instituted sanitation measures that included inspections, immediate case reporting, complaint investigations, evacuations, and disinfection of possessions and living quarters. These efforts kept an outbreak of cholera at bay, while cities without a public health system for monitoring and combating the disease fared far worse in the 1866 epidemic. Similar collective efforts, such as information sharing followed by immediate actions, have persisted in infectious disease control. However, as programs are getting bigger and bigger, the collaboration root of public health is getting weaker and weaker, especially in the area of chronic disease prevention. The erosion of the collaboration tradition of public health programs has caused major concerns in information sharing and interprogram disease surveillance and prevention.

The fragmentation of public health programs and lack of communication among them are rooted in organizational and funding stream fragmentation, which individuals can hardly overcome. In the process of drafting the Institute of Medicine (IOM) public health report (1988), the U.S. Committee for the Study of the Future of Public Health visited six states in the United States and identified several issues. In terms of public health organizations, the committee found that regardless of organization, health services were often fragmented along organizational lines, with almost no communication among the public health–related programs or organizations, such as mental health, social services, and public health services. The committee identified 10 barriers for effective public health actions, and 3 of them related to translating data to knowledge. They included disjointed decision making without necessary data and knowledge, organizational fragmentation, and disparate knowledge development across the full array of public health needs. To the greatest extent, these assessments are still true, and in some cases, they are even exacerbated.

As a result, when new questions or challenges arise that require collaboration, individual public health practitioners (e.g., surveillance specialists and epidemiologists) often do not have the time

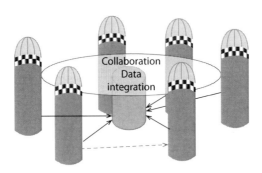

FIGURE 1.1   Data integration and program collaboration.

and energy to spend on them. Since the IOM report, tremendous progress has been made in integrating data at both the national and state levels. In addition, many individual programs and individuals from a program often take their own initiatives to collaborate and use integrated data to advance our knowledge in public health. In this book, we promote data integration to aid crosscutting program collaboration. Figure 1.1 provides a diagram to highlight our idea. Each silo in the figure represents a public health program, and the circle represents a crosscutting activity table. Data integration provides a common language among programs to share and act on.

Such a data integration approach is the key to evidence-based health disparity assessments. A key dimension in health disparities is racial disparities in health. The traditional program-based public health activities and reporting are based on each silo, such as women's health, the cardiovascular disease (CVD) program, the cancer prevention program, smoking and tobacco control, and home visitation. Health statistic reports within a public health agency are also organized by public health program or subject area. Examples include behavioral health, mental health, vital statistics, cancer, and radon reduction. If one wants to conduct a racial health disparity assessment, data have to either come from many programs or cut through many program areas. In addition to race or ethnicity, SES also cuts through many programs. As a disparity assessment expands to cover more program areas along a health disparity dimension, it is natural that the resultant report would lack depth and rely on programs to carry out in-depth assessments.

The focus on width and crosscutting in health disparity assessments fits well with chronic disease surveillance. Years after the epidemiologic transition in the United States, the Centers for Disease Control and Prevention (CDC) started to develop chronic disease indicators, which cover a wide range of public health areas from many datasets (CDC, 2004). Surveillance based on chronic disease indicators means pulling and manipulating data from different sources, programs, agencies, and organizations. In 2013, the expanded list of chronic disease indicators included 124 indicators and 201 individual measures from more than 20 data sources (CDC, 2015). Monitoring those indicators requires data integration, program and agency collaborations, and unprecedented infrastructure development. Analogously, the requirements for monitoring chronic disease indicators are also applicable to capacity building for health disparity assessments. In the rest of this chapter, we first review some of the efforts at the national and state levels and then summarize data integration efforts in Nebraska. Finally, we provide an outline of the book.

1.2   DATA INTEGRATION AT THE NATIONAL AND STATE LEVELS

At the national level, the National Center for Health Statistics (NCHS) has a data linkage program for linking interagency datasets. The National Health Interview Survey (NHIS), National Health and Nutrition Examination Survey (NHANES), and Longitudinal Study of Aging have been linked to the National Death Index (NDI) for mortality outcomes. These datasets have also been linked to Medicare enrollment and claims data and Social Security Administration (SSA), Retirement, Survivors, and Disability Insurance (RSDI), and Supplemental Security Income

(SSI) data. In addition, the National Institutes of Health (NIH) and National Science Foundation (NSF) founded surveys such as the Panel Study of Income Dynamics surveys and the Health and Retirement Survey that have both been linked to the Medicare claim data to obtain health care costs and diagnoses from respondents. Surveillance, Epidemiology, and End Results (SEER) cancer registry data have also been linked to Medicare records. In addition, individual respondents in most national surveys are also geocoded to the census tract unit to obtain neighborhood characteristics. Most federal linkage programs have broad public policy impacts and have generated countless academic publications, policy briefings, and data reports. Together, they demonstrate the fruitfulness of these collaborations.

At the state level, programs in data sharing and data integration have been uneven. Most states do not fund large-scale surveys. The Behavioral Risk Factor Surveillance System (BRFSS) is funded by the CDC. Respondents from the surveys cannot be identified or linked to other health claim and mortality data. However, some states use BRFSS as the basis to conduct sampling among children, such the Children's Health Assessment Survey (CHAS) in South Carolina, in which case both the parents' survey and children's health survey can then be linked. In addition, some state and city surveys, such as California Health Interview Survey (CHIS), have added geographic and environmental variables. Other than survey data, individual states have their own priorities, and some data are linked for case finding and ascertainment and health outcomes. Depending on need and sometimes leadership, some states had a robust data integration program for a few years, and then discontinued it.

South Carolina, for instance, had a vibrant data linkage program across many state agencies. The program was part of the federal initiative that provided funds to states with the intent to link administrative data from many sources to identify welfare recipients (Ver Ploeg et al., 2001). The program in South Carolina was wide ranging, with its impacts extending far beyond the original program design (Wheaton et al., 2012). The initial project attempted to check eligibility for participation in social services, mental health services, disability services, alcohol and drug abuse programs, educational programs, the criminal justice system, elderly services, housing programs, public safety programs, and disease, immunization, and child abuse registries in more than a dozen state government agencies. The integrated data were also designed to determine the prevalence of many diseases and conditions via hospital discharge data and emergency visit data for recipients of food stamps or Temporary Assistance to Needy Families (TANF). It turned out that the system was much more useful than originally designed. An example is identifying children with special care needs (CSCN), such as juvenile-onset diabetes, complex congenital heart disease, and genetic syndromes (trisomy 13, 18, or 21). The Children's Rehabilitative Services (CRS) in the Department of Health and Environmental Control provided services to CSCN. The CRS used International Classification of Diseases, Ninth Revision (ICD-9) codes from Medicaid, inpatient hospitalization, outpatient surgeries, emergency department visits, and the state government employee health plan to identify CSCN. In addition, it linked relevant data files from the Department of Mental Health, the Department of Disabilities and Special Needs, the Department of Vocational Rehabilitation, the Department of Education, and the Department of Social Services. By integrating all of these separate systems using the definition provided, an unduplicated list of more than 340,000 children with special health care needs was developed. This was in stark contrast to the 10,000 CSCN that the CRS served at the time. Although the linkage program in South Carolina generated many external benefits, the lack of a sustainable operational model among agencies eventually doomed its fate. The statewide program of health service data linkage and integration ceased due to recessions and other funding issues.

In most cases, funds are provided through an existing public health program for data linkage that is narrow in scope and practice. In Chapter 14, we review birth outcome–related linkage programs. Here, we review one of the recent programs: the Special Projects of National Significance Program's Systems Linkages and Access to Care for Populations at High Risk of HIV Infection. The initiative funded six demonstration states (Louisiana, Massachusetts, New York, North Carolina, Virginia, and Wisconsin) to design, implement, and evaluate innovative strategies to integrate different components of the public health system, such as surveillance, counseling, testing, and treatment, to create new and effective systems of linkages and retention in care for hard-to-reach HIV populations who have never been in care, have fallen out of care, or are at risk for falling out of care. The prison population is hard to reach, and the Louisiana project was designed to expand HIV testing to jails and prisons, so that prison and jail linkage coordinators and peer coordinators can be utilized to provide prerelease planning services to HIV-infected inmates and ensure they are linked with appropriate medical and social services before release. All of these pilot projects had a novel component that differed from that of the other states. A challenge is to identify a sustainable linkage operation targeting a specific population or program common to all the states.

Both the federal and state data linkage initiatives suggest that technologies for data linkage have matured enough that it would not be difficult to implement at any level. However, program- or project-based data linkage has been piecemeal, and it tends to generate few external benefits beyond the program's specific needs. For instance, the breast and cervical cancer screening program, or what used to be called the Every Woman Matters (EWM) program in Nebraska, is required to link the positive screening individuals to a cancer registry to verify their Medicaid care status. Program staff involved in data linkage can easily conduct a program evaluation project by incorporating a data linkage component, but the lack of communications between programs hinders program specialists to go beyond a program-mandated evaluation. However, the program epidemiologist is not trained or authorized to analyze cancer registry data, while the cancer epidemiologist or biostatistician is not required to analyze EWM program data. For this reason, both programs rarely go beyond their mandated requirements to conduct cross-program data analysis projects. There are many similar programs within a state public health agency, and they conduct routine and ad hoc data linkage, serving regulatory and other funding requirements, but they rarely go beyond mandated operations. Such smaller linkage operations are inefficient with limited externalities. If resources can be pooled together, then a bottom-up linkage operation can be sustained through the participation of all relevant programs. Such an approach is infrastructure-oriented data linkage. In addition, to serve the existing program data linkage needs in a business-as-usual mode, the infrastructure-based data linkage products can enhance program collaborations that could never be achieved through piecemeal data linkage through each silo program.

1.3   INFRASTRUCTURE APPROACH TO DATA INTEGRATION

Most data linkage initiatives above are project oriented. Even though they provide data integration and data sharing, they tend to focus on integrating a number of datasets to enhance one dataset. Alternatively, we could think of public health data integration from the infrastructure building perspective. This perspective is to find the best way to enhance data infrastructure for broad public health applications. Program-based data integration and enhancements tend to focus on a program's need. For instance, a social determinant project for the CVD program would only extract CVD-related diseases and outcomes from the hospital discharge data. Any data enhancements (e.g., geocoding patients to census tracts) and data integration (e.g., linking census tract SES variables) would be program specific. The infrastructure building perspective, in contrast, would geocode the entire hospital discharge data and link them with census tract variables, so that many programs would be able to use the hospital discharge data to extract program-specific diseases and outcomes, together with enhanced variables. A major goal is to level the public health practice field so that smaller or poorly funded programs can just as easily access and use enhanced data as the well-funded programs. At the same level of data integration platform, public health programs can use enhanced data for collaboration, and researchers can use the enhanced data for secondary data analysis.

To enhance public health data infrastructure, we felt that the guiding principle should be that once a program dataset is linked to another dataset, it should be good for crosscutting collaboration, rather than going back to the silo. Health disparity assessment is a crosscutting activity among programs; if we keep thinking about what needs to be done to enhance data elements in health disparity assessments, the eventual integrated data are likely to serve many programs. In a program-oriented data linkage project, a program dataset is often linked to another program's data for a project-specific need. As shown in Figure 1.1 (dashed line), a program-to-program request is sufficient, and it does not require communications with other programs. A health disparity–oriented data linkage project is more comprehensive; it almost always requires crosscutting activities and communications among many public health programs. Regardless of which disparity dimension (race or ethnicity, gender, SES, or rural–urban status) is investigated, a comprehensive report often requires data extraction from many existing programs, and comparisons are made for many diseases or conditions. In Figure 1.1, the solid lines represent an infrastructure enhancement project that requires linkable data from major public health programs. After data linkage and integration, all programs on the table can collaborate based on shared data resources and linked variables.

In order to have program data talk to each other, it is necessary to develop a master data index that creates an index sheet for all program data. Through the master index sheet or lookup table, a record in a program can then be found in all indexed programs. Such an index has no differences in how customers are indexed; for example, for a utility company, a customer could be indexed by the (1) location for the meter reading, (2) type of repairs and maintenance requirements, (3) billing records, and (4) repair history. Although it is relatively easy for a utility company to index a custom on initiation of the service, it is not straightforward to index public health participants through various programs after they have been in the program databases for years. In this book, we proposed a bottom-up approach of indexing public health participants.

The infrastructure enhancement approach fits well with health disparity assessment projects. Most states in the United States have a minority health or health disparity program. However, most administrative data are collected through specific funding streams or mandates that do not budget for extra data integration efforts. A state minority health program is a good example. Normally, a state minority health program is responsible for administering and evaluating some block grants that would improve the health status among minority participants. A state minority health program is also responsible for releasing reports on minority health status, in which data can be pulled from many other programs. However, some program data do not have race and ethnicity variables. In addition, pronounced health disparities are often observed along rural–urban status and SES, and reporting along these dimensions tends to fall on the shoulder of a minority health and health disparity program. Since few program data include income, SES is often derived through census tract information that requires placing program participants in a census tract or geocoding.

The infrastructure enhancement approach to data integration should be sustainable without outside funding. This is analogous to building a highway. At the initial stage, external funding may be needed to integrate the data backward and set up a baseline analysis. As the data or information highway is built, it should be sustained on its own. The service provided by the information highway should cover its maintenance costs. If the maintenance costs exceed potential benefits generated from data services, then the data infrastructure enhancement project is not considered a success. Integrating and linking major public health program data sounds daunting, but it requires relatively small efforts in terms of full-time equivalence (FTE). This is especially true once a master patient index is built and multiple program datasets are standardized. However, one of major infrastructure tasks that might not be sustainable is geocoding that matches patients' or participants' addresses to the geographic locations in terms of latitude and longitude, and then to a census enumeration unit, most often the census tract. For programs that have resources such as a cancer registry, geocoding can be done routinely. For programs that lack resources, geocoding is rarely included in their workflow. In Chapter 6, we propose a master address index to geocode all statewide addresses. Through address standardization in the long run, the bulk of geocoding would become cross-referencing existing master address files and incrementally geocoding new addresses.

In assessing data enhancements, we kept three criteria in mind. First, all datasets should have at least a race variable that is fairly complete. Most datasets within public health programs meet this requirement with variable data quality issues. Datasets that do not meet this requirements include hospital discharge data and the Crash Outcome Data Evaluation System (CODES); in these cases, we had to link other data to either populate or create a race variable. Second, all datasets should have an SES variable that may include income, wealth, occupational status, and census tract poverty status. Since most data do not have income information, we settled with geocoding all records and assigning them with poverty-level information (<5%, 5%–10%, 10%–15%, 15%–20%, and >20%) at the census tract level. All datasets should be able to assign individuals according to their county of residence to urban and rural areas. We simply used a county population of >100,000 as the criterion to distinguish between urban and rural counties. In addition, we also kept the greater metropolitan Omaha–Lincoln area separate as an urban area. We realized that Grand Island became a metropolitan county in 2010. However, most of our data were from before that year, when its population was <100,000, which could potentially trigger confidentiality concerns if we singled it out by the year 2010. Finally, if a program is willing to participate in data integration, all data should have a minimum set of identifiable information for data linkage and for creating an agency-wide master patient or program participant index.

This book is for both multi-program-based data analysis and research-oriented secondary data analysis. Given the ever-increasing data volume and data sophistication, public health data scientists are in great need. Although program collaboration requires a multiprogram team, it takes a multiprogram trained data scientist or practitioner to pull everything together. We therefore aim to nourish a new generation of public health data practitioners, who are not only knowledgeable and skillful for data integration, but also understand substantive issues of public health for multiple programs.

Although we could say that we used data primarily collected administratively for various programs, in reality, we used secondary data. As the primary purpose of data collection was to meet a program-specific mandate, program collaboration and disparity assessments were usually not in the database design. In addition, the original program data were not integrated either. One could argue that data analysis and disparity surveillance are part of mandates in many administrative data collections. If administrative data are used in research, then they are considered secondary data. If administrative data are used for public health surveillance, then they are exempt from the Health Insurance Portability and Accountability Act (HIPAA). In all data presentations, we attempted to meet the secondary data analysis guidelines even though our analyses were mostly at the surveillance level. Data were generated by data analysts, and a secondary dataset was then generated without names, addresses, date of birth, zip code, and county identifiers. In order to track individuals, a record ID was generated that can be linked to individuals, and this ID does not have any connection to existing record IDs or other personal identifiers used by any public health programs. To provide rural–urban status, county codes were used to generate the secondary data, but deleted for users who conducted data analysis. Likewise, most person-level data were geocoded to the census tract level, but census tract information was only used to generate poverty status. After poverty status from each census tract was attached to individuals, census tract IDs were deleted. In this way, even though our data analysis was borderline between data analysis and research, we did not violate any confidentiality clauses stipulated in HIPAA. All data used in this book were at the state level; the only identifications at the person level are categorical rural–urban status, poverty status, age group, sex, and racial category within a state boundary.

1.4 CHAPTER HIGHLIGHTS

The infrastructure approach to data enhancement is to encourage public health practitioners and researchers to use the data. This book has three sections: Section I, Chapters 2 through 7; Section II, Chapters 8 through 12; and Section III, Chapters 13 through 20. Creating secondary datasets and indexing multiple datasets are helpful, but there is also a workforce development piece, so the first section prepares data analysts, giving them the basic knowledge and skills for disparity assessments. Chapter 2 describes major health disparity dimensions. Chapters 3 through 5 are for data preparation, which includes the descriptions of public health datasets, data linkage, and indexing. Chapters 6 and 7 provide measurements and statistical models for health disparity assessments. For informatics managers, this section provides a bottom-up approach, creating master data index for public health program participants and a master address index for enhanced geocoding. The second section, to some extent, is a test drive of the public health information highway that we have created. We took the social determinants approach and used the hospital discharge data to systematically assess health disparities. A variety of methods are used to highlight health disparity by SES (Chapter 8), sex (Chapter 9), rural–urban (Chapter 10), and race (Chapter 11) categories. Finally, Chapters 13 through 20 provide case studies. Each case study is closely related to a chapter in Sections I and II.

In Chapter 2, we develop conceptual and measurement frameworks on health disparity. We discuss each health disparity dimension in turn—race or ethnicity, gender, SES, rurality, and limited English proficiency (LEP)—using empirical data from Nebraska. We omit disability because the American Community Survey routinely releases disability data at the county and census tract levels by major demographic and SES dimensions. We highlight issues related to multiple races and their potential impacts on the magnitude of health disparities. In the discussion of gender differences in health, we emphasize those due to social, occupational, and family positions. Based on the calculation of life table and other empirical evidence, we also point out the morbidity–mortality paradox by gender: women live longer than men, yet women live with a greater number of comorbidities and have worse self-reported health than men. With regard to SES disparities, we discuss the importance of using area-based measures, as most administrative data do not have educational level and personal income. For both rural and LEP disparities, lack of access to care is a significant issue.

In Chapter 3, we first introduce the use of administrative data. We argue that almost all administrative data belong to secondary data, at least to outside researchers. We describe important characteristics of administrative data, such as limited variables, varied data qualities, and lack of time stamps. We then attempt to set out some criteria for assessing public health datasets. Using the criteria, we describe major administrative datasets commonly used by the Division of Public Health at the Nebraska Department of Health and Human Services (DHHS). It is noted that the division has access to more than 100 datasets. Examples include vital records, cancer registries, trauma registries, hospital discharge data, mental health data, and Medicaid claims data. We highlight their main use and applications in public health and point out their limitations.

In Chapter 4, we provide an overview of the data linkage methodology from an initial linkage assessment to the interpreting of the linkage results. We point out that many public health datasets can be linked by a common identifier, and important information can be gained through data linkage. Furthermore, even though the social security numbers (SSNs) are available, we are not authorized to use them for general record linkage applications. In addition, some people do not have a SSN, and some SSNs are incorrect. Hence, it is important to develop knowledge and skills about probabilistic data linkage. We describe in detail the four steps for data linkage: (1) data preparation, (2) matching and merging, (3) manual review, and (4) verification. We also introduce some commonly used concepts in record linkage, such as matching rate, blocking, pass, weight, false positive, and false negative. Finally, we offer our views on training and privacy protection.

In Chapter 5, we provide a rationale for why it is necessary to integrate state public health data to improve data support efficiency and increase the value of existing datasets. To overcome current technical and organization barriers, we propose a bottom-up approach to data integration and building a master patient index (MPI) for public health program participants, as well as for patients under public health surveillance. An MPI within a state public health agency can increase operational efficiency and communications and collaboration among participating programs. Workforce development in data integration requires learning from real-world practices and lessons learned. In this chapter, we provide three examples of how to implement a bottom-up approach to build an MPI within the state public health agency. Based on examples and other practice-based data linkage projects, we can loosely connect more than a half-dozen data marts or program datasets. Each program was responsible for the maintenance of its data mart similar to its standard operation. The informatics unit maintains and updates the MPI by unifying a set of personal identifiers and providing a link to each program data mart. All data marts together are in a federation that follows the existing protocol for data sharing.

In Chapter 6, we highlight the importance of geocoding, a process of assigning individual addresses to their geographic location. We propose a master address index approach to sustain geocoding operations. Theoretically, each standardized address should only be coded once to a very high quality. After updating new addresses for a while, there will be a very moderate effort in geocoding each year within the public health agency. In reality, low-quality geocoding is easy and cheap; people might just take an easy route to geocoding and sacrifice quality. Since a major application of geocoding is to associate patient residential location to neighborhood SES variables, we also describe how to attach census tract data to geocoded coordinates. In addition, we introduce some exploratory spatial data analyses and spatial statistical methods.

In Chapter 7, we introduce study designs and common measurements for how to conduct health disparity surveillance. We stress that disparity surveillance is based on observational studies, and they often cannot meet the standards of experimental and cohort study designs. Although disparity surveillance may not need a research question, it does need a scope to frame surveillance questions. Measurement descriptions include rate, relative rate, odds ratio, rate ratios, standard mortality, and morbidity. In addition, we introduce the log-rate model or Poisson regression for incidence, Surfling regression for cyclic effects, and a quasi-experimental design for program evaluation.

In Chapter 8, we set up the analytical framework as well as empirical evidence for SES-based health disparity assessments in Chapters 9 through 11. We prefer the term *surveillance* to *assessment*, as we emphasize width rather than depth of disparity assessments. We first set out the log-rate modeling approach and some study design issues, and then list steps on how to conduct comorbidity surveillance along the SES gradients defined by census tract poverty levels in a typical range of <5%, 5%–10%, 10%–15%, 15%–20%, and >20%. Using incidence or record-based data, we present in sequence comorbidity, hospital procedure, hospital readmission, and 1-year mortality for about 40 diseases or conditions. With some exceptions for hospital procedures, most outcomes follow an SES gradient: poor neighborhoods tend to have a greater chance of hospitalization for most diseases and conditions.

In Chapter 9, we present another way of measuring disparities. We use age-standardized rates to assess gender differences in diseases and hospital procedures. In the context of the morbidity–mortality paradox in demography, we showed that hospitalization data in general do not support the paradox. What sets our analysis apart from most gender difference studies is that we systematically use neighborhood SES in gender disparity surveillance. We show that among the 10 leading causes of deaths, the hospital-based prevalence rates were higher in high-poverty neighborhoods than in low-poverty neighborhoods by both genders. In addition, sex also interacts with SES, which either ameliorates or exacerbates some sex differences in disease rates.

In Chapter 10, we present yet another way of measuring disparities. We use principal diagnoses to examine rural–urban differences in common hospitalizations. We use the same set of diagnoses used in Chapter 8. We show that urban residents are more likely to be hospitalized than rural residents. SES gradients in health are much stronger in urban areas than in rural areas. In addition, we introduce travel distance and rural hospital bypassing rates by diagnosis. Depending on poverty groups and diagnoses, rural hospital bypassing rates can be as low as 5% and as high as 72%. Traumatic brain injury, back problems, and cancer have the highest bypassing rates, while chronic obstructive pulmonary disease (COPD), bronchiectasis pneumonia, and asthma have the lowest bypassing rates. Finally, we present a case study of injury admissions by using E-code data.

In Chapter 11, we present yet another way to measure disparities. We conduct racial disparity surveillance by using patient or prevalence data. We use hospital discharge data by attaching race information from multiple public health and other databases. We first assess some uncertainties associated with different uses of multiple-race categories in the census data for the at-risk population. The first part of the chapter is devoted to methodological assessments. Our conservative calculations of standard hospitalization rates show that Blacks are twice as likely to be admitted to hospitals as Whites for most diagnoses. Decomposing race effects by neighborhood SES would enlarge the Black–White disparity for the midpoverty group, and White–other race disparities for the high-poverty groups. Our model-based estimates show that Blacks are not only more likely to be admitted, but also more likely to be readmitted to hospitals. While Blacks have higher mortality rates than Whites in major diseases, they are less likely to die in the hospital than White patients. Finally, we offer a case study in cardiac rehabilitation for acute myocardial infarction (AMI) patients. We find that Blacks and other races are much less likely to participate in an outpatient cardiac rehabilitation program than Whites. These effects are independent of many risk factors, such as age, sex, SES, in-hospital procedures, comorbidities, and place of residence.

In Chapter 12, we use hospital emergency visit data that are part of outpatient claim data. One motivation of this chapter is that we have not touched emergency department (ED) data, and a demonstration of using ED data is needed. Another motivation is that we have not really evaluated age-based disparities, which are often related to age-based population vulnerability. Since many population vulnerability studies deal with public health preparedness, we offer two evaluations: one is to evaluate population vulnerability by place of residence and age group for influenza-like illnesses (ILIs), and the other is to evaluate the impact of extreme weather on injury-related ED visits. We find that place of residence has little effect on ILIs, while the elderly group had a very different seasonal variation from that of the younger age groups, especially for the H1N1 flu in 2009. In the second part of the chapter, we experimented with a set of weather-related explanatory variables. We define extreme weather as being very cold (<20°F), very hot (>90°F), rain, or snow, and explored its potential associations with a set of external injuries. We found that extreme weather conditions relate to a half-dozen injury categories.

Chapter 13 is an application study of Chapter 4. Cancer registry data often lack complete chemotherapy and radiation therapy information. To conduct treatment disparity surveillance, we linked 2005–2009 Nebraska Cancer Registry data with Nebraska hospital discharge data. Due to the high quality of both datasets and the proposed linkage procedure, we had a linkage rate of 97%. We demonstrate the utility of the linked dataset in case finding, treatment update, and treatment surveillance. The results show that the linked dataset is likely to identify up to 5% of potential missed cases. We investigated the use of radiation therapy in treating colorectal and breast cancers as case-finding examples. The linked dataset found 12.5% and 14.9% more treatment cases of colorectal and breast cancer patients, respectively. When radiation therapy information is more complete, it can be used to assess disparity. In addition, comorbidity information can be derived for assessing survival.

Chapter 14, to some extent, it is an application of Chapter 5 for indexing. We review major birth certificate data-based data linkage projects. One common characteristic among these projects is funding, which is a double-edged sword. On the one hand, major funding sources provide opportunities for the state and cities to initiate data linkage projects. On the other hand, as funds dry up, the linked datasets are often left without caretakers to maintain, update, and so forth. We indicate that building a simple and low-cost mother index based on the birth certificate data is an easy task. Using the mother index for Nebraskans, we examine the relationship between first and later birth outcomes in terms of low birth weight (LBW) and preterm birth (PTB). We show that 20% of mothers have second LBW infants when their previous births are LBW. Likewise, 23% of mothers who have PTBs have PTBs the next time. To a lesser degree, these trends also repeat for third births. In addition, we show how a mother index might be used for creating longitudinal datasets based on Pregnancy Risk Assessment Monitoring System (PRAMS) and birth certificate data. We also provide a case study about rapid repeat births using the mother index.

Chapter 15 is an in-depth study of the geocoding in Chapter 6. It summarizes geocoding improvement experiments in the Nebraska Cancer Registry. An initial assessment of previous geocoding suggests that some proven geocoding procedures have not been followed, and the results are unacceptable. It is also suggested that the combination of a match score of 80 and a spelling sensitivity of 80 in the ArcGIS geocoder is sufficient for most geocoding purposes. Given the sizable number of unmatched addresses, the Google Maps geocoding service is used. A comparison of 1500 high-quality addresses that are matched by both Google Maps and ArcGIS geocoders shows that the location between the two in most cases is acceptable. The median distance between each pair of 1500 coded locations was 36.6 meters. It is concluded that by strictly following the major successful steps, including the use of address coding specification, Internet-based white pages for reverse address finding, and Internet-based geocoding, a 90% or even 95% match rate is achievable. The chapter also includes some recent developments, such as using local government and North American Association of Central Cancer Registries (NAACCR)–provided web services for geocoding, and provides some tips for managing geocoding projects.

Chapter 16 is an in-depth study of Chapter 9 on gender differences in health. It links stroke hospitalizations to community-based mortality records to examine sex differences in stroke case fatalities and associated prognosis factors. Hospital discharge data and death certificate data from January 2005 to December 2009 in Nebraska were linked. Multivariable logistic regression was used to estimate sex differences in 30-day mortality, and the Cox proportional hazard model was used to predict overall survival. A total of 15,806 patients was included. Females were more likely to die during the 30 days after stroke hospitalization. However, there was no significant difference in overall survival in the multivariate analysis that controlled for age, comorbidity, and rehabilitation factors. Females were more likely to have comorbidities such as atrial fibrillation, anemia, and heart failure, while males were more likely to have chronic kidney disease. In addition, males were more likely to receive rehabilitation services after stroke.

Chapter 17 is an in-depth and data-driven study of Chapter 10 on rural–urban differences in health by using AMI. AMI is one of the most studied hospitalization reasons in the United States, and despite some improvement in hospital-based outcomes, there are some inconsistent findings regarding place of residence–based outcomes versus hospital-based outcomes between rural and urban areas. Previous studies tend to study either rural–urban residence differences in health outcomes or rural–urban hospital differences in health outcomes. Based on the integrated public health data infrastructure, we can easily make a contrast between rural and urban patients, or between urban hospitals and rural hospitals. The former is a place of residence–based rural–urban difference, while the latter is a place of hospital–based rural–urban difference. Using 30-day readmission and out-of-hospital survival, the study examines rural–urban differences in both place of residence and place of hospital. The study linked 2005–2011 hospital discharge data to the 2005–2012 death index and outpatient data. In addition, distance to the hospital is calculated for each hospitalization. Hospital bypassing rates are also examined. It was found that both residence- and hospital-based measures yield similar results. We found that participating in a rehabilitation program is the most important predictor of out-of-hospital survival.

Chapter 18 is a systematic application of racial disparities to motor vehicle crash (MVC) injuries. To some extent, it is an in-depth application study of Chapter 11 on racial health disparities. The lack of race information for nonfatal motor vehicle crash injuries in the United States has limited the understanding of racial disparities in MVCs. In this chapter, we described a pilot surveillance project in Nebraska that linked crash reports and driver's license records to investigate racial disparities among nonfatal MVC injuries. The chapter first examines disparities based on police-reported injuries, and then repeats the analysis for hospital-based injuries. In both cases, Black drivers had a higher rate of injuries, especially nonsevere injuries. In addition, the project evaluated SES disparities by geocoding injured drivers. It showed that those living in poor neighborhoods tend to drive short distances, but have a high rate of injuries, independent of the observed disparities by race.

Chapter 19 can be viewed as a general application study of program evaluation. The motivation is twofold. On the one hand, lack of income measures is an important drawback of cancer registry data. Linking cancer screening program participants, who were all relatively low income and underinsured, provides an indirect way of measuring income effects. On the other hand, breast cancer screening programs funded by the CDC have been implemented for more than 25 years, but few studies have evaluated their mortality outcomes. We compare 3-year mortality and long-term survival outcomes using the linked Nebraska Cancer Registry—the breast screening data from the EWM program. The chapter employs two study designs: comparisons of (1) the full sample and (2) the geographically matched sample to the census tracts where the EWM participants reside. Compared to all non-EWM participants, women diagnosed through EWM were more likely to survive 3 years after diagnosis. However, this short-term survival advantage was a nonfactor in terms of long-term survival. These results hold for the geographically matched sample, too. In addition, there was no significant difference in resection surgery, radiation, and chemotherapy treatments among EWM participants and nonparticipants.

Chapter 20 attempts to open a new surveillance avenue by linking environmental exposure data and patient-level data in spatial disease surveillance. It uses Parkinson's disease because its etiology is largely unknown. We demonstrated both spatial cluster detection and case-control approaches linking patients to environmental exposures. The cluster detection approach is first to detect spatial clusters and then compare exposures within and outside the clusters. The case-control design assesses whether cases and controls could be differentiated by past exposures. We used 5 pesticides and 15 herbicides in 2005 as exposures. We selected cases from both the Nebraska Parkinson's Disease Registry and controls from hospital discharge data, all from 2007 to 2011. Although we made some interesting findings, we tend not to interpret them due to the methodological emphasis and uncertainties associated with long-term exposure to Parkinson's disease.

REFERENCES

CDC (Centers for Disease Control and Prevention). 2004. Indicators for chronic disease surveillance. *Morbidity and Mortality Weekly Report* 53(RR11): 1–6.

CDC (Centers for Disease Control and Prevention). 2015. Indicators for chronic disease surveillance—United States, 2013. *Morbidity and Mortality Weekly Report* 64(RR01): 1–15.

IOM (Institute of Medicine). 1988. *The Future of Public Health.* Washington, DC: National Academies Press.

Rosen, G. 1993. *A History of Public Health.* 2nd and exp. ed. Baltimore: Johns Hopkins University Press.

Ver Ploeg, M. ver, R.A. Moffitt, and C.F. Citro, eds. 2001. *Studies of Welfare Populations: Data Collection and Research Issues.* Washington, DC: National Academy Press.

Wheaton, L., C. Durham, and P.J. Loprest. 2012. TANF and related administrative data project: Final evaluation report. Washington, DC: Urban Institute, June 18. http://www.urban.org/research/publication/tanf-and-related-administrative-data-project-final-evaluation-report/view/full_report (accessed June 2015).

Section I

Conceptual, Analytical, and Data Preparations for Health Disparity Assessments

2 Common Population-Based Health Disparity Dimensions

2.1 INTRODUCTION

The U.S. government's Healthy People initiative first listed reducing health disparities as of one its Healthy People 2000 goals. Broadly speaking, health disparities are differences in health outcomes by population segments according to demographic characteristics, socioeconomic status (SES), and geographic environment. This definition includes two interrelated components: populations segmentation and health outcomes. The most commonly used population segments are race and ethnicity. The most commonly used SES measures are individual income and educational level. In the absence of or in addition to individual SES measures, area-level poverty concentration, such as the percent of population under the 100% poverty line, is often used. The most commonly used geographic disparity measure is rural–urban difference. In addition, gender, disability, and limited English proficiency (LEP) are also used to segment different population groups. The most commonly used health outcome measures for a state health agency can be grouped by behavior, health condition or morbidity, and mortality. Commonly used health behavior measures include physical activity, smoking, substance abuse, and clinic visits. Both mortality and morbidity measures are often based on the International Classification of Diseases, 9th and 10th Revisions (ICD-9 and ICD-10, respectively). Commonly used diseases and health conditions include heart disease, cancer, chronic respiratory disease, stroke, unintentional injury, diabetes, and obesity, and they are measured by rates of incidence, prevalence, and mortality. In this chapter, we describe and elaborate on these measures in detail and lay out the challenges of linking health disparity population and health indicator measures.

2.2 RACE AND ETHNICITY

Data on race and ethnicity, the most frequently used population segmentations for health disparity, are mostly collected by self-identification. The federal and most state governments in the United States have adopted the Office of Management and Budget (OMB) definitions of race and ethnicity (Wallman and Hodgdon, 1977). In 1997, the OMB issued revisions to the 1977 standards for classification of federal data on race and ethnicity, which expanded the four-race-category reporting system (OMB, 1977) to the five-category system (OMB, 1997). The five minimum categories for race data include American Indian or Alaska Native (AI/AN), Asian, Black or African American, Native Hawaiian or Other Pacific Islander (NHPI), and White; the first four categories are racial minorities, while the White category is the racial majority. Only Hispanic or Latino ethnic origin is required, which refers to a person of Cuban, Mexican, Puerto Rican, South or Central American, or other Spanish culture or origin, regardless of race. In this book, we use *Hispanic* and *Hispanic or Latino* interchangeably.

Race and ethnicity data are often used in statistical analysis, and they serve as biomarkers for unmeasured biological differences. In the absence of income and other SES variables, race is used as a proxy for SES. Regardless of purpose, race and ethnicity are most often used for health statistical reporting, where rates are calculated based on disease counts and at-risk population (Mays et al., 2003). However, it is not straightforward to calculate disease incidence and mortality rates by race, because cases are normally from a registry, a surveillance system, or vital statistical records, and it is necessary to pair case counts and at-risk populations by race consistently. In many cases, race categories from different datasets or from different periods are not identical, and concerns are often raised about how to make race categories consistent as numerators or cases and as denominators or at-risk populations.

The first concern is that different census years may have inconsistent race and ethnicity classifications, which require either backward fitting or forward bridging. Backward fitting is a process to find the smallest denominator from early years of data, when race categories were fewer, and then combine the same data from later years, which have more racial categories. Forward bridging is more difficult to implement, as it requires dividing a single racial category from an early year into two categories to be compatible with later years of data. For example, the 1990 census had four race categories according to the 1977 race reporting standards of the OMB. Since 2000, the census has been based on five race categories, because the former Asian and Pacific Islander category was divided into two categories: Asian and NHPI. It would be difficult to split the 1990 population for an area from four race categories into five, even though we have the 2000 census five-category race information for the area. Consequently, in order to pair racial data over the two periods, researchers often use the backward-fitting approach to combine the two racial categories of the 2000 census. In the above example, NHPI and Asian would be combined into the single category of Asian or Pacific Islander according to the 1990 census. Table 2.1 uses such an approach to provide an overall picture of cancer cases and at-risk population by race and ethnicity from 1991 to 2010. The incidence data have been collected by the Nebraska Cancer Registry in four race categories since 1991. The at-risk populations over four periods were provided by the Surveillance, Epidemiology, and End Results (SEER) program based on the 1990 race and ethnicity categories, which go back as far as 1970 at the county level. Note that backward fitting is not without problems. In the early 1990s, cancer patients tended not to indicate whether they were of Hispanic origin. Therefore, the 0.01% of "Other Hispanics" may not reflect the true percentage. The percentage for Other Hispanics in the 1991–1995 period would be misleading if it were compared with the rate in the 2006–2010 period. In addition, as more races or different ways of defining races are added, respondents are likely to be unsure or confused about which race to choose, which may partially explain the increase in missing race information from the cancer registry data.

There are two common methods of treating records with missing race information. One simple way is to delete them and then add weights to the remaining records with valid race information. In this way, the overall incidence rate accounts for all patients and therefore is precise. However, if

TABLE 2.1

Distribution (%) of Cancer Cases and At-Risk Population by Race and Ethnicity

Race/Ethnicity	At-Risk Population				Cancer Cases			
	1991–1995	1996–2000	2001–2005	2006–2010	1991–1995	1996–2000	2001–2005	2006–2010
NH-White	91.64	89.22	86.54	83.89	96	94.62	93.31	92.93
NH-Black	3.72	4.07	4.43	4.86	2.51	2.48	2.66	2.79
NH-American Indian	0.74	0.79	0.85	0.89	0.28	0.4	0.4	0.35
NH-Asian	0.92	1.26	1.62	1.85	0.2	0.42	0.47	0.66
Hispanic-White	2.81	4.43	6.2	7.69	0.45	0.99	1.34	1.42
Other Hispanic	0.17	0.23	0.36	0.82	0.01	0.07	0.12	0.14
Missing race					0.55	1.03	1.69	1.71
Total[a]	1,625,846	1,694,833	1,737,703	1,799,067	7,614	8,228	8,657	8,544

NH = non-Hispanic.

[a] Total is annualized.

those with missing race information are biased toward late or early stages of cancer, or particular cancer sites, the rates according to these dimensions are biased and should be reported with caution. Sometimes it is necessary to generate estimates for small areas, such as census tracts, by race or ethnicity, for which preserving observations in each area is critical for generating stable rates. In this case, the distribution of the at-risk population is often used to allocate race to records with missing race information according to small-area racial distribution. For example, if four patients in a census tract do not have race information, and the census tract population by race is distributed as 50% Whites, 25% Blacks, 25% Asian and Pacific Islanders, and no American Indians, then according to probability, two persons out of four (or 50%) would be allocated randomly to White, and one to Black and one to Asian. One important difference between the two methods is that the weighting method uses patient population information, while the race allocation method uses race information from the at-risk population.

The second concern is how to treat multiple races. Even though racial categories may be identical, the method of identifying races may differ. The OMB's 1997 revision of the race reporting standard allows multiple-race categories, resulting in 31 potential racial groups, ranging from a *one-race-only* group to a combination of four races. Table 2.2 summarizes how the 31 races are produced. For two or more races, the leftmost column or columns separated by a semicolon (;) represent one, two, or three multiple races that can be combined with each bold race category on the right. In the first line of the three-race panel, for instance, we can have three multiple races: (1) AI/AN; Asian; Black or African American, (2) AI/AN; Asian; NHPI, and (3) AI/AN; Asian; White.

Since many health statistics are reported with single or primary race categories, one has to combine multiple-race categories into a single-race category for the at-risk population. This process is often called race bridging, which generates bridged race categories. There are many race bridging methods that include both deterministic assignments and statistical allocations (Mays et al., 2003). One has to be careful about how to combine multiple races into a single race. For example, to determine the total White population, one can use the White-alone category, which produces the smallest number of White population. One the other hand, one can use White in combinations with all other racial groups, which produces the largest White population. Both SEER and the Centers

TABLE 2.2
Races and Multiple Race Combinations in the 2000 Census

One race alone	**AI/AN**	**Asian**	**Black**	**NHPI**	**White**
Two races	AI/AN;	**Asian**	**Black**	**NHPI**	**White**
	Asian;		**Black**	**NHPI**	**White**
	Black;			**NHPI**	**White**
	NHPI;				**White**
Three races	AI/AN;	Asian;	**Black**	NHPI	White
	AI/AN;		Black;	**NHPI**	White
	AI/AN;			NHPI;	**White**
		Asian;	Black;	**NHPI**	White
		Asian;		NHPI;	**White**
			Black;	NHPI;	**White**
Four races	AI/AN;	Asian;	Black;	**NHPI**	White
	AI/AN;	Asian;		NHPI;	**White**
	AI/AN;		Black;	NHPI;	**White**
		Asian;	Black;	NHPI;	**White**
	AI/AN;	Asian;	Black;	**NHPI**	

Note: A semicolon (;) separates each race group. A total of 31 races and multiple race combinations are in bold.

for Disease Control and Prevention (CDC) use a statistical method to provide at-risk population by age, sex, and bridged race categories at the county level. At the census tract level, however, there is no official at-risk population by bridged race categories, which forces people working with census tract data to come up with their own estimates. Since census tracts tend to have a smaller population size, small adjustments of at-risk populations by race can have a substantial impact in estimated disease rates by race.

The above two concerns are mainly about the consistency between data sources of the composition of the at-risk population. The third concern arises because many administrative data collected by state governments do not follow the federal data collection standards. These administrative data are often related to disease cases, for which the at-risk population often comes from the census. There are always some minor differences between the race categories used in data collection and the race categories in the census. Table 2.3 lists the distribution of injured drivers in Nebraska aged 19 years and older and the corresponding at-risk population (aged 19+ years) averaged over 2006–2010. There are at least three discrepancies. First, even though we tried to make these racial categories as compatible as possible, the American Community Survey (ACS) identifies drivers using only single-race categories, and the administrative database of driver's licenses identifies drivers by primary race. Despite these discrepancies, the proportion of White population was still greater in the ACS data than in the driver's license data. Second, the Other race category means other multiple races in the ACS, but it means that the driver's license applicants were unable to identify themselves among the four categories above in the driver's license data. Third, the driver's license data included Hispanic as a racial category that an individual could choose when completing an application for a driver's license, but Hispanic is clearly not a race category in the ACS.

If we use injured drivers in motor vehicle crashes as cases, and drivers from the driver's license database as the at-risk population, we would have few problems, because they come from the same database. If we do not like Other as a category, we can simply group Others and Hispanic, or even missing, as Other. If we were to use the ACS population as the at-risk population, then we would not group Hispanic cases with the Other race group. In this example, the results for the Black and White races in the at-risk population are unlikely to differ significantly between the driver's license data and the ACS data, but others can.

Even when the state health statistics follow the federal standards, completely classifying all cases into identifiable racial groups is almost impossible due to missing values and multiple-race parents. If the child self-identifies his or her race after age 18, the situation can be worse. As an example, Table 2.4 shows the mother's race category in the row denoted by m for Nebraska babies born between 2005 and 2012, and the father's race category in the column denoted by f. Among the 221,411 babies, only 155,584, or 70%, had a mother and father of the same race (Table 2.4). In addition, 25,931 mothers and 23,611 fathers identified themselves as multiple races either by choosing "other" or by checking several race self-identification boxes on the birth certificate. Except in the

TABLE 2.3
Driver's License Data and ACS Data

	Injured Drivers	Total Drivers	ACS Population
White	84.14	87.43	90.29
Black	7.43	4.31	3.97
AI/AN	0.96	0.77	0.72
Asian/NHPI	1.6	2.04	1.67
Other	3.05	4.02	3.35
Hispanic	2.81	0.95	
Missing	0.01	0.48	
Total	8,942	1,305,680	1,319,164

TABLE 2.4

Race by Birthing Mothers and Fathers in Nebraska: 2005–2012

m/f	White	Black	AI/AN	Asian	NHPI	2+ Race	Missing	Total
White	145,215	3,566	819	743	59	5,941	16,979	173,322
	(83.8%)	(2.1%)	(0.5%)	(0.4%)	(0%)	(3.4%)	(9.8%)	
Black	636	6,257	56	26	5	291	5,716	12,987
	(4.9%)	(48.2%)	(0.4%)	(0.2%)	(0%)	(2.2%)	(44.0%)	
AI/AN	559	120	800	14	10	402	1,752	3,657
	(15.3%)	(3.3%)	(21.9%)	(0.4%)	(0.3%)	(11.0%)	(47.9%)	
Asian	1,149	110	8	3,188	6	135	268	4,864
	(23.6%)	(2.3%)	(0.2%)	(65.5%)	(0.1%)	(2.8%)	(5.5%)	
NHPI	70	16	1	12	124	38	53	314
	(22.3%)	(5.1%)	(0.3%)	(3.8%)	(39.5%)	(12.1%)	(16.9%)	
2+ race	3,244	659	170	91	43	16,780	4,944	25,931
	(12.5%)	(2.5%)	(0.7%)	(0.4%)	(0.2%)	(64.7%)	(19.1%)	
Missing	63	7	1	0	0	24	241	336
	(18.8%)	(2.1%)	(0.3%)	(0%)	(0%)	(7.1%)	(71.7%)	
Total	150,936	10,735	1,855	4,074	247	23,611	29,953	221,411

Note: Row percentages are in parentheses.

case of missing values, which cannot be classified, a baby's race can be classified by three methods: (1) conservatively (e.g., when both the mother and the father are Black, the baby is Black); (2) moderately (e.g., when either the mother or the father is Black, the baby is Black); and (3) liberally (e.g., identifying all the mothers and fathers in the two or more race [2+ race] category that are Black plus any other racial groups and adding those cases to method 2). Method 1 generates 6,257 Black births, and methods 2 and 3 generate 17,465 and 19,329 Black births, respectively. If we use low birth weight (LBW) as an indicator for birth outcomes, method 1 corresponds to a 12.58% LBW, and methods 2 and 3 correspond to an 11.92% and 11.70% LBWs, respectively. The rate of LBW for the whole state from 2005 to 2012 was 7.30%. To improve birth outcomes, it is better to cast a wider net by using method 3, which captures the largest number of Black births.

Finally, there is an issue of how to distinguish or combine racial and ethnic categories. Normally, Hispanic Whites are excluded from minority health statistics, but minority population reports often include all Hispanics, regardless of race, as a single category. Care must be taken when estimating disease and health condition burdens for Hispanics. Younger age groups tend to have a larger proportion of Hispanics. Even though non-White Hispanics have a slightly lower overall cancer burden and a slower rate of increase relative to the population's increase (Table 2.1), when age is adjusted, they may have a greater burden, a point that will be discussed in later chapters. In the birth certificate data, we also found that mothers who checked "other" or multiple race boxes, or who did not answer race questions, were more likely to be Hispanics. For example, among the 25,931 mothers in the Other or multiple-race category in Table 2.4, 85.6% were Hispanics, and among the 336 mothers in the missing race category, 71.73% were Hispanics. The overall Hispanic percentage for all the mothers was 14.7%. This propensity may be due to respondents' confusion about the difference between race and ethnicity. If this same discrepancy was also true in cancer data for the missing race category, the cancer burden could be heavier for Hispanics than for non-Hispanics. Moreover, Hispanics are the fastest-growing population in Nebraska, and they tend to be more mobile than other ethnic groups due to age, immigration status, and farm- and construction-dominated occupations. These characteristics suggest that (1) their exposures to environment and neighborhood characteristics are less likely to be captured than for other ethnic groups, and (2) community- or neighborhood-based public health interventions need to adapt to this mobile population.

2.3 GENDER

Health disparities by gender can be studied or measured from at least four perspectives: demographic, gender inequality in economic status and employment opportunities, access to care, and sexual orientation. Although gender is easily accessible in most data, variables from the above perspectives may not always be available. Below, we touch briefly on these perspectives.

It has long been established that women live longer than men, which can be measured in age-specific mortality and life expectancy. Life expectancy at birth between men and women is about 5 years apart in the United States, and the number is fairly consistent among states. Life expectancy at the state level can be obtained from the CDC, which maintains life tables for all states, updated periodically. According to the CDC tables for 1999–2001 (http://www.cdc.gov/nchs/data/dvs/lewk4_nebraska.pdf), the life expectancies in Nebraska for males and females are 76.00 and 80.78 years, respectively, and for the 2009–2011 period are 77.65 and 82.00 years, respectively. To calculate a life table, it is necessary to have age-specific deaths and populations, and the rest of the table can be derived from these numbers (Plane and Rogerson, 1994). Table 2.5 provides life tables for males and females based on Nebraska data for 2009–2011. The column heads in Table 2.5 are described as follows:

D: Number of age-specific deaths in a population during a year or averaged annually.
P: Observed age-specific populations, which is normally from the Census Bureau estimates. When multiple years are used, the midyear population estimate is often used. Sometimes, the first age group also comes from the number of real births.
n: Number of years in the age interval (e.g., $n = 1$ for age 0–1, $n = 4$ for age 1–4, $n = 5$ for age 5–9, $n = \infty$ for age 85+)
$_nM_x$: Age-specific mortality rate for persons aged x to $x + n$. When the midyear population is used, the annualized mortality rate is D/P.
$_nq_x$: Probability of dying between age x and $x + n$:

$$_nq_x = \frac{2n \times {}_nM_x}{2 + n \times {}_nM_x}$$

l_x: Number of survivors at age x each year from the stationary population starting at 100,000:

$$l_0 = 100{,}000; \; l_{x+n} = l_x \times \left(1 - {}_nq_x\right)$$

$_nd_x$: Number of deaths each year between age x and $x + n$ in the stationary population:

$$_nd_x = l_x \times {}_nq_x$$

$_nL_x$: Number of person-years lived between age x and $x + n$. Since 100,000 births are added each year, and all individuals follow age-specific mortality rates, $_nL_x$ reflects the age-specific size of the stationary population:

$$_1L_0 = 0.3l_0 \times 0.7l_1; \quad _nL_x = \frac{n}{2}\left(l_x \times l_{x=n}\right); \quad _\infty L_x = \frac{l_x}{_\infty M_x}$$

TABLE 2.5
Life Tables for Nebraska Based on 2009–2011 Population (P) and Death (D) by Gender

Age	D	P	$_nM_x$	$_nq_x$	l_x	$_nd_x$	$_nL_x$	T_x	e_x
Male									
0–1	228	13,372	0.0057	0.0057	100,000	566	99,498	7,764,578	77.65
1–4	38	54,196	0.0002	0.0009	99,434	93	397,512	7,665,080	77.09
5–9	30	65,846	0.0002	0.0008	99,342	75	496,472	7,267,568	73.16
10–14	25	62,692	0.0001	0.0007	99,266	66	496,158	6,771,096	68.21
15–19	116	65,982	0.0006	0.0029	99,200	290	495,418	6,274,938	63.26
20–24	192	66,997	0.001	0.0048	98,910	471	493,400	5,779,520	58.43
25–29	193	66,094	0.001	0.0049	98,439	478	491,060	5,286,121	53.7
30–34	208	59,484	0.0012	0.0058	97,961	569	488,423	4,795,061	48.95
35–39	247	55,772	0.0015	0.0074	97,391	716	485,334	4,306,638	44.22
40–44	340	55,934	0.002	0.0101	96,675	975	481,195	3,821,305	39.53
45–49	624	63,968	0.0033	0.0161	95,700	1,545	475,024	3,340,109	34.9
50–54	1,038	65,349	0.0053	0.0262	94,155	2,462	465,039	2,865,086	30.43
55–59	1,315	58,464	0.0075	0.0368	91,693	3,377	450,378	2,400,046	26.17
60–64	1,671	47,486	0.0117	0.0571	88,316	5,039	429,589	1,949,668	22.08
65–69	1,760	33,211	0.0177	0.0847	83,277	7,057	399,519	1,520,080	18.25
70–74	2,078	25,220	0.0275	0.129	76,220	9,832	357,994	1,120,560	14.7
75–79	2,677	20,170	0.0442	0.2003	66,388	13,300	300,618	762,566	11.49
80–84	3,347	15,289	0.073	0.3108	53,088	16,498	226,092	461,949	8.7
85+	6,007	12,907	0.1551	1	36,590	36,590	235,857	235,857	6.45
Female									
0–1	195	12,707	0.0051	0.0051	100,000	509	99,548	8,200,752	82.01
1–4	42	51,515	0.0003	0.0011	99,491	108	397,700	8,101,203	81.43
5–9	19	63,003	0.0001	0.0005	99,383	50	496,757	7,703,503	77.51
10–14	25	60,046	0.0001	0.0007	99,333	69	496,483	7,206,746	72.55
15–19	63	62,772	0.0003	0.0017	99,264	166	495,986	6,710,263	67.6
20–24	80	63,716	0.0004	0.0021	99,098	207	494,984	6,214,277	62.71
25–29	78	63,349	0.0004	0.0021	98,891	203	493,973	5,719,293	57.83
30–34	104	57,358	0.0006	0.003	98,688	298	492,718	5,225,320	52.95
35–39	156	53,986	0.001	0.0048	98,390	473	490,880	4,732,602	48.1
40–44	230	54,605	0.0014	0.007	97,917	685	488,056	4,241,722	43.32
45–49	472	64,058	0.0025	0.0122	97,232	1,188	483,489	3,753,666	38.61
50–54	580	65,155	0.003	0.0147	96,045	1,415	476,926	3,270,177	34.05
55–59	827	60,003	0.0046	0.0227	94,629	2,150	467,998	2,793,251	29.52
60–64	1,040	48,907	0.0071	0.0349	92,479	3,223	454,726	2,325,253	25.14
65–69	1,253	35,929	0.0116	0.0566	89,256	5,048	434,217	1,870,527	20.96
70–74	1,596	29,281	0.0182	0.0871	84,209	7,337	403,802	1,436,310	17.06
75–79	2,396	26,131	0.0306	0.1426	76,872	10,959	358,552	1,032,509	13.43
80–84	3,548	22,522	0.0525	0.2333	65,913	15,380	292,885	673,957	10.22
85+	10,608	26,665	0.1326	1	50,533	50,533	381,072	381,072	7.54

T_x: Total number of person-years lived after age x. Since the size of each age-specific stationary population is $_nL_x$, the total size of the stationary population at T_0 is the sum of column L. In general,

$$T_x = \sum {}_nL_x$$

e_x: Life expectancy at age x, calculated by dividing the number of person-years lived beyond age x by the number of persons reaching age x:

$$e_x = \frac{T_x}{l_x}$$

For example, the average remaining life expectancy for those age $25x$ is $5,286,121/98,439 = 53.70$.

A life table is often used by demographers and actuaries to calculate survival ratios. For example, the probability that a woman aged 55 years will live to her 65th birthday is $l_{f65}/l_{f55} = 89,256/94,629 = 0.943$. Here the f subscript before age indicates female. This simple ratio calculation has fairly wide application. If a woman had breast cancer at age 55, her prognosis of 10-year survival is 90%, which is close to the population-based survival probability. The lack of informal support by a spouse can also be calculated using the life table. The chance that a 55-year-old wife dies within 10 years while her 60-year-old husband survives for 15 years can be calculated as the joint probability $[1 - l_{f70}/l_{f55})]*l_{m75}/l_{m60})] = [1 - 89,256/94,629]*(66,388/88,316) = 0.083$ or a very small probability.

To systematically take age-specific survival rates (i.e., an inverse relationship between age and survival rate) into consideration for chronic diseases, it is necessary to calculate the relative survival rate, which is the ratio of the proportion of observed survivors in a cohort of disease-specific patients to the proportion of expected survivors in a comparable set of disease-absent individuals (Brown et al., 1993). In other words, relative survival measures survival probability for a given disease, such as cancer, in the absence of other causes of death. In this regard, relative survival is a hypothetical measure adjusting for the background risk of death. For a state-level relative survival calculation, the background mortality is often obtained from a state life table. To calculate relative survival, it is necessary to divide the observed disease-specific survival during a period of time by the expected survival for the total population. Rutherford et al. (2012) have compared a number of methods of calculating relative survival.

Since the survival calculation is quite cumbersome, age-specific mortality can simply be calculated, so that age-specific mortality ratios can be graphed. Figure 2.1 plots relative mortality

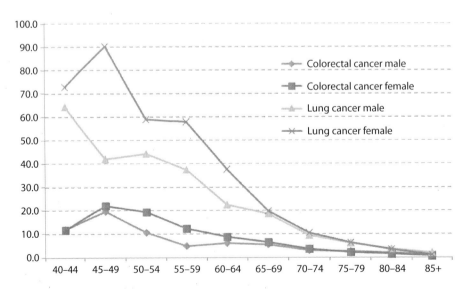

FIGURE 2.1   Age-specific relative mortality ratios of colorectal and lung cancer patients in relation to general populations.

ratios for colorectal and lung cancer for those aged 40 and above, where females had higher relative mortality ratios between ages 45 and 65. After age 70, males and females had almost identical relative mortality ratios. From a demographic perspective, a possible explanation for these ratios is that females are generally expected to outlive males in each age group until age 85; thus, they contribute more than males to denominators. This expectation and the ensuing calculation exacerbate female and male relative mortality ratios for both cancers. It is also worth pointing out that the primary explanation for diminished disparities in older populations is that the least healthy individuals are no longer in the population, and mortality will eventually converge between men and women. Such an explanation by age also applies to other dimensions of disparities.

Since disease-specific mortality may not always be available, for example, for a rare disease, a general life table can be used to calculate relative survival and relative mortality ratios. To calculate age-adjusted survival rates for a particular disease for a state, the state-specific life table should be used.

Many factors contribute to gender differences in mortality. If biological factors and natural risk-taking behavior factors could be excluded, the life table would provide a reference point for adjusting behavior or disease-specific mortality by gender. In Nebraska, behavioral factors disproportionally harmful to men include cigarette smoking, drinking, gun violence, and heavy labor occupational hazards. Diseases related to these factors include lung cancer, unintentional injuries, and homicide. Women experience higher rates of arthritis, asthma, and depression (Table 2.6). Likewise, females are more likely to have a physician, but more often experience difficulties in seeing a physician due to cost. There are also a number of paradoxical facts about men's and women's health that fit the morbidity–mortality paradox: women live longer than men, yet women live with a greater number of comorbidities and have worse self-reported health than men.

Even though gender disparities in health status are many and obvious, reducing these differences requires a greater understanding of structural factors. Women tend to be unemployed, and the unemployed are less healthy. However, there is also a selection effect, the so-called healthy worker effect, in which unhealthy people are disproportionately unlikely to be selected for employment (Ross and Mirowsky, 1995). Women tend to bear more of the family burden of providing physical

TABLE 2.6
Sex Differences in Behavioral and Disease Risk Factors from BRFSS 2006–2010

Males at Greater Risk	Male	Female*
Had 5+ drinks on one occasion in the past month	24.8	13.4
Have driven when had too much to drink	8.5	3.4
Current smoker	19.9	17
Respondents without a personal physician	20.3	10.6
Told they had coronary heart disease by a health professional	5	2.9
Having 5+ servings per day of fruits and vegetables	17.7	26.9
Told they had high blood pressure by a health professional	27.7	23
Told they had a myocardial infarction (MI) by a health professional	5.1	2.4
Never get emotional support	5.2	2.3
Females at Greater Risk		
Could not see a physician due to cost	8.4	12.6
Not physically well 10± days in the past month	8.8	11.4
Told they had a depressive disorder	10.9	20.9
Not mentally well for 10± days in the past month	8	12.5
Told by a health professional they have arthritis	23.7	27.2
Currently have asthma	6.4	9

* All the male–female differences are significant at $p < 0.05$.

and emotional care. Moreover, married women are associated with better health than unmarried or divorced women, because marriage provides material advantages and social support (Elstad, 1996). State health agencies have access to many datasets (e.g., hospital discharge data and Behavioral Risk Factor Surveillance System [BRFSS] survey data), in which gender is the most complete variable. With or without data linkage, most datasets can support many analyses of gender in health. Through linkage between hospital discharge data and mortality data, public health practitioners can also assess follow-up care and health outcome indicators, such as rehabilitation visits after stroke and community-based mortality. For structural inequality and family-based data analyses, most health statistic datasets have some shortcomings that cannot be easily overcome. Marriage can take place anywhere, and the state marriage certificate system rarely captures out-of-state marriages. In addition, divorce is not in the public health data system, at least in the initial registration. Consequently, the data linkage system within a state can find only those who married at a given time point. It is known that hospital discharge data do not include marital status. Although marriage certificate data could be linked to hospital discharge data to assess whether married men are more likely to survive a heart attack or a stroke, incomplete certification information in the marriage certificate system would yield a low-quality study at the population level.

2.4 SOCIOECONOMIC STATUS

In the discussion of gender disparities in health in Section 2.3, we mentioned some relevant structural factors, such as employment status, occupational status, and family roles. These gender disparities, broadly speaking, can all be associated with socioeconomic position or SES (the two terms are used interchangeably). SES can be conceptualized as the social standing or class of an individual or group. Even though the term *standing* or *status* suggests a one-dimensional scale, SES can be measured by education, income or poverty, occupation, and housing, or by combinations of those dimensions. For this reason, SES is often used to measure two-dimensional space.

Health disparities due to socioeconomic conditions were noted more than 150 years ago. In his Census of Boston report in 1846, Lemuel Shattuck linked individual infant mortality to maternal mortality and related the linked records to poor neighborhood conditions (Rosen, 1958). In New York City in 1890, more than 20% of dwellings experienced a tuberculosis (TB)-related death, and the rate was much higher in poorer neighborhoods (Rosner, 1995). After World War II, it was expected that improved overall socioeconomic conditions, especially the introduction of universal health care systems in European counties, would greatly reduce SES-related health disparities. If basic health conditions are related to an income threshold, and if people in an affluent society all live above the threshold, their health conditions would be equalized. This reasoning also dictated the early practice of using a threshold indicator, such as above versus below poverty level. Here, poverty is a control variable to contrast health and health behaviors. When it is introduced, it is expected that health disparities due to poverty will disappear, and in the early years of health disparities research, this variable was almost exclusively placed at an individual level. Although an individual poverty threshold would often reduce some poverty effects on health disparities, many studies found that as an indicator variable, such a threshold explained only a small portion of health disparities.

Threshold measures suggest that there are many SES pathways leading to a better or worse health status. Broader SES concepts and measures of SES are needed to capture various aspects of SES disparities in health status. To understand the multiple pathways linking SES and health, it is necessary to broaden SES to socioeconomic positions that relate an individual's socioeconomic characteristics, such as income, education, wealth and occupation, social environment or social determinants, and life-course exposure and programming, to that individual's health and health behaviors. Furthermore, in order to understand the effect or multiple effects of each socioeconomic position on health, it is necessary to use different measures of SES at different levels of analysis (e.g., individual, family, neighborhood, legal, and regulatory environments).

2.4.1 OCCUPATION

Occupation, income, and education are all associated with health outcomes. In the early years of health disparities research, occupation was most often used to highlight disparities. For instance, it was reported that during the industrialization period, manual laborers and farm laborers had worse standard mortality than people in white-collar occupations. In addition, some occupations were directly related to diseases, such as chimney sweeps having excessive cancer (Gustavsson et al., 1988) and coal miners in Appalachia having excessive black lung disease (Friedl, 1982). However, occupational status misses a significant segment of the population: the unemployed and those not in the labor force often have health deficits. As occupations have shifted to more service-related occupations, occupations in which workers have direct contact with natural and man-made hazards have become fewer. Consequently, occupation disparities in health have become less apparent than disparities related to education or income measures.

2.4.2 INCOME AND EDUCATION

Income and education represent two interrelated but distinct SES measures. To show how income level and educational attainment might be related, we combined the years 2011 and 2012 and used weighted samples from the Nebraska BRFSS that had both variables. Our interest was current use of mammography, which was defined as having had a mammogram within the past 2 years. We cross-tabulated education and income and presented the rate of breast cancer screening using mammography among women aged ≥40 years (Table 2.7). First, the aggregated income effects, shown as percentages in the last row, were comparable to the aggregated education effects (last column). The diagonals, which show the combined income and education effect, had the steepest gradient in current use of mammography. The lowest income and education categories together had the lowest screening rate (30.3%), while the highest income and education categories together yielded the highest screening rate (57.6%). Controlling for educational attainment would give us an income gradient (in screening) steeper than the educational attainment gradient we would get if we controlled for income level.

Even though income inequalities have a simple conceptualization, the income variable is often not available at the individual or patient level. When the variable is available, mostly from survey data, it suffers from a high percentage of missing values. For instance, the BRFSS 2011–2012 data had 13.4% missing values for the income variable, while the educational attainment had <0.16% missing values. The former should be treated, while the latter values can be ignored in most analyses. To overcome the lack of income variables, studies of income inequality and health have often used aggregated income variables, such as median family or household income and percent of people under the federal poverty guideline from a geographic unit. The aggregated measures usually

TABLE 2.7

Having a Mammogram in the Past 2 Years: Income and Education

Education/Income	<15	15–24.999	25–49.999	>50	Column Total (%)
Less than high school	30.3	37.8	30.4	49.6	32.7
High school	37.0	48.2	48.4	52.3	44.8
Some college	37.8	45.3	48.4	48.8	45.1
College degree	41.8	46.4	53.6	57.6	53.5
Row total (%)	36.5	46.2	50.0	53.9	46.6
Total N	8,692	7,939	4,051	4,987	25,669

Source: Nebraska Department of Health and Human Services. Nebraska BRFSS 2011–2012, weighted by age, sex, and race or ethnicity.

Income levels are in 1,000.

follow census geography—state, county, census tract, and census block group—with a lower unit being nested within an upper unit. For instance, a census tract has to be within a county, and it cannot cross the boundary line. However, many studies also use zip codes, which do not follow census geography. Early work in this area used an area income variable as either a control or a proxy measure for individual income. However, marked health inequalities among different geographic units, as large as a country and as small as a neighborhood with different income levels, led to the conceptualization of social environments and health. It is believed that the social environment influences an individual's behavior through social cohesion, social capital, neighborhood resources for health promotion, and community hygiene (Berkman and Kawachi, 2000, p. 7). These effects are cumulative, especially for children during the critical stages of their early development. Malnutrition, substandard housing, witnessing violence, and early childhood traumatic experiences make the individual vulnerable or resistant to various diseases in adulthood.

There have been many measures of neighborhood SES and other environmental variables, but no single variable can capture the many different dimensions of the social environment. Social capital and community cohesion are often measured jointly through community-based surveys. Neighborhood resources can be measured by the accessibility of playgrounds, parks, and healthy groceries or food. However, a measure of the individual social environment that is appropriate from one perspective may be inappropriate from another perspective. This problem is exacerbated when two locales have no social environment factors in common. A lakeside community and a mountain community have quite different built environments. Retirement neighborhoods, such as those around Sun City, Arizona, have quite different demographic and economic characteristics than a typical suburban community. For this reason, composite measures have been developed.

The Townsend Deprivation Index is an early index developed in the United Kingdom. It is based on four population census ward variables (a census ward in the UK has about 2000 people): households without a car, households not owner occupied, overcrowded households (more than one person per room), and persons unemployed. All measures are standardized, with the latter two also being log-transformed. Although the index is widely applied in different countries, the measure cannot be compared between countries or even within a country. In the UK, the index is less applicable in rural areas, because rural wards are more socially heterogeneous than urban wards, and some elements of the index do not vary much in rural areas. For example, car ownership is more or less a necessity for rural residents due to travel distance and lack of public transportation. This fact suggests that geographical accessibility to essential services, such as food and primary care physicians, may be more meaningful than census-based measures. However, because accessibility measures are not usually included in census data, indices that include accessibility (e.g., the Index of Multiple Deprivation) have not been as widely used as the Townsend index. In the United States, both unemployment and overcrowding measures are also not very effective. The former varies widely between years and among different regions, and the latter cannot be directly applied due to different housing conditions in different regions. In addition, many scholars have attempted to develop composite SES indicators by following a multivariate analysis approach, such as factor or principal component analysis (Hightower, 1978; Steenland et al., 2004; Havard et al., 2008; Fukuda et al., 2004). However, due to lack of meaningful and comparable explanations for a number of derived factor dimensions, such an approach has very limited application in the United States. Consequently, most composite indices have not been used frequently in the United States for neighborhood social epidemiology analyses.

In the United States, some consensus was reached after the Public Health Disparities Geocoding Project was conducted at Harvard University. The project aimed to determine appropriate socioeconomic measures and geography levels for monitoring socioeconomic disparities in various health outcomes. Based primarily on birth outcome data from Massachusetts and Rhode Island, poverty level at the census tract level was determined to be the most suitable measure for a variety of area-based disparity assessments. The recommended cutoff values are 5%, 10%, 15%, and 20% of a population. A neighborhood in a census tract that has 5% of people living under the poverty level is considered well off, but a neighborhood in a census tract with more than 20% of people living under the poverty level is likely in disrepair, and residents may be suffering many health deficits.

Even though the single measure of poverty rate is straightforward, some cautionary notes are necessary. First, poverty guidelines for the 48 contiguous states and the District of Columbia are the same, while the cost of living can be quite different between coastal states and rural states. However, for a regional study with a number of adjacent states, this difference poses little problem. Second, although the poverty rate at the county level is updated annually by the Census Bureau, the rate at the census tract level is not normally readily available or linkable to administrative data, which tend to be based on zip codes or other administrative units. For this reason, large-scale, neighborhood-level data analysis of social environments is not common in the United States.

Using the standard five poverty levels to categorize each census tract, Figure 2.2 displays age- and sex-adjusted odds ratios for lung cancer incidence between 2005 and 2009 according to the five poverty levels of the ACS 2005–2009. The reference category is poverty level below 5%. As can be seen, lung cancer incidence follows a gradient: those living in poorer neighborhoods are much more likely to be diagnosed with lung cancer, due perhaps to more people in poor neighborhoods being smokers and to poor neighborhoods tending to have worse indoor and outdoor air quality.

We will discuss how to calculate these odds ratios and other measures in Chapter 7, but for now, let's take a look at another group of conditions, called avoidable hospitalizations, according to poverty level. Avoidable hospitalizations are those hospital visits that could be avoided if adequate care were provided in an ambulatory setting, for example, for conditions such as hypertension and diabetes. Since poorer people rely more on emergency room visits than on primary care, it is expected that poorer neighborhoods would have a higher avoidable hospitalization rate. Indeed, compared to a neighborhood with 5% of the people under the poverty level, those from neighborhoods with 20% or more people under the poverty level are 1.5 times more likely to be hospitalized due to a lack of primary care.

Poverty data at the county level are released by the Census Bureau every year. At the census tract level, the data were released every 10 years between 1970 and 2000. In the 2010 census, questions about income and other SES variables were no longer asked. The Census Bureau instead collects SES data through the annual ACS. At the end of 2010, the 5-year combined ACS sample of 2005–2009 was released. Since the annual ACS sample was about 1/40, the 5-year sample would at most be 5/40, or 12%, which is much smaller than the 16% long-form sample from the 2000 census. For this reason, a margin of error accompanies every census ACS variable. This uncertainty should be considered when using ACS data. Researchers should also be aware that the poverty population

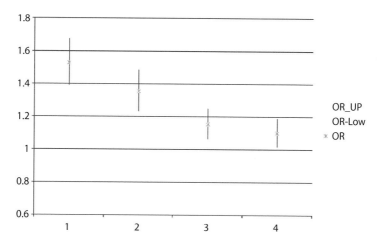

FIGURE 2.2 Odds ratios of all stages of lung cancer by neighborhood poverty level.

differs slightly from the census population because the estimate of poverty population excludes those residing in institutions and group quarters, while a decennial census includes them. According to the Census Bureau, poverty status was determined for all people except institutionalized people, people in military group quarters, people in college dormitories, and unrelated individuals under 15 years old. These groups were excluded from the numerator and denominator when calculating poverty rates. For this reason, the poverty population by age group or other demographics will always be smaller than the census population.

Another neighborhood measure, preferred by economists, is the Gini index, which measures the observed distribution of income among individuals within an area unit against the equal distribution. In contrast to absolute poverty, we can treat the Gini index as a relative income distribution within a neighborhood. If a neighborhood includes individuals with very low income and social position, as well as individuals who enjoy greater wealth, the income inequality, even in an affluent neighborhood, can cause health disparities. The Gini index is the most commonly used inequality measure. A Lorenz curve plots the cumulative percentages of total income received against the cumulative number of recipients, starting with the poorest individual (Figure 2.3). The Gini index measures the area between the Lorenz curve and a hypothetical line of absolute equality, expressed as a percentage of the maximum area under the line. Thus, a Gini index of 0 represents perfect equality, while an index of 100 implies perfect inequality.

In Figure 2.4, the Lorenz curve plots the cumulative income share on the vertical axis against the distribution of the population on the horizontal axis. In this example, 40% of the population obtains around 20% of the total income. If each individual had the same income, that is, total equality, the income distribution curve would be the straight line in the graph—the line of total equality. The Gini coefficient is calculated as area A divided by the sum of areas A and B. If income is distributed completely equally, then the Lorenz curve and the line of total equality are merged and the Gini coefficient is zero. If one individual receives all the income, the Lorenz curve would pass through the points (0, 0), (100, 0), and (100, 100), and surfaces A and B would be similar, leading to a value of 1 for the Gini coefficient.

The Gini index is much more difficult to calculate than a simple poverty percentage or other simple inequality measure, such as the index of dissimilarity. This difficulty is especially true at the census tract level, as it requires, ideally, personal or household income for each individual or household. However, this disadvantage has now become trivial, because the ACS, which will be described in more detail in Chapter 6, provides the Gini index measures at the census tract level. Another disadvantage of the Gini index is that it is not additive across neighborhoods. In other words, the total Gini value of a city is not equal to the sum of the Gini indices from all neighborhoods. Finally,

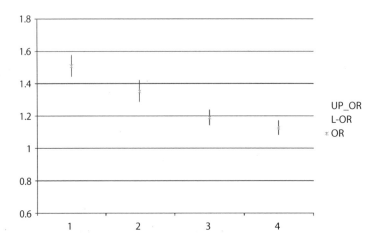

FIGURE 2.3   Odds ratios of avoidable hospitalization by neighborhood poverty level.

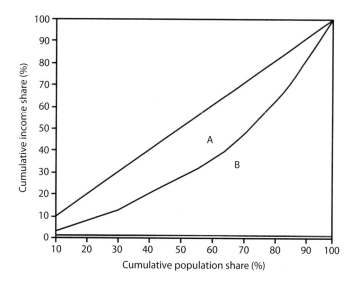

FIGURE 2.4 Lorenz curve of income distribution.

the Gini index has no direct implications in income. A city can have a very high income level, but its Gini index can be very high too. Hence, unlike the absolute poverty measure, a high Gini index at the census tract level may not always be related to adverse health outcomes.

In general, however, both the Gini index and poverty level measures can provide good group comparisons of health outcomes. Table 2.8 cross-tabulates two relationships: the Gini index versus LBW and poverty versus LBW. The data are based on census tracts for Douglas County, where Omaha is located. Both the Gini index and the poverty data are based on the ACS 2006–2010 pooled data. In the context of exploratory analysis, we use the quintiles for the Gini index cutoff points because we do not have empirical and theoretical justifications otherwise. Since the poverty cutoff points are already established, we opted to use the standard percentages of 5, 10, 15, and 20 as cutoff points for poverty level. The results of the LBW rates are shown in each cell, with the column and row averages at each margin. As can be seen, both the Gini index and the poverty rate categories can pick out LBW gradients. The average LBW rates according to the Gini index categories are from 0.063 to 0.090, while the average rates according to the poverty levels are from 0.066 to 0.10. Depending on conceptual frameworks, both measures can effectively show the effects of SES on LBW.

TABLE 2.8
LBW Rates by Poverty and Gini Index (Omaha, 2005–2011)

Poverty/Gini	<0.32	0.32–0.38	0.38–0.42	0.42–0.45	>0.45	Average	Total Population
<5	0.060	0.073	0.075	0.067	0.070	0.066	14,163
5–10	0.066	0.072	0.072	0.073	0.068	0.070	12,859
10–15	0.067	0.080	0.080	0.072	0.083	0.076	7,959
15–20	0.074	0.077	0.086	0.094	0.063	0.082	5,198
>20	No values	0.093	0.101	0.102	0.101	0.100	17,921
Average	0.063	0.079	0.086	0.087	0.090	0.080	
Total population	13,706	11,030	12,266	12,167	8,931		58,100

2.5 OTHER DIMENSIONS OF HEALTH DISPARITIES

In the remainder of the chapter, we briefly discuss rural–urban disparities and LEP health disparities; both dimensions share access to care issues. We omit disability because the census data and ACS data routinely release disability data at county and census tract levels by major demographic and SES dimensions. In addition, BRFSS also includes a disability question, so we decided not to deal with this dimension in this book.

2.5.1 Rural–Urban Health Disparities

Two distinct disparities issues in rural–urban health are often studied: the lack of access to care, especially to primary care physicians, and rural-specific risk factors. Factors of rural areas in the United States include the following:

- Rural residents generally lack access to health care due to distance, lack of employer-sponsored insurance, and low health literacy on preventive care (Hartley et al., 1994).
- Rural residents are more likely than urban residents to be on supplemental health insurance due to injury and other health care coverage.
- Rural areas have a disproportionate number of motor vehicle crash injuries, especially severe injuries or death.
- Rural residents face particular occupational health hazards, such as agricultural chemical poisoning and contamination, and agricultural injuries.
- Rural residents are much more likely than urban residents to die from unintentional injuries other than motor vehicle crash injuries.
- Rural residents are more likely than urban residents to be smokers and alcohol abusers and less likely to wear a seat belt.
- Rural residents, especially women, exercise less than urban residents and are more likely to be obese (Wilcox et al., 2000).
- Rural residents, especially the rural elderly, tend to have poorer health than the urban elderly, but the gaps have been narrowing in recent decades (Mainous and Kohrs, 1995; Morgan, 2002; Hartley, 2004).

There are several definitions of rural areas, and most are county based. The U.S. Census Bureau uses urbanized areas (50,000 people or more) and urban clusters (>2,500 and <50,000 people) to define urban areas and considers all other areas rural. However, since this definition does not follow city or county boundaries, it is rarely used in public health analyses that rely on administrative data. The OMB uses counties as a basis for rural–urban designation. Prior to 2003, only metropolitan and nonmetropolitan counties were used. Since 2003, the definition has included micropolitan counties. A metro area contains a core urban area of 50,000 or more population, and a micro area contains an urban core of at least 10,000 but less than 50,000 population. If a county is not a metro or a micro county, then it is a rural county. Besides these common definitions, rural–urban continuum codes, which include nine county-based categories, are also used quite often. Essentially, the OMB metro counties are further divided into three metro categories, and the nonmetro counties are further divided into six nonmetro categories. Further divisions are based on population size and adjacency to metro counties. The rural–urban continuum codes were originally developed in 1974, and they were called Beale codes in the early days. They have been updated each decennial since (1983, 1993, 2003, 2013), and they can be downloaded from the U.S. Department of Agriculture (USDA) Economic Research Service (ERS) website.

There is also a rural–urban definition based on census tracts: rural–urban commuting area codes (RUCAs). RUCAs consider not only the population size of a community, but also its relationship to large urban centers. Starting in 2010, the decennial census no longer collects commuting data.

Therefore, commuting patterns come from the ACS. The classification contains two levels. Whole numbers (1–10) delineate metropolitan, micropolitan, small town, and rural commuting areas based on the size and direction of the primary (largest) commuting flows. These 10 codes are further subdivided based on secondary commuting flows, providing flexibility in combining levels to meet varying definitional needs and preferences. Descriptions of the codes are found within the data files and also in the documentation. Some federal programs have identified areas with a RUCA code of 4 and above as nonmetropolitan. The Office of Rural Health Policy accepts the RUCA methodology in determining rural eligibility for its programs. The ERS has an Excel file on its website whereby an applicant can link to a table of RUCA codes (2010) by state census tracts or all U.S. census tracts.

In Nebraska, county-based definitions are most frequently used to identify rural and urban areas. In addition to metro and nonmetro categories, the three categories of metropolitan, micropolitan, and rural have recently become more common. In Table 2.9, we used the three-category definition. The metropolitan areas are nine counties, two of which have a city of 50,000 or more residents and seven of which are metropolitan "outlying" counties. Micropolitan areas consist of the 10 Nebraska counties that are not metropolitan and have at least one city of 10,000 or more residents. Rural areas include all of the remaining 74 counties in Nebraska.

At this point, there is no overall rural disadvantage in health outcomes and risk factors, judging from the BRFSS and mortality data. The 5-year mortalities from 2006 to 2010 are comparable among the three types of residents, with an age-adjusted death rate in Nebraska of 736.0 deaths per 100,000 population. The rates for metropolitan, micropolitan, and rural areas are 734.5, 748.6, and 736.0, respectively. Rural residents are at greater risk for mortality from heart disease, unintentional injury overall, and motor vehicle crashes, whereas urban residents are more likely to die from cancer overall, lung cancer, homicide, and drug abuse.

Table 2.9 lists some factors selected from the Nebraska BRFSS from 2008 to 2010. As can be seen, rural residents had a higher rate of no health care coverage: 11.32 in rural areas versus 16.24 in micropolitan areas. In addition, rural respondents (7.4%) were more likely than metropolitan respondents (6.0%) to say they never get the social and emotional support they need. Even though rural residents were less likely to have a personal physician, they tended to have similar rates of

TABLE 2.9
Rural and Urban Differences in Selected Risk Factors from BRFSS 2008–2013

Access Factors	Metropolitan		Micropolitan		Rural	
	%	SE	%	SE	%	SE
No health care coverage	14.58	0.77	16.24	0.99	11.32	0.61
No personal physician	17.82	0.4	15.77	0.36	14.22	0.25
Had a routine checkup past year	60.47	0.47	57.56	0.44	58.63	0.33
Visited a dentist < 2 years	84.39	0.6	79.07	0.59	77.85	0.46
Last didital rectum exam < 2 years	50.55	1.19	44.64	0.97	43.04	0.8
Had a sigmoidoscopy/colonoscopy[a]	62.98	0.76	52.84	0.61	50.58	0.5
Had prostrate-specific antigen (PSA) test < 2 years	52.02	1.21	52.4	1	53.74	0.83
Behavioral Factors						
Current smoker	19.05	0.38	18.53	0.36	16.7	0.26
Had drinks in past month[b]	6.45	0.23	5.52	0.22	5.73	0.17
Had too much to drink while driving	5.5	0.46	5.77	0.46	6.53	0.35
Always use seat belts	79.62	0.4	64.24	0.47	56.05	0.36

[a] Age 50+.

[b] Males age 60+ and females age 30.

routine checkups as urban residents. Even though rural residents had a lower colorectal cancer screening rate, their prostate cancer screening rate was comparable to that of urban residents.

For general health status, urban residents were less likely than rural residents to report poor general health and lack of social and emotional support. Micropolitan (13.4%) and rural (12.9%) residents were more likely than metropolitan residents (10.9%) to state that their health was "fair" or "poor." However, the difference was <2%. In addition, rural residents were generally disadvantaged in smoking status, drinking, drunk driving, and use of a seat belt.

There are a number of rural health agencies, centers, and rural health associations at both the national and the state level. However, most of these organizations tend to address health care needs through health care services (e.g., primary care physicians, critical care hospitals, and emergency medical services). The Federal Office of Rural Health Policy and its sponsored centers, state offices of rural health, national and state rural health associations, and Agency for Healthcare Research and Quality (AHRQ) all tend to focus on health care access, quality of care, and telemedicine. Very few organizations specifically serve rural public health needs. Consequently, there has been little change in behavioral and lifestyle risk factors among rural residents, although some measures of health status have improved through greater access to care.

Finally, pertinent to the classification used in this book, it is necessary to mention a 1990 urban–rural classification that classifies metropolitan counties with a central city of 50,000 or more or a county population greater than 100,000. We found that this scheme is convenient in Nebraska because only three counties were qualified as metropolitan or urban counties in 2000 and 2010. In addition, 100,000 is a criterion for area units that can be considered for public use without confidentiality concerns. This confidentiality protection scheme was used by the CDC's national mortality and natality files and census public use microdata areas (PUMAs). In later chapters, we use this scheme for ease of presentation and for minimizing confidentiality concerns.

2.5.2 LEP DISPARITY

LEP is a recently added disparity dimension for contrasting LEP individuals with others. Individuals whose primary language is not English and who have a limited ability to read, speak, write, or understand English are considered LEP. Executive Order 13166, "Improving Access to Services for Persons with Limited English Proficiency," requires Federal agencies to examine the services they provide, identify any need for services to LEP persons, and develop and implement a system to provide those services so persons with LEP can have meaningful access to them. Further legal framework for language access in health care is reviewed by Chen et al. (2007), and a review of LEP on clinical research factors is provided by Bustillos (2009).

Lack of access to care and preventive medical services among immigrants and LEP individuals has been an issue for a long time. However, before Executive Order 13166 in 2000, there were few studies that explicitly documented and assessed access to care and health status among LEP individuals. Kandula et al. (2004) reviewed 10 leading health indicators among immigrants in the United States. The 10 indicators are physical activity, overweight and obesity, tobacco use, substance abuse, responsible sexual behavior, mental health, injury and violence, environmental quality, immunizations, and access to health care. The study found that (1) data about the health of immigrants were limited and often did not distinguish between the U.S. and foreign born, (2) differences in culture and language both inhibit providers' understanding of issues and limit immigrants' ability to take full advantage of disease prevention and health intervention services, and (3) immigrants were substantially disadvantaged for access to care. In the same year, Yeo (2004) reviewed how language barriers contribute to health disparities among ethnic and racial minorities in the United States. It was found that language barriers are associated with lack of awareness about health care benefits, such as Medicaid eligibility, longer visit time per clinic visit, less frequent clinic visits, less understanding of the physician's explanations, more lab tests, more emergency room visits, and less satisfaction with health services. It was also pointed out that it is not clear if these negative consequences are related to health outcomes or disease incidences.

Studies of health status and risk factors among LEP individuals are more recent. Data at the national level are mainly from two sources. One is the Behavioral Risk Factor Surveillance System (BRFSS), which is an annual national survey conducted by each state. The BRFSS has a Spanish-language instrument that can be used to indicate language barriers among Hispanics and Latinos. Studies employing this strategy include DuBard and Gizlice (2008) and Linn et al. (2010). The other is Medical Expenditure Panel Survey (MEPS), which also has a Spanish-language instrument that can be used to indicate language barriers among Hispanics and Latinos. Studies using MEPS include Cheng et al. (2007) and Brach and Chevarley (2008). Data sources at the state level are rare, and the California Health Interview Survey (CHIS) is a valuable source. CHIS covers all the counties in California, and it includes five languages for non-English-speaking adults (Spanish, Mandarin, Cantonese, Vietnamese, and Korean). Studies using the CHIS for LEP or language barriers to health care studies are numerous, and they include Gee and Ponce (2009), Ponce et al. (2006), and Maslanda et al. (2011). California also leads the nation in improving access to language services for LEP individuals (Moreland et al., 2012). These studies consistently found that LEP individuals are disadvantaged in access to health care and preventive care services, and they tend to be worse off in health conditions (e.g., self-rated health status, arthritis, and high blood pressure) and behavioral risk factors (e.g., physical activity and fruit and vegetable intake).

Although the foreign-born and LEP populations account for a small proportion of the Nebraska total population, they are the fastest-growing populations in Nebraska. For example, the foreign-born individuals in Nebraska increased from 28,198 in 1990 to 74,638 in 2000 and to 106,298 in 2010 (with the 2010 numbers coming from the ACS. In addition, the U.S. census surveys routinely asked questions about languages spoken at home. For those who were not "speaking English only," the census also asked, How well do you speak English? with one of the following four answers to be selected: "very well," "well," "not well," and "not at all." According to the census, about 2.4% of the population age 5+ in Nebraska spoke English less than "very well" in 1990, and in 2010, this percentage increased to 4.4% (Table 2.10).

Even though recent census releases tend to group "speaking English less than very well" as a single category (implicitly referring to them as LEP individuals), the guidance issued by the U.S. Department of Justice on discrimination against LEP persons (DOJ, 2002) uses speaking English "not well" or "not at all" to define LEP individuals. For this reason, we tend to refer to LEP individuals as those who speak a language other than English at home and who speak English "not well" or

TABLE 2.10

Language Ability by Age and Language Spoken at Home in Nebraska

Censuses	Age	Speaks English Only	Speaks Non-English Language at Home			
			Very Well	Well	Not Well	Not at All
Census	5–17	298,450	7,933	10,911	6,065	113
1990[a,b]	18+	1,090,582	39,687	11,745	6,369	815
Census	5–17	304,915	17,124	6,398	3,835	780
2000[c]	18+	1,164,131	50,758	18,882	18,733	9144
ACS	5–17	283,013	29,965	7,370	3,204	369
2006–2010[d]	18+	1,224,692	58,546	23,686	25,711	13,159

[a] http://www.census.gov/hhes/socdemo/language/data/census/table2.txt.

[b] http://www.census.gov/hhes/socdemo/language/data/census/table3.txt.

[c] STF4 Age by Language Spoken at Home (PCT038) for Nebraska.

[d] ACS Table B16004 for Nebraska.

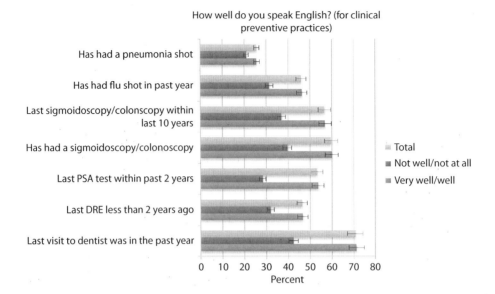

FIGURE 2.5 Some indicators of LEP population access to health care. DRE, digital rectal examination. Since very few LEP persons were sampled, the total and the well/very well population are almost identical. (Data from Nebraska Department of Health and Human Services. Nebraska Behavioral Risk Factor Surveillance System (BRFSS) 2008–2010. Nebraska DHHS.)

"not at all." Using this definition, we find that Nebraska LEP adults increased rapidly from 18,929 in 1990 to 46,759 in 2000 and to 62,556 in 2010 (Table 2.10). In other words, 4.6% of adults in Nebraska were considered LEP individuals according to the ACS of 2006–2010.

At this point, with the exception of disability from the ACS data, BRFSS is the only source to study risk factors among LEP individuals at the state level. The BRFSS has a Spanish-language instrument that can be used to indicate language barriers among Hispanics and Latinos. However, the standard BRFSS survey does not have questions identifying the LEP population. From 2008 to 2010, a language ability question was added to the Nebraska BRFSS: "How well do you speak English?" The four response categories were "very well," "well," "not well," and "not at all." Based on this question, we can examine a number of health care accessibility dimensions. Figure 2.5 shows that those who could not speak English well or not at all have deficits in access to the pneumonia shot, flu shot, colon cancer screening, prostate cancer screening, and dental care. Because of limited data sources for LEP, we will not study it in later chapters.

2.6   CHAPTER SUMMARY

In this chapter, we reviewed major health disparity dimensions (race, gender, SES, rural–urban, and LEP) and provided empirical evidence whenever possible by using data from Nebraska. We noted that race classification has become more complicated since the 2000 census. It affects the race-based, at-risk population at all levels, but most critically below the county level, because the CDC routinely provides a race bridge file above the county level.

Gender-based health differences can be due to either biological reasons, which are not health disparities, or social, occupation, and family positions, which are gender health disparities. There is also a long-known phenomenon often labeled the morbidity–mortality paradox. Women live longer than men, yet women live with a greater number of comorbidities and have worse self-reported health than men. Because of life expectancy at birth for males and females, one needs to be aware of adjusted survival, for which a state-specific life table is often used.

Within SES disparities, both educational level and personal income are not available for most datasets, except some survey data and birth certificate data. To perform a more uniformed assessment of health disparity, census tract–based SES measures seem more realistic. In other counties, employment, housing, and other factors were used to derive a deprivation index. In the United States, the census tract–based poverty level and Gini index are readily available; a more suitable measure was the poverty level, with the recommended cutoff values being 5%, 10%, 15%, and 20% of a population. A neighborhood in a census tract that has 5% of people living under the poverty level is considered well off, and a neighborhood with more than 20% of its people under the poverty line is considered very poor.

Finally, we describe some measures and empirical findings about rural–urban differences and LEP and non-LEP differences, mainly along the line of access to health care. For both rural and LEP people, lack of access to care is a significant issue.

REFERENCES

Berkman, L.F. and I. Kawachi, eds. 2000. *Social Epidemiology*. New York: Oxford University Press.

Brach, C. and F.M. Chevarley. 2008. Demographics and health care access and utilization of limited-English-proficient and English-proficient Hispanics. Research Findings No. 28. Rockville, MD: Agency for Healthcare Research and Quality, February. http://meps.ahrq.gov/mepsweb/data_files/publications//rf28/rf28.pdf.

Brown, B.W., C. Brauner, and M.C. Minnotte. 1993. Noncancer deaths in white adult cancer patients. *Journal of the National Cancer Institute* 85(12): 979–987.

Bustillos, D. 2009. Limited English proficiency and disparities in clinical research. *Journal of Law, Medicine and Ethics* 37: 28–37.

Chen, A.H., M.K. Youdelman, and J. Brooks. 2007. The legal framework for language access in healthcare settings: Title VI and beyond. *Journal of General Internal Medicine* 22(2): 362–367.

Cheng, E.M., A. Chen, and W. Cunningham. 2007. Primary language and receipt of recommended health care among Hispanics in the United States. *Journal of General Internal Medicine* 22(Suppl. 2): 283–288.

DOJ (Department of Justice). 2002. Guidance to federal financial assistance recipients regarding Title VI prohibition against national origin discrimination affecting limited English proficient persons. Washington, DC: DOJ, June 18. http://www.usdoj.gov/crt/cor/lep/DOJFinLEPFRJun182002.htm.

DuBard, C.A. and Z. Gizlice. 2008. Language spoken and differences in health status, access to care, and receipt of preventive services among US Hispanics. *American Journal of Public Health* 98(11): 2021–2028.

Elstad, J.I. 1996. Inequalities in health related to women's marital, parental, and employment status: A comparison between the early 70s and the late 80s, Norway. *Social Science and Medicine* 42(1): 75–89.

Friedl, J. 1982. Explanatory models of black lung: Understanding the health-related behavior of Appalachian coal miners. *Culture Medicine and Psychiatry* 6(1): 3–10.

Fukuda, Y., K. Nakamura, and T. Takano. 2004. Municipal socioeconomic status and mortality in Japan: Sex and age differences, and trends in 1973–1998. *Social Science and Medicine* 59(12): 2435–2445.

Gee, G.C. and N. Ponce. 2010. Associations between racial discrimination, limited English proficiency, and health-related quality of life among 6 Asian ethnic groups in California. *American Journal of Public Health* 100(5): 888–895.

Gustavsson, P., A. Gustavsson, and C. Hogstedt. 1988. Excess of cancer in Swedish chimney sweeps. *British Journal of Industrial Medicine* 45(11): 777–781.

Hartley, D., L. Quam, and N. Lurie. 1994. Urban and rural differences in health insurance and access to care. *Journal of Rural Health* 10(2): 98–108.

Hartley, D.A. 2004. Rural health disparities, population health, and rural culture. *American Journal of Public Health* 94: 1675–1678.

Havard, S., S. Deguen, J. Bodin, K. Louis, O. Laurent, and D. Bard. 2008. A small-area index of socioeconomic deprivation to capture health inequalities in France. *Social Science and Medicine* 67(12): 2007–2016.

Hightower, W.L. 1978. Development of an index of health utilizing factor analysis. *Medical Care* 16(3): 245–255.

Kandula, N.R., M. Kersey, and N. Lurie. 2004. Assuring the health of immigrants: What the leading health indicators tell us. *Annual Review of Public Health* 25: 357–376.

Linn, S.T., J.M. Guralnik, and K.V. Patel. 2010. Disparities in influenza vaccine coverage in the United States, 2008. *Journal of the American Geriatric Society* 58(7): 1333–1340.

Mainous, A.G., III, and F.P. Kohrs. 1995. A comparison of health status between rural and urban adults. *Journal of Community Health* 20: 423–431.

Masland, M.C., S.H. Kang, and Y. Ma. 2011. Association between limited English proficiency and understanding prescription labels among five ethnic groups in California. *Ethnicity and Health* 16(2): 125–144.

Mays, V.M., N.A. Ponce, D.L. Washington, and S.D. Cochran. 2003. Classification of race and ethnicity: Implications for public health. *Annual Review of Public Health* 24: 83–110.

Moreland, C.J., D. Ritley, J.A. Rainwater, and P.S. Romano. 2012. State policies and language access in California's HMOs: Public reporting and regulation of HMOs' language services. *Journal of Health Care for the Poor and Underserved* 23(1): 474–498.

Morgan, A. 2002. A national call to action: CDC's 2001 urban and rural health chart book. *Journal of Rural Health* 18: 382–383.

Nebraska Department of Health and Human Services. 2011–2012. Nebraska Behavioral Risk Factor Surveillance System (BRFSS). Nebraska DHHS.

OMB (Office of Management and Budget). 1977. Race and ethnic standards for federal statistics and administrative reporting. OMB Directive 15. Washington, DC: OMB, May 12. http://wonder.cdc.gov/wonder/help/populations/bridged-race/Directive15.html (accessed August 13, 2014).

OMB (Office of Management and Budget). 1997. Revisions to the standards for the classification of federal data on race and ethnicity. Washington, DC: OMB, October 30. http://www.whitehouse.gov/omb/fedreg/1997standards.html (accessed February 11, 2013).

Plane, D.A. and P.A. Rogerson. 1994. *The Geographical Analysis of Population: With Applications to Planning and Business.* New York: Wiley.

Ponce, N.A., R.D. Hays, and W.E. Cunningham. 2006. Linguistic disparities in health care access and health status among older adults. *Journal of General Internal Medicine* 21(7): 786–791.

Rosen, G. 1958. *A History of Public Health.* New York: MD Publications.

Rosner, D. 1995. *Hives of Sickness: Public Health and Epidemics in New York City.* New Brunswick, NJ: Rutgers University Press.

Ross, C.E. and J. Mirowsky. 1995. Does employment affect health? *Journal of Health and Social Behavior* 36: 230–243.

Rutherford, M.J., P.W. Dickman, and P.D. Lambert. 2012. Comparison of methods for calculating relative survival in population-based studies. *Cancer Epidemiology* 36(1): 16–21.

Steenland, K., J. Henley, E. Calle, and M. Thun. 2004. Individual- and area-level socioeconomic status variables as predictors of mortality in a cohort of 179,383 persons. *American Journal of Epidemiology* 159(11): 1047–1056.

Wallman, K.K. and J. Hodgdon. 1977. Race and ethnic standards for federal statistics and administrative reporting. *Statistical Reporter/Office of Federal Statistical Policy and Standards* 1977(77–110): 450–454.

Wilcox, S., C. Castro, A.C. King, R. Housemann, and R.C. Brownson. 2000. Determinants of leisure time physical activity in rural compared with urban older and ethnically diverse women in the United States. *Journal of Epidemiology and Community Health* 54(9): 667–672.

Yeo, S. 2004. Language barriers and access to care. *Annual Review of Nursing Research* 22: 59–73.

3 Common Public Health Data in a State Health Department

3.1 INTRODUCTION

Many public health core functions are based on collection and assessments. Public health programs usually service at least one public health core function, and many of them are inherently data-intensive. Complete, accurate, comprehensive, and timely information is the backbone of public health policy, practice, and research. Policy makers, program managers, providers, and the general public need to be able to easily access up-to-date information in order to develop effective policy and practice. With few exceptions, most public health data from the state public health agency are administrative data, which can be categorized as (1) program-based record systems (e.g., vital records), (2) health provider–based systems (e.g., hospital discharge data [HDD]), and (3) payer-based systems (e.g., Medicare and Kaiser Permanente). In this chapter, we introduce mostly program-based data.

There are numerous data sources that can be used for public health practices, and they are not limited to public health agencies. Public health agencies are authorized and responsible for a variety of administrative and surveillance data collection, storage, analysis and interpretation, and distribution, such as vital records, reportable disease registries, surveys, and medical facility utilization statistics reports. A large amount of data collected by other governmental agencies, nongovernmental agencies, and private organizations provide valuable information such as education, motor vehicle registration, criminal record, agricultural accidents, and medical and health associations. The advantages of using administrative data include

1. It is economical because they are collected by someone already.
2. They are most important because most data collections were mandated or funded by state and federal agencies due to their significance to public health or lifesaving areas.
3. Most data can be used for time series and trend analyses due to many years of data collection.

Program-based public health datasets have some limitations. First, their geographic and population coverages vary widely. Some program data serve the entire population within the jurisdiction, while others are for targeted services or a population segment. Vital statistics are population based, while Women, Infants, and Children (WIC) program data are for a segment of the population. Sometimes, a dataset may start with a region or segment of the population, and then as it expands, it can cover the entire population. An example is emergency medical services (EMS) data that started in major cities (Omaha and Lincoln) in Nebraska and then expanded to include the entire state. The children immunization registry also started with a number of local health departments. Each local health department may not cover the entire school system. The registry now covers almost the entire state and captures nearly 90% of children.

Second, data items or variable coverage may not be sufficient for research needs. For instance, demographic variables included in the Nebraska Parkinson's Disease Registry are age and sex only. There are no other risk factors, such as smoking and occupational status. There are also no disease severity, functional limitation, and comorbidity information.

Third, data quality varies, and people should not expect a clean dataset from many public health programs. Data analysts within the data-producing unit need to be aware of data quality issues and systematically assess them prior to any data analysis. However, it is often that the primary use of a public health dataset is limited to a few data items. If one looks at those data items, the quality may be very good. When data are used by another program internally or externally for secondary purposes, some data quality issues may emerge. Maintaining high-quality program data rests with the program that produces the data, but a program may not have the scope and appreciation for enforcing data quality beyond their primary use. Again, for the Nebraska Parkinson's Disease Registry, high-quality data items are the names and contact information for patients, physicians, and pharmacists. Diagnosis date provided by a physician had more than 25% missing values. If a researcher made a data request without knowing this information, a time series analysis would not be useful. For this reason, it is important to ask about missing values for important variables. In many cases, further data manipulation would improve data quality. For instance, if the diagnosis date was missing, it could be constructed from the Parkinson's disease form submission date or receiving date from a physician, pharmacist, or individual patient. Since the administrative data are not for research, no one would purposefully construct such a composite diagnosis date. Only when data are internally analyzed can such a variable be created.

Karr et al. (2006) outlined three data quality measures: process, the data, and user-friendliness. *Process* refers to the generation, assembly, description, and maintenance of data, such as reliability, metadata, security, and confidentiality, and they are not easy to quantify. Assessing the process often requires accessing the data. The data are the core of the data quality measures. At the record level, they include accuracy, completeness, consistency, and validity. At the database level, they include identifiability and joinability. User-friendliness refers to how users may access and manipulate the data, such as accessibility, integrity, interpretability, rectifiability, and timeliness. Data users often face significant hurdles in terms of data awareness, accessibility, and utilization due to different infrastructures and organizational settings and the disparate nature of the many data sources. Users, including researchers, policy makers, and public health practitioners, usually do not always know what information is available.

From the user's perspective, Love et al. (2008) evaluated multistate HDD, and the authors used four criteria: accessibility, reliability and consistency, analytic utility, and enhancement potential. Accessibility included if the user can access individual-level data, how the state regulations govern public health data use, and data acquisition cost. Data reliability and consistency include how the data are coded. If the data are coded by certified professionals, they are more reliable than data coded by untrained data coders. In terms of HDD, E-code data are less reliable than billing and procedure data, as the latter were coded by certified professionals. For Parkinson's disease, diagnoses from neurologists are more reliable than those from generalists. Data analytical utility includes potential statistical power, the time lag for accessing time-sensitive data, geographic subunit, such as zip code, and availability. Data enhancement potential includes if the data are linkable to other data and can be geocoded to include areal socioeconomic status (SES) information.

To increase user awareness, a series of activities have been conducted to develop a public health data inventory that involves internal public health programs and external partners to explore and identify data sources and follow surveys and interviews to relevant sources of information. A data inventory is a comprehensive list of public health relevant datasets that can be disseminated widely and easily accessed. The webpage of the data center also serves as a bridge to lead users to the appropriate data sources for access and utilization. The inventories have been regularly updated and reviewed and play a key role in promoting awareness, access, and use of relevant information to address various needs. In this chapter, we describe a number of important datasets from the Nebraska Pubic Health Data Inventory. We provide some data quality and usability indicators whenever possible, such as accessibility, reliability, analytic utility, and enhancement potential. We tried to organize datasets first by program data and then by survey data.

3.2 HOSPITAL DISCHARGE DATA

There are two types of hospital discharge records collected from hospitals: inpatient and outpatient data. The outpatient data include those from the emergency department (ED). The Nebraska Department of Health and Human Services (DHHS) normally receives data from the Nebraska Hospital Association (NHA) annually, approximately every October. Hospital discharge records contain information on the date of admission; date of discharge; patient's age, gender, and county of residence; and primary and secondary International Classification of Diseases, Ninth Revision, Clinical Modification (ICD-9-CM) diagnosis codes. Data are used to examine important topics of interest in public health and for a variety of activities by governmental, scientific, academic, and commercial institutions. Each year, more than 200,000 Nebraska residents are discharged from hospitals. Outpatient data are normally available from the NHA upon request.

Information on each hospital discharge is reported from acute care hospitals in Nebraska to the Nebraska Association of Hospitals and Health Systems (NAHHS). The information is reported by hospitals on the UB-04 inpatient and outpatient data form. The data have been collected annually since 1975; however, the geographic coverage of hospitals has only become complete since 2004. Race and ethnicity are not consistently collected because they are not a core uniform billing data element. Effort has been made to add race information through data linkage, but it is very time-consuming. Inpatient data have been geocoded to the census tract level since 2005. HDD are probably the most commonly used data source for public health surveillance, local health status assessment, and planning. They are population-based data primarily reflecting nonfatal conditions.

Our general impression is that HDD have good accessibility, high reliability and consistency, very good analytic utility, and great enhancement potential.

3.3 NEBRASKA CANCER REGISTRY DATA

The purpose of the registry is to gather data that describe how many Nebraska residents are diagnosed with cancer, what types of cancer they have, what type of treatment they receive, and the time and quality of survival after diagnosis. These data are put to a variety of uses both inside and outside of the Nebraska DHHS. The agency closely monitors records from year to year to determine trends that are developing and to see how Nebraska's cancer experience compares to that of the rest of the nation. Cancer clusters either by small areas or by occupations are also occasionally investigated. The agency also uses these data to help plan and evaluate cancer control programs within the agency.

There are about 12,000 new cancer cases each year. With proper data requests, the registry has furnished information to many individuals, institutions, and organizations. In addition to sponsor organizations, such as the Centers for Disease Control and Prevention (CDC) and the North American Association of Central Cancer Registries (NAACCR), common users of the registry data include the University of Nebraska Medical Center, the National Cancer Institute, and the American Cancer Society.

Cancer registry data have been collected monthly from hospitals, clinics, and physicians since 1987. Data quality is generally excellent, and the Nebraska Cancer Registry (NCR) has maintained the NAACCR gold standard since 1995. The data are population based for ongoing surveillance, which includes personal identifiers such as name, residential address, and census tract numbers. However, for most users, they would not need identifiable information. Descriptions of data items or variables are available from the NAACCR website. All state cancer registries follow the same data formats, especially those from non–Surveillance, Epidemiology, and End Results (SEER) states. For SEER states, such as Connecticut, New Jersey, Iowa, and New Mexico, the SEER standard or data format is also used. Commonly used variables are around incidence reporting. Variables about treatment and comorbidity are available, but their qualities are not good.

Cancer registry data have been one of the most frequently requested datasets from outside of the DHHS. Researchers have used it to study incidence, staging trends, and treatment compliance between rural and urban areas. Our general impression is that NCR has very good accessibility, high reliability and consistency, moderate analytic utility, and great enhancement potential. NCR has been linked to HDD, death certificate data, and other datasets, and after the linkage, the analytic utility has increased markedly.

3.4   CRASH OUTCOME DATA EVALUATION SYSTEM

The Crash Outcome Data Evaluation System (CODES) follows a collaborative approach to obtaining medical and financial outcome information related to motor vehicle crashes for highway safety and injury control decision making. CODES links motor vehicle crash records with other medical records in Nebraska, including hospital discharge, EMS records, and death certificate data. Linkage is conducted annually and now provided by request. The database can accommodate special requests. Data are used to better understand the factors contributing to motor vehicle crashes and their associated injury outcomes. The data are population based for ongoing surveillance.

Data are collected by the Nebraska Department of Roads (crash data), the Nebraska Department of Motor Vehicles (driver's license data), the NHA (HDD), and the Nebraska DHHS (EMS and death data). The data have been linked annually since 1996. Usually there is a 1-year lag. There are more than 30,000 motor vehicle crashes each year, with 15,000–18,000 individuals sustaining some injuries and about 300 deaths. Crash site locations are geocoded, and some injured patients are also geocoded to their residential locations.

Our general impression is that CODES has moderate accessibility, fine reliability and consistency, low analytic utility, and great enhancement potential.

3.5   NTR DATA

The Nebraska Trauma Registry (NTR) was established and started to collect data statewide in 2005. Data from 2007 onward tend to have a better quality in terms of geographic coverage and data item completeness. The purpose of having the statewide trauma registry is for ongoing surveillance. The data system also makes the delivery of trauma care cost-effective, reduces the incidence of inappropriate or inadequate trauma care, prevents unnecessary suffering, and reduces the personal and societal burden resulting from trauma.

These data are collected from state-designated trauma centers (hospitals) and hospitals that voluntarily submit information. Data are submitted on a monthly basis by those designated hospitals. Data collected by the trauma registry include patient demographics, prehospital information, injury information, ED and acute care information, hospital procedures and diagnosis, and patient outcome information. The data have been collected monthly since 2005, but the statewide coverage started in 2007. From 2008 to 2013, there were on average more than 3500 in-state trauma cases each year.

There are nearly 200 variables in the trauma registry, which include demographics (age, sex, and race), primary external cause of injury (E-code 1), ED admission- and discharge-related variables (ED arrival date, ED arrival time, ED length of stay, and disposition at ED discharge), hospital admission-related variables, transfer hospital, expression and motor disability at discharge, feeding disability at discharge, locomotion disability at discharge, zip code of injury, injury state, time of injury, primary payer, injury severity score, patient address, and transport mode. However, users need to ask about data quality before making a request. Blood alcohol content is a common variable, but its item missing values are very high, as it is most often tested after a motor vehicle crash injury.

Our general impression is that NTR has good accessibility, moderate reliability and consistency, moderate analytic utility, and great enhancement potential through geocoding and linkage to HDD.

3.6 TRAUMATIC BRAIN AND SPINAL CORD INJURY REGISTRY

Traumatic brain injury (TBI) is a major public health problem in Nebraska and the United States. Inpatient discharges from acute care hospitals in Nebraska who have a diagnosis code indicating a TBI are reported to the Nebraska DHHS. These data sources could be utilized in the future to expand TBI surveillance in the state and can provide additional information on contributing factors for TBI, such as race and ethnicity of individuals treated for TBI.

State law (§81-653 to §81-662) requires that any hospital, rehabilitation center, psychologist, or physician report the following information about the person sustaining the injury to the Nebraska DHHS: name, social security number (if known), date of birth, gender, residence, date of injury, final diagnosis or classification of injury, cause of injury, place where injury occurred, identification of the reporting source, and dispensation upon discharge. Diagnosis and treatment information are collected from the patient's medical record. The data have been collected monthly since 2004. In addition, the NHA provides a subset of HDD records of TBI patients.

3.7 NEBRASKA PARKINSON'S DISEASE REGISTRY

Parkinson's disease (PD) is a chronic motor system disorder. As a result of 1996 legislation, Nebraska became the first and only state to create a population-based Parkinson's disease registry. A registry provides information about the incidence of the disease in the populace. The purpose of the Parkinson's disease registry is to provide a central data bank of accurate, historical, and current information for research purposes. The data are population based for ongoing surveillance. The database has both confirmed and suspected cases. Except from 2004–2006, the NPDR has been collecting new cases each year since 1997. In the 4 years between 2011 and 2014, there were on average 2100 PD cases reported to the NPDR each year; among them, 503 were confirmed cases.

State statute requires that pharmacies and physicians report data directly to the Nebraska DHHS. Pharmacies report information semiannually about patients who received a Parkinson's disease drug that is being tracked as part of the registry. Physicians are required to report information about patients who are newly diagnosed with Parkinson's disease within 60 days of diagnosis, or they are to complete a patient confirmation form for patients submitted by pharmacies or patients who self-report to the registry that they have Parkinson's disease. Patients are also identified for inclusion in the registry using state vital records. Registry data since 1997 are available. There are only two demographic variables: age and sex.

Our general impression is that NPDR has good accessibility, low reliability and consistency, low analytic utility, and moderate enhancement potential.

3.8 NEBRASKA STATE IMMUNIZATION INFORMATION SYSTEM

The Nebraska State Immunization Information System (NESIIS) is a secure, statewide, web-based system that was developed to connect and share immunization information among public clinics, private provider offices, local health departments, schools, hospitals, and other health care facilities that administer immunizations. The data can be used to analyze trends in immunizations, identify pockets of needs, produce reminder and recall notifications, and recommend future immunization needs. It should be noted that not all providers submit data to the system because it is not mandatory.

NESIIS is populated with birth records from the Office of Vital Records. Birth doses of hepatitis B are populated from the Newborn Hearing Screening System (which is part of the vital records database). Real-time data collection started in 2001.

3.9 EMERGENCY MEDICAL SERVICES

Nebraska EMS and trauma rules and regulations provide the basis for the Nebraska data elements contained in the Nebraska EMS database. The mission of the EMS program is to strengthen emergency care through cooperative partnerships to promote the well-being of the citizens of Nebraska. The data are patient care records for ongoing surveillance.

All basic and advanced life support services are required to collect a patient care record for each response that the service makes. The current EMS regulation requires EMS services to report their data to DHHS within 72 hours of each incident. The data have been collected quarterly since 2000. Based on 2005–2009 data, there were, on average, about 240,000 EMS runs each year; among them, about 80,000 were transported to hospitals.

3.10   ENHANCED HIV/AIDS REPORTING SYSTEM

The Enhanced HIV/AIDS Reporting System (eHARS) is a browser-based application provided by the CDC. Nebraska's HIV Surveillance Program uses eHARS to collect, manage, and report Nebraska's HIV/AIDS cases' surveillance data to the CDC. The data are population based for ongoing surveillance.

All information is entered and stored in eHARS. All information obtained from reported cases is entered into a case report, and this information is then entered into eHARS. The data have been collected from 1983.

The eHARS database is not available for sharing or use by the Division of Public Health (DHHS). All data requests must be made to the HIV surveillance coordinator, who will then fulfill the data requests following the program's confidentiality and release of data policies.

To protect the privacy of persons with HIV in Nebraska, the Nebraska DHHS will only release total numbers of HIV/AIDS cases by any breakdown if the cell size is three or greater. The exception to this is when providing statewide data. HIV/AIDS data by county will only be given if the cumulative number of reported cases is five or more. Selected variables will only be given for counties having 25 or more cases, and then only when the cell size is >3.

3.11   NEBRASKA EMERGENCY ROOM SYNDROMIC SURVEILLANCE

The Syndromic Surveillance Program is designed and implemented to facilitate public health rapid detection of and response to unusual outbreaks of illness and injuries that may be the result of bioterrorism, outbreaks of infectious disease, or other public health threats and emergencies. The data are hospital based for ongoing surveillance.

Electronic health record data are collected from all emergency room encounters in all participating facilities. Data elements include date of visit, patient gender, zip code, diagnosis codes, chief complaint and triage notes, date of birth, and additional clinical information, such as vital signs and influenza test results, as available. Data have been collected in real time or near real time since 2010.

The system has been used to study influenza, motor vehicle crash injuries, and emerging infectious diseases.

3.12   BEHAVIORAL RISK FACTOR SURVEILLANCE SYSTEM

The Nebraska Behavioral Risk Factor Surveillance System (BRFSS) has been conducting surveys annually since 1986 for the purpose of collecting data on the prevalence of major health risk factors among adults living in the state. The surveillance system is based on a research design developed by the CDC and used in all 50 states, the District of Columbia, and three U.S. territories. The system aims to target health education and risk reduction activities throughout Nebraska in order to lower rates of premature death and disability.

Data are available from 1998 onward based on the landline survey. Data have been collected through both landline and cell phones since 2010 from randomly selected Nebraskans aged 18 and older. The survey is conducted by the Nebraska DHHS. The number of respondents in the last 10 years ranged from 10,000 to 22,000. In most years since 2005, the BRFSS in Nebraska was representative at the local health department level. The data have zip codes, but people should not use them as an area unit due to inconsistent sampling. Between 2006 and 2010, there was also a minority oversample of more than 5000 respondents. However, due to a different weighting scheme, the minority sample cannot be combined with the general sample.

Our general impression is that BRFSS has good accessibility, high reliability and consistency, moderate good analytic utility, and low enhancement potential.

3.13   VITAL RECORDS

The Vital Records Office mainly keeps three vital statistic records:

Birth records (vital—birth). All Nebraska birth certificate information is collected at the hospital and entered directly into the vital records electronic registration system at Nebraska DHHS. Home births are entered by health statistics staff. Out-of-state resident births are received electronically through the State and Territorial Exchange of Vital Events (STEVE) or via paper and entered into the system. Birth records have been registered since 1904. Birth certificates have been filed since 1912. Digital records can be traced back to 1970 with main birth outcomes, such as birth weight, infant death, mother's age at birth, race and ethnicity, and educational level. More complete identifiable information started in 1989. Most urban births since 2005 have been geocoded to the census tract level. In 2012, there was a total of 25,939 live births among Nebraska women.

Death records (vital—death). Records are collected from funeral directors, physicians, physician assistants, county attorneys or their designated representatives, and coroners. Death certificates have been filed since 1904. Electronic registration began in 2006. Out-of-state resident deaths are received electronically through STEVE or via paper and entered into the system. Demographics such as age, sex, race, and ethnicity, together with a list of underlying causes of deaths (ICD-10 coding), are available. Annual compiling certificate data are usually available with about a 6-month time lag. Most urban deaths since 2005 have been geocoded to the census tract level. A total of 15,654 deaths occurred among Nebraska residents in 2012.

Marriage records (vital—marriage). All marriage certificate information is gathered from the county clerk when the bride and groom apply for the marriage license. After the couple is married, the license is transmitted both electronically and by paper to the Nebraska DHHS Office of Vital Records, where the information is verified. Marriage certificates have been filed since 1909. A total of 12,376 marriage certificates were issued in 2012.

Our general impression is that vital records have good accessibility, moderate reliability and consistency, moderate analytic utility, and good enhancement potential.

3.14   BIRTH DEFECT REGISTRY

The Nebraska Birth Defect Registry (NBDR) is a statewide, population-based data system. It provides data for the purpose of initiating and conducting investigations of the causes, mortality, methods of prevention, treatment, and cure of birth defects and allied diseases. Data are collected monthly from physicians, hospitals, and persons in attendance at births. Such reports are required to be submitted to the department no later than the 10th day of the succeeding month after the birth. When either parent objects to furnishing information relating to the medical and health condition of a live-born child because of a conflict with his or her religion, such information shall not be required. However, not all hospitals report on time, and there is also long time delay until when data are entered into the system. A total of 2285 birth defects were diagnosed among 1213 children born to Nebraska women in 2012.

Our general impression is that NBR has good accessibility, moderate reliability and consistency, low analytic utility, and moderate enhancement potential.

3.15   NATIONAL ELECTRONIC DISEASE SURVEILLANCE SYSTEM

The Nebraska DHHS and local health departments have access to the state's Electronic Disease Surveillance System (NEDSS), which is currently used to capture and document information regarding laboratory reports and human exposures of public health importance. The data are laboratory reports of reportable diseases for ongoing surveillance.

State regulations dictate what conditions are reportable. Any physician or laboratory that has a Nebraska resident who has been diagnosed with a reportable condition must provide that laboratory test to the state. Physicians who identify a cluster of cases or a new or novel condition must report the cluster or novel condition to the state. Data collected include name, address, age, date of birth, laboratory performing the lab test, physician information, and lab test results. Data are received daily.

The data have been collected since 2005. Laboratory reports and outbreak information are collected for conditions and diseases, including, but not limited to, rabies and animal-to-human transmission diseases, food- and waterborne diseases, hepatitis, chronic hepatitis C, HIV, sexually transmitted diseases (STDs), vaccine-preventable diseases, tuberculosis, and antimicrobial resistance. The NEDSS can also be adapted to address emerging public health issues such as Ebola. Local health departments have access to NEDSS, which allows the secure exchange of information between the state and local health departments. However, due to the sensitive issues related to the data system, the data are rarely released to public health researchers.

3.16   NEBRASKA NEWBORN SCREENING

Newborn screening is conducted for every baby born in Nebraska, which includes a set of blood tests that detect conditions that could be harmful to a child. Eight conditions are screened: biotinidase deficiency, congenital primary hypothyroidism, congenital adrenal hyperplasia, cystic fibrosis, galactosemia, hemoglobinopathies, medium-chain acyl-CoA dehydrogenase deficiency (MCAD), and phenylketonuria (PKU). The data are population based, with more than 97% of newborns being screened each year. All newborns who receive a screen are in this database. Babies who die before a screen is collected may not be entered in the database. Filter paper blood specimens collected at the hospital have demographic information recorded from the medical record. These are sent to the contracted laboratory that enters all these data into the system. When test results are completed, they are added to the individual record.

3.17   PREGNANCY RISK ASSESSMENT MONITORING SYSTEM

The Pregnancy Risk Assessment Monitoring System (PRAMS) is an ongoing population-based risk factor surveillance system (survey) designed to identify and monitor selected maternal behaviors and experiences that occur before, during, and shortly after pregnancy. A random sample, stratified on the maternal race and ethnicity of about 10% of new mothers across the state who recently delivered a live birth, is selected from the birth certificate registry monthly to receive the survey. The goal of PRAMS is to reduce infant illness and death. PRAMS influences health care systems and maternal behaviors that affect health during and right after pregnancy. The data are population based for ongoing surveillance.

Each month, a stratified sample of approximately 200 mothers is randomly selected from recent birth certificates. PRAMS participants are first contacted by mail, and if results are not received, then they are contacted by PRAMS phone interviewers. The response rates have been around 80%. Hence, there are nearly 2,000 respondents each year. The data have been collected annually since 2000. Normally, researchers cannot access data with identifiable information, so the linkage potential is moderate.

Our general impression is that PRAMS has moderate accessibility, high reliability and consistency, moderate analytic utility, and moderate enhancement potential.

3.18   NEBRASKA ADULT TOBACCO SURVEY AND SOCIAL CLIMATE SURVEY

The Nebraska Adult Tobacco Survey (ATS) (in 2000, the Social Climate Survey [SCS] was also included) is a population-based ongoing telephone survey that the state developed. The survey provides tobacco-related information on the Nebraska adult population not captured through the BRFSS. The survey collects data on tobacco use, cessation, exposure to secondhand smoke, smoke-free policies, attitudes and beliefs about tobacco use, tobacco control and prevention, knowledge and awareness about tobacco and health, tobacco control and prevention media efforts, and youth access to tobacco products. The data are population based for ongoing surveillance. Data are collected from a random sample using random digit dialing techniques. The data have been collected annually since 2002.

3.19   NEBRASKA WIC PROGRAM

Nebraska WIC is the special supplemental nutrition program for women, infants, and children. The WIC system is used to enroll WIC participants; store and manage participant records; track participant health-related data; issue WIC food instruments; issue, process, and reconcile food instrument redemption data; reconcile and report monthly participation and financial data; manage vendor data and the vendor cost containment system; manage the WIC infant formula rebate contract and billings; maintain U.S. Department of Agriculture (USDA) participant characteristics data and reporting; maintain USDA integrity profile data and reporting; and produce ad hoc reports. The data are program level for ongoing program management and surveillance.

Data are collected from anthropometric and biochemical data and from self-reported WIC client records; self-reporting WIC retailers, with some data verified by state and local WIC staff for application purposes; program accountability and financial information; various tools used to collect client data, including the health assessment form, diet survey form, and computer system screens; retailer information collected from program applications and on-site visits; and financial and accountability information from computer system and processing contractors. Data are not available for individual clients or vendors or in a format that may identify individuals. Aggregate data are available from standard reports.

3.20   PREGNANCY NUTRITION SURVEILLANCE SYSTEM

The Pregnancy Nutrition Surveillance System (PNSS) is a national program-based public health surveillance system that monitors risk factors associated with infant mortality and poor birth outcomes among low-income pregnant women that participate in federally funded public health programs. In Nebraska, all of the data are collected through the Nebraska WIC program in coordination with the CDC. WIC state and local agencies utilize PNSS information to identify priority areas for education and targeting of services, monitor progress toward program objectives, and evaluate program effectiveness. The data are program level for ongoing surveillance.

During the prenatal clinic visit, demographic and maternal health and behavioral data are collected, and at the postpartum clinic visit, infant health data describing the birth outcome are obtained. Each woman contributes one record representing one pregnancy. The PNSS record that includes both prenatal and postpartum data is collected in the clinic and aggregated at the contributor or state level and then submitted to the CDC on a quarterly basis. PNSS data have been collected annually since 1983. In 2012, the CDC discontinued PNSS. The final data year available is 2011. Historical data are available from the data contact.

3.21   PEDIATRIC NUTRITION SURVEILLANCE SYSTEM

The Pediatric Nutrition Surveillance System (PedNSS) is a national public health surveillance system that describes the nutritional status of low-income infants and children who attend federally funded child health and nutrition programs. In Nebraska, all of the data are collected through the Nebraska WIC program in coordination with the CDC. WIC state and local agencies utilize PedNSS information to identify priority areas for education and targeting of services, monitor progress toward program objectives, and evaluate program effectiveness. The data are program level for ongoing surveillance.

Data are collected at the clinic level and then aggregated at the state level and submitted to the CDC for analysis. When multiple visit records are submitted for a child during the reporting period, the CDC creates a unique child record following specific selection criteria that may contain some data from all available records. The CDC then calculates the nutrition-related indices and sends each contributor agency a series of annual tables that summarize the prevalence and trends of nutrition-related indicators by age of child and race and ethnicity. PedNSS data have been collected annually since 1982. In 2012, the CDC discontinued PedNSS. The final data year available is 2011. Historical data are available from the data contact.

3.22   YOUTH RISK BEHAVIOR SURVEY

The Youth Risk Behavior Survey (YRBS) is a health behavior–related survey of Nebraska public high school students. It is part of the National Youth Risk Behavioral Surveillance System, created in 1990 by the CDC. The YRBS identifies and monitors priority health risk behaviors that are established during youth and result in sickness, disability, death, and social problems among youth and adults. It is the only school-based surveillance system to monitor the six priority health risk behaviors identified by the CDC among school-aged youth.

Survey topics include intentional injuries and violence, physical activity, nutrition, tobacco use, alcohol and substance use, HIV/AIDS and human sexuality, and asthma. The YRBS is conducted biennially by the Nebraska Department of Education among a representative sample of public high school students. The YRBS is the only survey in Nebraska that provides state-level estimates for high school students across a variety of important health areas, making it extremely important to the public health work in the state.

It provides information to assist communities, schools, districts, and state agencies to plan effective health prevention strategies for school-aged youth. It also provides critical data needed for local, state, and national grants to access funding to address health priorities in youth. In addition, the national coordination of the survey allows for Nebraska to compare its survey results to the results from other states and the nation as a whole. The Nebraska Department of Education, in collaboration with the Nebraska DHHS and schools, has conducted the YRBS biennially since 1991.

3.23   NEBRASKA YOUTH TOBACCO SURVEY

The Youth Tobacco Survey (YTS) is an ongoing population-based, school-based survey. The survey provides tobacco-related information not captured through the YRBS and Nebraska Risk and Protective Factor Student Survey (NRPFSS). The information helps the state and communities plan effective youth tobacco prevention initiatives. YTS uses a two-stage probability sample design that is proportional to school enrollment. The YTS data were collected from Nebraska public middle schools and high schools in 2000, 2006, 2008, 2010, 2013, and 2015 biennially (2015 data were only from high schools). Data are generally accessible to public health researchers, but they cannot be enhanced.

3.24 NEBRASKA RISK AND PROTECTIVE FACTOR STUDENT SURVEY

The NRPFSS is part of the Nebraska Student Health and Risk Prevention (SHARP) Surveillance System, which includes the coordinated administration of three school-based student health surveys in Nebraska. The NRPFSS is designed to measure adolescent substance abuse, delinquent behaviors, gambling, and the risk and protective factors that predict adolescent problem behaviors. The survey targets all public and nonpublic school students in grades 6, 8, 10, and 12 in Nebraska.

The survey has been administered six times in Nebraska, with the first administration occurring during the fall of 2003. All participating school districts get a personalized report of their results. The NRPFSS is primarily intended to provide local-level data to support local health planning and evaluation. Data are collected, processed, and reported by the Bureau of Sociological Research at the University of Nebraska–Lincoln as a contractor for the Nebraska DHHS.

The NRPFSS, which is based on nationally validated surveys, is specific to Nebraska and not part of a national surveillance system. All Nebraska schools, public and private, with a grade 6, 8, 10, or 12 are eligible and can choose to participate. In participating schools, all students in eligible grades are asked to complete the survey. Data are collected by paper-and-pencil administration in the classroom. Due to the voluntary nature of the survey, participation consistency and geographic coverage may not be ideal.

The survey has been administered during the fall of 2003, 2005, 2007, 2010, 2012, and 2014. Participating school districts get a report with their results. In addition, reports are generated at the county level (for counties that met the reporting requirements), as well as for substance abuse prevention coalitions, local health departments, behavioral health regions, and the state overall. Data are generally accessible to public health researchers, but they cannot be enhanced.

REFERENCES

Karr, A.F., A.P. Sanil, and D.L. Banks. 2006. Data quality: A statistical perspective. *Statical Methodology* 3: 137–173.

Love, D., B. Rudolph, and G.H. Shah. 2008. Lessons learned in using hospital discharge data for state and national public health surveillance: Implications for Centers for Disease Control and Prevention tracking program. *Journal of Public Health Management Practice* 14(6): 533–542.

4 Data Linkage to Gain Additional Information

4.1 INTRODUCTION

Let's start with a hypothetical event. One hot summer evening, John crashed his car into a telephone pole on a state highway, broke his collarbone, and severely injured his shoulder. A state trooper was the first on the scene, and tests showed that John's blood alcohol content was 1.7. An ambulance arrived a few minutes later and took John to the emergency room (ER) at a local hospital. In this example, the state trooper had critical information about the blood test, while emergency medical services (EMS) had in-vehicle treatment information, and the hospital had information about the type of injury, initial (ER) treatment, surgical procedures during hospitalization, and later rehabilitation treatments. From the point of view of preventing driving under the influence of alcohol, we want to know if John had any behavioral problem, such as overdrinking, and if he would adhere to rehab treatment after the accident. Since each government or health service unit (motor vehicle administration, EMS, and hospital) maintains its own data, and there was no data sharing agreement except for trauma patients, we would not routinely be able to link these data. Fortunately, the U.S. Federal Highway Administration, together with some state highway administrations, funded the Crash Outcome Data Evaluation System (CODES), which links the three administrative datasets. In addition, we could also link to other data. It turned out that John had previously sought behavioral treatment and been admitted to a hospital for alcohol problems. It also turned out that John had participated only in the court-directed alcohol treatment program, while he had opted out of a physical therapy program for his injuries. Although we knew from the hospital data linkage that John had private insurance, we suspected that he might not have sufficient coverage due to either co-pay or insufficient income. We then geocoded his address and found that he lived in a poor neighborhood or census tract with about 21% of individuals living below the federal poverty line.

This example highlights the importance of linking multiple datasets to gain additional information. We linked John's records to his own personal information in hospitals and other settings. We also linked his records to the neighborhood he lived in to gain some insight about his socioeconomic status. Although our record linkage definition is broadly defined, including linking (1) person-specific data from multiple data sources and (2) contextual data that refer to the same person, we limit our discussion to the linkage of personal records.

Most public health data in a state public health agency are collected administratively to fulfill legislative and federal grant mandates. Data can be arranged by a number of topics or themes. At the broader end, cancer registry and trauma registry data are collected for general disease surveillance and analysis. These data tend to have many variables or data items covering a wide range of topical areas. At the other end of the data spectrum, administrative data are collected for very specific purposes, with limited data items and designated use. Examples include cancer screening data for disadvantaged groups (e.g., from the Every Woman Matters program), death records, and infant screening records. Regardless of the breadth of data, items linking one dataset to another would likely add another dimension for data analysis to improve disease surveillance and program management.

Many programs in a state public health agency require data linkage as part of their operations. In Nebraska, the Nebraska Cancer Registry is routinely linked to death certificates to ascertain vital status for cancer patients. The breast cancer screening program routinely links records of those patients with a positive screening result to the cancer registry and Medicaid data to track cancer care insurance. The CODES program routinely links the police crash report, EMS, and hospital E-code data to trace injured individuals from the crash scene throughout the health care system and identify the causes and outcomes resulting from motor vehicle crashes. The Women, Infants, and Children (WIC) program is required to link eligible mothers to other social services. In addition, there are often ad hoc data requests linking infant screening, childhood immunization, and refugees to corresponding downstream service data.

Advocacy of record linkage in public health can be traced back to at least 1946, when Dr. Halbert Dunn (1946) wrote a commentary in the *American Journal of Public Health*. At the time, Dunn was the chief of the national Office of Vital Statistics of the U.S. Federal Security Agency. He envisioned a book of life created by data linkage from birth to death using primarily birth certificate numbers for all the states. The linked data would then become the basis for linking to other records from hospital care, insurance payment, pension, social security, and other administrative data. He coined the term of Life Record Index to capture his linkage strategy and thought that the overall linkage strategy would be relatively easy to design and implement. While the unique birth certificate number system was recommended and later implemented in all states, its linkage to other data systems was rarely seen until the late 1980s, when computerized linkage become less computationally intensive. In this regard, the numbering system developed in the early years made it possible to retrospectively link birth, death, and marriage records within the vital statistic system going back many years.

Datasets linked through a unique identifier are deterministic linked data. However, not many datasets have a unique identifier, in which case multiple record identifiers, such as name and date of birth (DOB), are necessary for linking two datasets. Some datasets even have a unique person identifier, such as a social security number (SSN). The use of the SSN is rather restrictive and, in most cases, is not allowed for linkage outside of the authorized purpose. For example, a mother's SSN is available from the birth certificate, but it can only be used for welfare verification. If one wants to link multiple births for birth outcomes, the SSN cannot be used. In addition, some people do not have an SSN, and some SSNs are incorrect. Hence, a better practice is to design a probability linkage procedure even if an SSN is available.

In a probability linkage process, personal identifiers other than the SSN are used even though a deterministic identifier, such as an SSN, is available. The earliest probabilistic linkage method is from Newcombe et al.'s (1959) work published in *Science*. Newcombe and colleagues in Canada were interested in determining genetic disease passed from parents to children, and they used birth and marriage certificate data in British Columbia to develop a probability linkage method based on the computer technology of the time. Twenty-three years later, Newcombe and colleagues summarized their statistical methods of using name to link personal records (Newcombe et al., 1992). Later, Jaro (1995) and Cook et al. (2001) provided an introductory summary of linkage methods and steps, while Roos and Wajda (1991) and Hser and Evans (2008) shared some lessons learned from multiple data linkage projects.

In the rest of the chapter, we first describe essential elements of probability data linkage and then provide an example to lead a complete linkage process. Since many health departments may not carry out data linkage by themselves, we also provide some guidelines about managing a data linkage project and protecting data confidentiality. Finally, we provide a summary of the chapter.

4.2 DATA LINKAGE ESSENTIALS

4.2.1 DATA FILE ASSESSMENTS FOR DATA LINKAGE

Data linkage has two dimensions: identifying multiple appearances of individuals within a dataset and among different datasets. Although linking multiple records in a single dataset, such as a mother with multiple births from a birth registry, is routine, most data linkage projects are associated with integrating records across datasets. In the latter case, record linkage involves at least two files. Newcombe et al.'s (1959) early work actually included both dimensions. Suppose that file B contains individual birth records and file A contains marriage certificate records. Each file consists of a large number of fields or variables pertaining to characteristics of each file. Linkage can be carried out when at least one field on file A is available on file B. For example, when a mother's first and last names and DOB are deemed appropriate fields for matching, both files A and B should have at least these fields. If file A has j records and file B has k records, a simple matching algorithm is (1) start from record 1 ($j - (j - 1)$) in A and search through all records in B for matches, (2) move to record 2 ($j - (j - 2)$) and go through all records in B for matches, and (3) repeat 2 until the end of records in A. If each record in A found only one match in B, and all records in A found matches, then $j < k$, and vice versa. Even though counting matched pairs seems trivial, it comes in handy when assessing matching strategies and diagnosing potential matching problems, as shown below:

1. Total record match pairs versus total unique person matching pairs. This matching process generates $j \times k$ pairs of records for potential linkage at the record base. The above algorithm works more efficiently when it links unique persons first and then populates all linked records. Hence, matching or linking in practice is most often according to unique person identified by matching fields. Care must be taken when a linked record is populated to multiple records of the same person. For example, a cancer patient had two records, one for breast cancer and the other for colon cancer. Linking the patient to hospital discharge data can identify multiple hospitalization records. A person-to-person unique record pair would say only that the cancer patient is found in the hospital discharge data. To pair hospitalization events, hospital admission date, diagnosis date, or hospital treatment date should be used.

2. Preliminary assessment of the appropriateness of matching fields. Selecting matching variables or fields is not necessarily a straightforward process. Obviously, if two fields, such as age and sex, cannot uniquely identify an individual person or entity, then the two fields are not appropriate. However, when two sets of variables can both be used as identification fields, choosing which set to use can be done quantitatively. Let j_{u1} represent the number of unique values in A for the first set of selected match fields (e.g., first name, last name, DOB) and k_{u1} represent the number of unique corresponding values in B. Let j_{u2} represent the number of unique values in A for the second set of selected match fields in A (e.g., first name, last name, and birth year) and k_{u2} represent the number of unique corresponding values in B. If the product of $j_{u1} \times k_{u1}$ is greater than the product of $j_{u2} \times k_{u2}$, then the first set of matching fields is better than the second set. To put it another way, if age and sex were to be used as match fields in the second set, the product would be much smaller than the product of $j_{u1} \times k_{u1}$. Data analysts should be aware of this issue, because it is not always better to use name and DOB. For datasets based on hospitalization within a hospital, hospital-based identifiers, such as admission date, discharge date, and last name, may be a better set of matching fields than a set based on first name, last name, and DOB.

3. Preliminary matching rate assessment. Based on a better set of matching fields, an initial assessment can be performed based on a quick deterministic linkage operation on unique persons from two files. If the number of a deterministically linked pair is large relative to the number of unique person records in either A or B (J_u or k_u), then the potential linkage rate for probabilistic linkage is likely to be high. This operation can also be used for assessing data quality for matching fields, as spelling, typos, and informal names can also affect the eventual linkage ranges. The rationale behind this method is simple. Suppose we are interested in the matching rate for file A. Matching files A and B will generate an indicator variable of matching status with values of match (m) and unmatch (u) in file A. This variable will then separate all the matched and unmatched records into a matched set (M) and an unmatched set (U), with the number of matched sets in M being n_m and the number of unmatched sets in U being n_u. When n_m is large, $n_m/(n_m + n_u)$ is also large.

4.2.2 Dealing with Large Dataset Blocking

When two matching files are very large, such as more than 100,000 records each, matching based on multiple fields is very time-consuming. If the two files can be partitioned into smaller segments, the matching process speeds up substantially. Blocking is a way to partition both files into mutually exclusive and exhaustive subsets, so that a match within each pair of subsets can be performed sequentially. Using age as a blocking variable, with the top age as 100, would potentially yield 100 age-specific subsets. The first subset is age <1, and the second is age 1–2, and the last subset is age 99–100. These age segmentations are called blocks, and they allow record matching within each block. Assuming that each age has 1000 records in each file and that files A and B each have 100,000 records, the number of search pairs without blocking would be 100,000*100,000, or 10 billion. With 100 blocks, the search pairs would be 1000*1000, or 1 million, which is 10,000 times smaller. The matching speed of the blocked datasets would be potentially 10,000 times faster than the matching speed without blocking.

However, if age in file A has a systematic error of 1 year older than age in file B (e.g., due to a coding error of the beginning age), then there would be no match between the two files if each age is a block. If each 10-year age group is a block, then there would be 10 segments of 1 year that would not match. Therefore, a record matching design should include at least two blocking variables, so that fields not matched by one blocking scheme will be matched by the other. A computer run based on a blocking scheme is often called a pass, and multiple passes are needed to improve matching speed and matching rates. This blocking and pass taxonomy is implemented in Link Plus, a public domain software supported by the Centers for Disease Control and Prevention (CDC). Other linkage software packages may not use the term *pass* to describe a computer run based on a blocking scheme. And even if the software uses the term *one pass* or *pass 1*, it may not mean that one pass is based on one blocking variable. In fact, most linkage software allows a number of blocking variables in a single pass, and at least two passes are required.

The next question is, which variables should be used for blocking? Since improving speed is a major concern, we suggest the following properties:

1. A blocking variable should meet at least two criteria: having as many categories as possible and having as few errors as possible. As a rule of thumb, a blocking variable with less than 10 categories would not be desirable. In this regard, sex would not be a good blocking variable. Even though addresses tend to be numerous, using addresses as a blocking variable is messy and not recommended.

2. An ideal blocking variable should be able to divide a file more or less evenly. Birth year would not meet this criterion because events that people are interested in matching, such as birth, death, and hospitalization, tend to be skewed by age distribution. In this regard, birth month (1–12) without year and birth date (1–31) without month and year are good blocking variables.

3. An ideal blocking variable should have a one-to-one correspondence between files A and B. For this reason, last name is rarely used as a single blocking variable in the first pass, as a lot of last names in file A would not be found in file B.

4. A blocking variable can be created by combining a number of variables. For example, a combination of last name, birth month, and sex could be used as a blocking variable in the first pass, and another combination of variables with at least one variable difference from the first pass could be used as the blocking variable in the second pass, and so on.

5. As a rule of thumb, one should not go beyond five passes when designing blocking variables and blocking schemes, and the less dependence among those passes, the better the blocking design. The main reason for this rule is that if you have more than five passes, and the first two passes are heavily dependent, then the blocking scheme might not efficiently catch the matched records in each pass. As each pass is made, the number of unmatched

pairs should be reduced. When a blocking scheme has five passes, the number of additional unmatched pairs should be stabilized in the last two passes. In mathematical terms, the fields with the highest weights make the best blocking fields.

4.2.3 DETERMINING MATCHING QUALITY: WEIGHTS

When selecting matching variables, it is expected that some variables are more important than others in determining matching pairs, even though all the selected variables contribute somewhat to the matching process. For a common set of matching variables, such as first name, last name, DOB, and sex, common sense suggests that last name is more important than first name, and DOB is more important than sex. The contribution of each match field can be measured by its weight in a given matching scheme.

To understand weight formally, it is necessary to introduce two more notations: p_m for the field-specific probability of matched pairs and p_u for the field-specific probability of unmatched pairs. Calculating p_m is a straightforward process because the matched set is much smaller than the unmatched set. For example, if the sex field has 10% discrepancies, then the p_m for the field would be 0.9, which is 1 minus the error rate of 0.1. The more reliable a field, the greater the p_m. Since p_u is the probability that a field agrees given that the record pair being examined is an unmatched pair, its denominator is the number of possible unmatched pairs, which tends to be a much bigger number than for matched pairs. For example, if files A and B contain 1000 unique person records, there are at most 1 million possible record pairs, but only a maximum of 1,000 matches, leaving 999,000 unmatched pairs. Even though P_u = 1,000/999,000 = 1/999, it is very close to 0.001(= 1,000/1,000,000). P_u is therefore essentially the probability that the field agrees at random when the file size is large. In Fellegi and Sunter's (1969) statistical treatment, the weight for a field is defined as the log2 of the ratio of p_m and p_u, which means that the larger the weight, the more important the field in determining a matched pair.

To show how a relationship between matched and unmatched pairs translates into actual values, we examined the match between first name and SSN fields. Assume that sex has a 10% discrepancy rate or error rate and SSN has a 40% error rate. The p_m for sex is 0.9. The u probability is 0.5 in situations with an equal number of males and females. Thus, the weight for sex is log2(rn/u) = ln(rn/u)/ln(2) = ln(0.9/05) = 0.85. Assume that the probability of chance agreement for a seven-digit SSN is 1 in 10 million. Given that the p_m for SSN is 0.6 (1 – 0.4), the weight for SSN is log(0.6/0.0000001) = 22.51. Thus, the weight for a match on sex is 0.85, and for a match on SSN, it is 22.51. Since the sex field intuitively weighs less than the SNN field, this weighting calculation reflects common sense.

In most software packages, a final or composite weight is calculated by combining all field-specific weights. If a field agrees in the pair being compared, the agreement weight, as computed above, is used. If a field disagrees in the pair being compared, the disagreement weight is computed as log[(1 – rn)/(1 – u)]. This results in field disagreements receiving negative weights. Thus, agreements add to the composite weight and disagreements subtract from the composite weight, and p_m must always be greater than p_u. Similar to field-specific weight, the higher the score, the greater the agreement is.

Although the final matching weight or score is a continuous variable, a cutoff point must be chosen, so that matched pairs and unmatched pairs can be divided into two sets. A good practice is to select 20 pairs around an expected cutoff point and then do some manual checking to empirically determine the cutoff score. However, the location of an expected cutoff score usually depends on file size. Based on the above example of 1000 unique person records in files A and B, potentially all records in A could be matched. According to Cook et al. (2001), the calculation of the cutoff score requires three steps. The first step is to calculate the weighting score based on the desired confidence interval. If we want the p_m to be 95%, then the corresponding weight in the absence of file size is log2(0.95/(1 – 0.95)), or 4.24. The second step is to calculate the –log2(p_u); in this case, it is log2(1/999) = −9.96. The third step is to add the values from steps 1 and 2 together, which equals 14.21. One should therefore move inspecting matching values empirically around weight score 14. For the statistical theory behind this calculation, interested readers can refer to Fellegi and Sunter (1969), where they also took attribute value distribution into consideration. For instance, Smith is a common last name, and its attribute value weight is smaller than that of a rare family name, such as Dankworth.

Note that as confidence level increases, the step 1 value increases. As the number of records increase in a file, the step 2 value increases. Both would call for an increase in the composite weight cutoff score. Since the final weight score is a function of matching fields and blocking, the greater the weight demands, the more discriminative information that is needed to determine matched pairs. In other words, as the file size increases, the demand for the number and the quality of linkage variables increases. Cook et al. (2001) provide some empirical relationships between file size and linkage variables.

4.2.4 Postmatching Evaluation

We have now decided matched and matched pairs by selecting an appropriate weight score. However, among the matched pairs, there are unavoidably some record pairs that are not true links among linked pairs, and some record pairs that are true links among the unmatched pairs. To evaluate postmatching quality, we need to define some concepts. According to Blakely and Salmond (2002), it is necessary to distinguish match and link, the former in a deterministic sense and the latter in a probabilistic sense. Table 4.1 lists the combinations of match and link.

Based on Table 4.1, we can define positive predictive value (PPV) as $PPV = a(a + b)$. PPV is the proportion of linked pairs that are deemed true links, and it is the most commonly used indicator for postmatching assessment. If all linked records in one file are true matches, then $PPV = 1$. If 50% of linked records are true matches, then $PPV = 0.5$. For example, in hospital discharge data, there is a variable called discharge status. For a patient who died in a hospital, the discharge status of 20 is used. When mortality data are linked to hospital discharge data and the linked person died either 30 days before or after the discharge date, then it might be a false positive. As we will see later,

---

TABLE 4.1

Combination of Probabilistic Link Outcomes vs. Real-World Match Outcomes

	Match	Nonmatch
Linked	a (true positive)	b (true negative)
Unlinked	c (false positive)	d (false negative)

Note: Match: A pair of records from files A and B refer to the same real-world entity.

Nonmatch: A record in file A is not in file B for the same real-world entity.

Link: A pair of records that are accepted for the same real-world entity.

Unlinked: A pair of records that are not accepted for the same real-world entity.

a: True positive link—an actual match was accepted as a link.

b: True negative link—a nonmatch was accepted as a link.

c: False positive link—a nonmatch was accepted as a link.

d: False negative link—an actual match was not accepted as a link.

PPV is often calculated from samples rather than from complete records. On the other hand, false negative is not so easy to determine. In cancer registry operations, one is often involved in updating the vital status by using the National Death Index (NDI). NDI uses the SSN as the primary linkage variable. When the SSN and DOB are matched, this is often a true link. However, it can be rejected if the name fields do not match. After using other information, such as marriage certificate or naturalization certificate, we might find out that the person changed his or her name, but insurance and other billings were kept under the original name. In this situation, we have a false negative link and we could accept the link again in the data system. For example, in hospital discharge data, there is a variable called discharge status. When we find a lot of false negative, we can calculate negative predictive value (NPV) as $NPV = d(c + d)$. In practice, however, we normally do not have a lot of cases, and NPV is therefore rarely used.

4.3   CASE STUDY: A COMPLETE LINKAGE PROCESS

The above introduction of data linkage concepts also points the importance of to a record linkage process. As with most data analytical processes, a data linkage process should start with prematching data standardization, data blocking, record matching, and postmatching review. In the hypothetical example of two data files below, we highlight some issues in matching names. Given that DOB is matched, we list a number of issues with name matches. In Table 4.2, the first three columns after DOB are for the first, middle, and last names for file A, and the last three columns are the corresponding three name fields for file B. The first issue is that first and middle names are often not consistent. Due to changes in family name, people often use a middle name to indicate a maiden name. In the absence of a maiden name field in a database, using a middle name field can be

TABLE 4.2
Hypothetical Data Linkage Table from Two Files

No.	Sex	dob_m	Fname	midName	LastName	Fname	midName	mom_lnm
1	1	19581021	Ryyando	Robertven	Reader	Ryyando	Robertven	Reader
2	1	19660525	Daniel	Smith	Miller	Daniel	Allen	Miller
3	1	19631001	Jason	Smith	Miller	Jason	Allen	Miller
4	1	19710514	Jeremy	Lee	Miller	Jeremy	Eugene	Miller
5	**1**	**19250300**	**Armando**		**Fernandez**	**Armando**		**Fernández**
6	1	19581020	Ryyando	Robertven	Reker	Ryyando	Robertven	Reader
7	1	19320701	Georgine	Ray	Smith	Georgine	Ray	Smith
8	1	19541105	Douglas	Russells	Wade	Dog	Russell	Wade
9	1	19590419	Rendell		Cassell	Rendell		Cassell
10	1	19470208	Bradley	Mountain	Wegner	Bradley	Mountain	Wegner
11	2	19641223	Jean	Wade	Raskto	Jean	W	Raskto
12	2	19581000	Amy	Glenn	Almansorion	Amy	Glen	Farhranm
13	2	19320418						Crawford
14	2	19381212	Marcia	Veronica	Perez Mejia	Marcie	Veronica	Perez-Mejia
15	2	19640905	Melanie	Beth	Hovland	Melanie	Beth	Hovland
16	2	19370524	Teresa		Rathjen	Teresa		Rathjen
17	**2**	**19510822**	**Amelia**		**Hernandez**	**Amilya**		**Larson-Hernandez**
18	2	19691221	Alicia	Marie	Adams	Alicia	Marie	Adams
19	2	19570520	Joanettia	Loise	Miller	Joanettia	Louise	Miller
20	2	19661214	Marie	Carolyn	Miller	Mary	Carolyn	Miller
21	2	19700811	Jamie	Becker	Lindburg	Jamie	Becker	Lindburg
22	2	19520103	Lennett	Kahn	Livingston	Lennett	Kahn	Livingston

quite messy. In the second record, Daniel Smith Miller and Daniel Allen Miller could well be the same person. Second, some names come with special characters. In the fifth record, Fernandez and Fernández refer to the same last name, but they are spelled slightly differently when a Spanish spelling option is available for one data file but not for another data file. Third, the first name could be slightly different due to convenience, as shown in record 17, where Amilia and Amilya could be the same first name. Record 17 also shows that the last name could be added in addition to the original. The above example in Table 4.2 relates only to name. There are issues related to dates, facilities, and events. In the following, we systematically go through a complete linkage process.

4.3.1 Data Standardization

Even though weight and blocking are the most studied topics, data standardization or prematching data cleaning is the most time-consuming and important step.

1. Overall data quality assessment. Before starting to clean data, the data quality in terms of accuracy and completeness within each file, and the data consistency and timeliness between two files, must be assessed. It is not easy to check data accuracy and completeness without external data sources. However, it should be possible to assess field-specific value distribution and missing values. For example, if the admission date is disproportionally greater for the first day of each month, the first day of the month would be less trustworthy than other days of the month. A good variable to check for females before marriage is their maiden name for data linkage. However, if 40% of maiden names are missing, then the value of that variable will be lower than if only 5% were missing. Most often, data values in one field are consistent, but it is still a good practice to check their distribution to assess their consistency. Between two files, the same data fields may have a different format. For example, sex is often coded as 1 or 2, with 1 being male and 2 being female. However, if one field is numeric and the other is a character string, then the match would be difficult. In addition, some databases also use F and M to refer to female and male, respectively. When data are in flat files and have not been integrated over the years, data values may not be consistent within a field. For those reasons, data and format consistency should always be checked in a data linkage process.

2. Date format. Date can be entered in the U.S. format [MM DD, YYYY] or in a format commonly used by other countries [DD MM YYYY], in which day and month are swapped. An easy way to check date format is that a day value should be less than 31, and a month value should be less than 12. For example, 1375 persons were born on the first day of the month in a 5-year sample of Nebraska registry data. The average number of persons born on days 2–30 of the month was 1176 persons. Thus, there were 199 more persons than average born on the first day of the month than during the remainder of the month. This distribution could be due to human error or computer defaults, rather than to chance. By looking at the values in the month, day, and year fields, most problems associated with an eight-digit date field can be resolved. In addition, the position of YYYY can also be changed, as happened to North American registry data before and after 2012.

3. Data cleaning. When data quality issues are found, they must be resolved. A guiding principle is to standardize matching fields as much as possible in terms of names or numbers. For instance, Spanish names often have special characters that are not consistent between different databases. It is a good practice to remove the special characters even though some linkage software can handle them. In Table 4.2, Amelia and Amelya are the same person, but with a slightly different spelling. In Chinese, Yan Zhang can refer to more than four different names, such as 張嫣 (Zhang, Yān), 張妍 (Zhang, Yán), 張兗 (Zhang, Yǎn), and 張艳 (Zhang, Yàn).

   Deduplication that removes and keeps track of duplicate records is also part of data cleaning. It not only increases linkage speed, but also increases linkage consistency. Suppose that records a1 and a2 are for the same person with slightly different first-name spellings. Without deduplication, a1 is matched to b1 and a2 is matched to b2, contradicting intuition that the same known person in file A should not be matched to a different person in file B. This problem is also referred to as transitive closure. Finally, women often change their names. If the maiden name is available either as part of the last name or as a separate field, it can be included as a matching field. If only one data file has the maiden name field, then one can create a pseudorecord that indexes with the original record. If either the original or pseudorecord is linked, one can then determine a true linkage pair later.

4. If a linkage or deduplication process includes address fields or other fields that are prone to irregular entries, field parsing or segmentation is often necessary. Normally, apartment numbers should be segmented out to a separate field. A zip + 4 field should be used only with a five-digit zip code.

4.3.2 DESIGN BLOCKING VARIABLES

An ideal blocking variable should have high attribute data quality (for a correct classifying block) that can generate as many evenly distributed groups as possible. Assuming that Table 4.2 has eight matching variables, DOB, first name, middle name, last name, street address, city, state, and zip code, the last name could be used as a blocking variable. However, many records have Miller as the last name, defeating the purpose of blocking efficiency. The middle name could also be used as a blocking variable, but it has a number of missing values. Hence, the first name would be more appropriate for blocking, as it is more complete (no missing values) than the middle name and is distributed more evenly than the last name.

4.3.3 MATCHING

Before choosing matching variables, both files should be carefully examined for high-quality attribute values. It should not be taken for granted that personal identifiers are the best choice. When both sets of data have an event-based identifier, such as admission date, it should be included even for just a subset of records. An example is data from a cancer registry that can be linked to hospital discharge data. Even though the purpose might be to obtain comorbidity for individual cancer patients, both cancer registry data and hospital discharge data have event-based information. The date of surgery is often available, and it may correspond to admission date, procedure date, and so forth. The bottom line is that the data analyst should be knowledgeable about the two files to be linked. This knowledge acquisition should be started before requesting linkage variables and other variables for linkage purposes.

In probabilistic linkage, obviously not every character in a name will be matched. A linkage system normally employs a phonic encoding system to handle the most common English spellings and nicknames. Soundex, Phonex, Phonix, and the New York State Identification and Intelligence System (NYSIIS) are commonly available for linkage software. Soundex is the oldest phonetic encoding algorithm that converts English names into numeric codes according to a set of rules. Phonex extends Soundex by preprocessing a dozen characters, such as *s*, *kn*, and *ph*. Phonix further extends Phonex by including hundreds of preprocessing rules. The NYSIIS encoding algorithm was developed by considering the other three systems, and it is generally equal to or slightly better than the Soundex and Phonex encoding methods. In other words, when Soundex and NYSIIS are both available for name indexing or blocking, NYSIIS is the preferred choice.

String comparison algorithms are normally not explicitly available for data analysts to choose in a linkage software package, and a detailed discussion is beyond the scope of this chapter. However, a brief description of how string comparison algorithms work and how to choose a preferred algorithm may be helpful. Common algorithms include the edit distance; the Smith–Waterman edit distance; and Q-gram-based, Jero–Winkler, and Monge–Elkan string comparison methods.

The edit distance method finds the smallest number of edits that would make two strings identical, and the Smith–Waterman method is an extension, where the edit distance incorporates white space and dashes and other ways of expressing compound names. The Q-gram-based method splits a string into a number of substrings based on a number of characters (q) and then compares substrings in relationship to the total number of substrings. The above string comparison methods can be for any strings (e.g., address), while the Jero–Winkler method was designed specifically for the comparison of names. Similar to the Q-gram-based method, the Jero–Winkler method uses a moving window of predefined length to extract and compare substring similarity, such as substring agreement or number of transpositions (e.g., pi vs. ip). The Monge–Elkan string comparison is used mainly for a string value that contains several words, such as address matching. SoftTFIDF is another long string or composite string comparison method. Since most people do not know these matching algorithms, some software packages, such as Link Plus, use layperson's terms to correspond to some of the algorithms. Data analysts should follow the match method in layperson's terms when selecting a method.

4.3.4 POSTMATCHING EVALUATION

We have already mentioned that a common method to classify linked and unlinked pairs is based on the composite weighting score that sums the individual attribute weights. In practice, unlinked pairs are those below a threshold value that are predetermined to be unmatchable or highly questionable. The linked pairs are those that can be divided into matches and potential matches. The first task of postmatching evaluation is to determine the cutoff weight or score that distinguishes these two classes.

The threshold is normally determined manually. Usually, a sample of 20 or 30 records from each side of an expected cutoff value is evaluated. This evaluation can move upward or downward, depending on the initial evaluation. If the sampled records above the threshold have more mismatches (>5%), the cutoff score can be moved higher. On the other hand, if the mismatches below the threshold are <5%, the score can be moved further down. The goal is to find the cutoff score that minimizes the probability of mismatches above the score. It should be pointed out that the composite or summary weighting score masks individual weight, which in some cases is important to know. For example, if the weights assigned to first and last names are 1 and 2, respectively, a female who changed her last name would be weighted more heavily if her current last name in file A does not match her previous last name in file B.

Finally, at least two different people should conduct a manual review. The sample size for a manual review must be determined at the outset. At the 5% confidence interval, a sample of 400 is sufficient for any files having greater than 50,000 matched pairs. Alternatively, a sample calculator can be used to determine the exact sample size for review. Although reviewers do not need to know statistics, they should be able to pay attention to detail and determine similarity based on a set of attributes. In general, a manual review should have more information than what is already there from matching variables. For instance, addresses may not be used for linkage, but they should be provided for manual review if possible. When linking the cancer registry data to the hospital discharge data, the event date, such as the admission date for surgery in the cancer registry data, may be available. When this variable is not used in linkage, due to various reasons, it can be used for manual review. If the linkage quality is acceptable after the review, then the PPV can be calculated. Otherwise, the cutoff point must be fine-tuned and the review conducted again.

Besides the above manual review process, which is often conducted by linkage specialists, program managers can also conduct some review. At the managerial level, one needs to think about a big picture—we call it a broader or macroreview. For example, death records are often linked to cancer data. We could use pancreatic cancer to conduct some reviews by race or by year. If for race A a high proportion (75%) of patients died within a year, while a substantially low proportion (40%) of patients of another race died during within a year, this would raise a flag. Time is another good macrovariable for conducting a review. Hospital discharge data are often linked to death certificate data to update 1-year mortality. As a manager, a mortality rate comparison between 2 and 3 years before and after a cutoff time could yield useful information. If the 3-year mortality rate for acute myocardial infarction (AMI) is 8% between 2004 and 2006 and 15% between 2008 and 2010, then this should raise a flag for the matching quality of the early years. In some situations, one may also want to compare the state mortality rate with the national rate. If the state rate is too high, such as 5% more than the national rate, then a flag should be placed. In any cases, the managerial review should be timely and broad in nature. If a flag is raised, then the matching and technical review processes can be probed.

As mentioned above, data linkage should have a sponsor who has a keen interest in the linked product. For this reason, one of best quality controls at the managerial level is to encourage using linked data. A study of time trends could be used for quality control, but they can also be a stand-alone study with substantive interest. One of the main purposes for linking death records to hospital discharge data is to study patient survivals beyond hospitals. After initial linkage, one should encourage program data analysts to use data. Although the product may not be final, all the variables are unlikely to change, and data analysts can become familiar with newly added variables to the hospital discharge data and start some initial assessments. If there are some data quality issues, such as the death date is before the admission date, or a large number of in-hospital deaths due to a particular disease (i.e., with the discharge status being "expire") cannot be found in the linked death records, the data analyst can report them immediately—before the final linkage product.

4.4   OTHER ISSUES IN RECORD LINKAGE

4.4.1   TRAINING

Training is always important, as it sets the standard within the organization. Most data analysts who have biostatistics or informatics degrees do not have real-world experience with record linkage. Even if someone has record linkage experience, it is still a good idea to go through training. Protocols for training and linkage practices should be established, and a practice linkage dataset should be compiled that reflects major organizational needs. The purpose of establishing protocols is to reduce postlinkage regrets due to lack of experience or lack of an overall picture of what is to be achieved through data linkage. The established linkage protocol should then be carefully reviewed by experienced linkage experts within the organization. When an organization does not have an expert, an outside consultant can be helpful. All data analysts who are expected to conduct data linkage should meet the minimum standard, so that the practice within an organization can be standardized.

Training requires a project leader to design several linkage operations typically found in an organization. In a public health agency, there are typically two types of linkage: person based and event based. Linking mothers from birth certificate data to women from marriage data is person-based linkage. Linking motor vehicle crash injury data to EMS data and hospital discharge data is event-based linkage. Sometimes data may have both event and person identifiers, even though the purpose may be for person-based linkage. Designing slightly different training modules to go with different types of linkage projects is essential for a successful linkage operation.

Different types of data linkage operations require different training datasets. Organizations should develop some datasets for training. Alternatively, many public domain datasets can be downloaded. Link Plus comes with a number of datasets for practice. Data analysts should use these training datasets to go through the complete linkage process described in Section 4.3.

4.4.2   PRIVACY

Privacy concerns in a public health agency often arise because of the potential for inappropriate data use, data storage, and data sharing. When linkage is conducted within a single organization that is under the same privacy protocol, it is normally not an issue to request identifiable and other attribute data for record linkage. Even in this situation, however, some protocols are necessary. For instance, an informatics support unit may provide data linkage and conduct linked data analyses for multiple programs. However, the data linkage support unit should not share the linked data among the programs within the agency. In fact, the informatics support unit should not own any data. The separation of data support from programs increases professionalism in data linkage.

One of the major privacy issues related to record linkage in public health is consent to linkage. Although the public health agency has the authority to link person-to-person data without individual consent, a privacy issue can arise in other ways. Sometimes record linkage is conducted in another organizational setting, and a linkage protocol that satisfies privacy requirements from both organizations must be drafted, reviewed, and agreed upon. One of the most frequently encountered privacy issues is nondisclosure. Some linkages require data sharing between agencies, and when this occurs, certain laws and policies concerning disclosure and consent are relevant. Notably, the Health Insurance Portability and Accountability Act (HIPAA) may apply to other agencies or organizations, even though the public health agency is exempted. In such a case, the public health agency has to work with other agencies or organizations to develop a proper protocol.

Another privacy issue is reengineering risks for reidentification. Some datasets are linked using a universal ID number that is not supposed to be known by users. Secondary datasets are supposed to strip explicit identifiers when released to researchers. However, few people in a public health agency are trained for privacy protection, so some reidentification of some data subjects could occur through deductive reasoning. These problems already exist in nonlinked datasets. Linked datasets provide more information about an individual and are more susceptible to deductive reasoning. An example of this type of privacy issue is the linking of patients with a rare disease to census tracts: it would be relatively easy to identify the patients; thus, the risk of reidentification might be increased significantly. Sometimes a small risk of reengineering exists. When a linked dataset is requested, the public health agency has to rely on the institutional review board (IRB), which is often part of the data requester's organization. The potential harms to data subjects are then weighed against the potential benefits to all. Even though two relatively innocuous datasets might be cleared by the IRB separately, the heightened sensitivity of the data due to data linkage calls for further scrutiny.

Finally, as in any person-specific datasets, data security is crucial to protecting linked data. Besides the fact that a linked dataset may be more detailed than its components, data transfer between programs, and data transportation between agencies or organizations, calls for encryption and other secured data transfer protocols and practice. When data are to be shared with another agency or organization, or even a researcher, written agreements can help ensure that security measures are in place. For releasing data to researchers without the safeguard of an IRB review, it is normally not safe to use common data suppression methods, such as requiring at least three or five subjects within a data cell. When a researcher wants to use data with greater detail, an approved IRB is required. In addition, a data use agreement from the public health agency side is needed to set contractual conditions about data access, security safeguards, and prohibition of reidentifying data subjects.

4.5   CHAPTER SUMMARY

This chapter has provided an overview of data linkage methodology. It first highlighted the importance of data linkage in public health practice and introduced deterministic and probabilistic link concepts. It stressed the importance of probabilistic record linkage by pointing out that (1) even though the SSN is available in many datasets, it is not authorized to use in general record linkage applications, and (2) some people do not have a SSN and some SSNs are incorrect. It then went through a complete data linkage process that included an initial data evaluation, the performance of a data linkage, and a manual review. Related concepts such as matching rate, blocking, pass, weight, false positive, and false negative were introduced. In addition, we provided a manager's perspective for data linkage that included training and privacy protection. Even though weight and blocking are the most studied topics, data standardization or prematching data cleaning is the most time-consuming and important step. Normally, the format of data fields would be a major issue. Therefore, measures such as data quality checking, deduplication, parsing, and the standardization of the linkage variables should be taken into account to solve the problem.

REFERENCES

Blakely, T. and C. Salmond. 2002. Probabilistic record linkage and a method to calculate the positive predictive value. *International Journal of Epidemiology* 31(6): 1246–1252.

Cook, L.J., L.M. Olson, and J.M. Dean. 2001. Probabilistic record linkage: Relationships between file sizes, identifiers, and match weights. *Methods of Information in Medicine* 40: 196–203.

Dunn, H.L. 1946. Record linkage. *American Journal of Public Health* 36: 1412–1416.

Fellegi, I.P. and A.B. Sunter. 1969. A theory for record linkage. *Journal of the American Statistical Association* 64(328): 1183–1210.

Hser, Y. and E. Evans. 2008. Cross-system data linkage for treatment outcome evaluation: Lessons learned from the California Treatment Outcome Project. *Evaluation and Program Planning* 31(2): 125–135.

Jaro, M.A. 1995. Probabilistic linkage of large public health data files. *Statistics in Medicine* 14(5–7): 491–498.

Newcombe, H.B., M.E. Fair, and P. Lalonde. 1992. The use of names for linking personal records. *Journal of the American Statistical Association* 87(420): 1193–1204.

Newcombe, H.B., J.M. Kennedy, S.J. Axford, and A.P. James. 1959. Automatic linkage of vital records. *Science* 130(3381): 954–959.

Roos, L.L. and A. Wajda. 1991. Record linkage strategies. Part I: Estimating information and evaluating approaches. *Methods of Information in Medicine* 30(2): 117–123.

5 Indexing Multiple Datasets
A Bottom-Up Approach to Data Warehousing

5.1 INTRODUCTION

In Chapter 4, we introduced data linkage, which is the necessary condition for integrating different datasets at the record level. However, informatics support professionals do not own data. Although they can access sufficient information for data linkage, the data products are returned to data linkage requesters, normally a program or researchers. Even in a data warehouse, where all program data are stored in data servers, a data use protocol is often used to guide access of multiple data. In the past, there were very few data linkage operations within the state public health agency. However, as more and more programs require data linkage, overlapping and repeated data linkage for the same dataset become more and more common. Consequently, an integrated approach is needed to streamline data linkage and reduce project redundancy.

In this chapter, we propose a data infrastructure approach that treats each data linkage project as a piece of the whole. We also offer strategies that integrate data linkage processes and multiple datasets within a state or local health department. Some of the main concepts about data index and integration are from *Data Warehousing Fundamentals for IT Professionals* (Ponniah, 2008), but we tried to avoid technical details while maintaining a level that data managers and informatics administrators or decision makers could understand.

The remainder of the chapter is organized as follows. To motivate the integrated approach, we first lay out a typical data request process in a health department (Section 5.2). Duplicated and overlapping data requests are common; the challenge is how to strategically treat them (Ziegler and Dittrich, 2007). We believe that building an enterprise-based master patient index (MPI) in a data warehouse is an appropriate solution. With the intention of building an agency-wide data warehouse, we discuss the requirements for the top-down and bottom-up approaches. We conclude that the bottom-up approach is more appropriate for loosely connected and program-based public health departments. In Section 5.3, we present our experiences and lessons learned about the bottom-up experiment. In Section 5.4, we sketch the main ideas about data integration using the MPI and loosely joint programs and their datasets. In the data warehouse language, they are data marts in a data federation. Finally, we provide a summary and offer some concluding remarks.

5.2 TOP-DOWN AND BOTTOM-UP APPROACHES TO DATA INTEGRATION

5.2.1 Case for Data Integration

In a state health department, there are many data requests. Most are furnished by individual program data analysts or surveillance specialists. Data requests that require linking data from one program to another tend to be handled by informatics or statistics units within the department. Below we use a hypothetical data linkage request to state public health program A (e.g., motor vehicle crash injury outcomes) to illustrate the traditional protocol and practice for furnishing a data request (Figure 5.1).

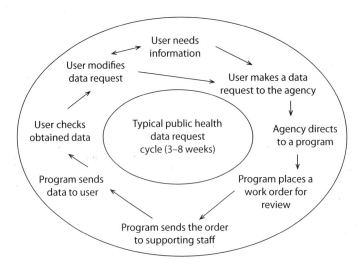

FIGURE 5.1   Data request cycle.

1. Program A receives a data request from either an internal or external user for linking dataset A to dataset D for the years 2000–2005. Even though a user made the request to the public health agency, it is often program A that is finally identified by the agency.
2. After some administrative reviews, a program A staffer may make data linkage a pair-to-pair data integration. But most often, a program A staffer furnishes the data integration request to an informatics support unit for linking data A to data D for years 2000–2005.
3. The informatics support unit is assigned a work order to link datasets A and D for the period of 2000–2005.
4. Program A receives the linked data labeled "linked data A to D years 2000 to 2005," or "L-A2Dy2000t05" for a short file name.
5. Program A delivers the linked file to the data requester.
6. If the data requester finds some discrepancy between what was requested and what was delivered, he or she may make an updated or revised request.

In the above scenario, program A is the data owner or steward for dataset A, which is an operational data store that keeps all the records and transactions at a point of time. Likewise, dataset D is owned by entity D, which can be either a program or an organization. Dataset D can be an operational data store or an extract from an operational database. A data request can be internal or external. An internal data request comes from an employee within the public health agency, which can request and view individual identifiable data for public health practice purposes. An external request comes from outside of the public health agency to a program within the agency, which in turn may ask for an external institutional review board (IRB) or legal review within the public health agency. In any case, the owners of datasets A and B have to grant permission for use of the data. Let's assume that dataset D is hospital discharge data (HDD) owned by organization D, which contributed to a data warehouse with the condition that after data linkage, personal identifiers must be removed before releasing data D to any programs.

Now, let's make the example more concrete. Dataset A is owned by the state Crash Outcome Data Evaluation System (CODES) program, and it is to be linked to the HDD, or dataset D. In a typical informatics support operation, linking dataset A to dataset D requires a cross-reference table that goes between datasets A and B. However, such a table may not be designed with repeated uses in mind. In addition, linking operational data stores A to D is not the only operation. In fact, there could be many datasets routinely linked to dataset D. Examples include trauma registry data, cancer registry data, and emergency medical services (EMS) to hospital data.

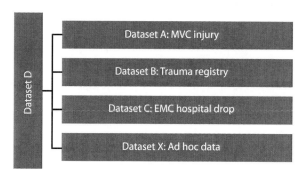

FIGURE 5.2   Data linkage requests.

Figure 5.2 shows four datasets—A, B, C, and X—to be linked to D. In a typical linkage operation, the informatics specialist would link data A to D, B to D, C to D, and X to D. In the probability linkage context, this operation has a number of problems. For example, one request could be for the 2000–2005 time period, while another could be for 2000–2006. The two requests could be 1 year extra for the latter, with 5 years overlapping. Since identifiers such as names have to be removed before releasing the files, they are treated as two separate data linkage projects. Most often, datasets A–C are related to each other. Trauma registry records (dataset B) and EMS records are often found in motor vehicle crash records. When all three datasets (A, B, and C) are separately linked to dataset D, a concern is often raised about whether the record from D is linked to the same person found in the other three datasets. This concern is also labeled interlinkage consistency or validity. While interlinkage operations are likely to be consistent in high-quality linkage operations, there are advantages in harmonizing datasets A, B, and C, so that a patient only needs to be linked once for the three. If one patient is linked to another database, there is no interlinkage operation to dataset D, thus avoiding potential inconsistency. The informatics support unit would avoid repeated linkage operations, such as overlapping years or interrelated databases, with essentially the same type of data linkage request.

On a broader spectrum, state and local public health agencies often discuss how to enhance interprogram collaboration, because in the past it has been known to be one of deficiency, as identified not only by the Institute of Medicine (IOM), but also by stakeholders, among others. One of the collaboration areas is integrated data sharing among programs to more effectively work with program recipients. Data linkage provides an impetus for moving forward with the discussion and eventual implementation of an agency-wide MPI. We borrow the term *MPI* to represent public health service registrants who are in various public health programs, similar to patients in multiple practice-based community health centers (Devoe et al., 2011). Some registrants may be patients, but most are people who receive some public health services. In this regard, MPI can also mean master (public health) participant index.

Let's take a newborn baby boy, John, and his mother, Alice, as an example, and walk through their life stages. When the newborn receives his birth certificate, the infant and the mother become registrants of the vital statistics record system, with their first names registered as John and Alice, respectively. When John receives infant screening, he becomes a registrant of the infant screening program. If Alice, together with John, is deemed eligible for the Special Supplemental Nutrition Program for Women, Infants, and Children (WIC), both of them become registrants of the WIC program. If Alice happens to be a refugee, she may also be registered in the Medicaid program, at least in the initial settlement period. Alice may also be screened for breast cancer if she meets income and age eligibility requirements for the Every Woman Matters program (or men and women's health program). As John receives his immunization shots, he is likely to be registered in a statewide child immunization system. As John goes through his life cycle, he will get married and obtain a marriage certificate, which will be recorded in the same vital records system as his birth. If he is admitted to a hospital within the state, his hospital records also become available to the state public health agency.

Can we track John or Alice through multiple datasets within the public health system? The answer is yes, and MPI is the key. There are many disease- or condition-specific programs, such as birth defects, early childhood disability, alcohol abuse and other behavioral health conditions, reportable infectious diseases, and cancer. Almost all program databases contain basic demographic data to uniquely identify a patient or registrant and enable association of his or her record. However, interprogram linkage is often problematic. Due to different data formats, even the same set of demographic variables from two closely related programs cannot easily identify the same person. There is a need to build an MPI that serves multiple programs. As each program makes a request to perform data linkage, we could potentially build an accurate MPI as the key to locating and linking program records in the statewide public health information system.

5.2.2 TOP-DOWN APPROACH TO DATA INTEGRATION

The question then becomes how to keep track of multiple datasets within a public health agency that may be integrated at the person or record level. One traditional data model is the enterprise data warehouse, which can be used within a health care organization. This traditional approach has two characteristics: the wholesale approach, which centralizes all datasets within an IT or informatics unit, and the top-down analytical approach, which looks at the whole picture when integrating program data. The centralized data warehouse essentially takes over most, if not all, program reporting functionalities and broadens many analytical functionalities that might not be available for individual programs.

Figure 5.3 displays a top-down approach to setting up an enterprise data warehouse. All program datasets are considered data sources, and they can be birth, cancer, and death records. All variables from data sources, regardless of confidentialities, are available for data integration. Identifiable information such as name, date of birth, and social security number (SSN) are available in the process of extracting, transforming, and loading (ETL). Sometimes, a dataset maintained by an organization outside of an agency (e.g., Nebraska HDD) has already gone through the ETL process, and we can only extract necessary variables from it. All the source data are integrated and stored in the enterprise data warehouse, and they can be used as if they are in individual programs. A data mart refers to a composite data warehouse, where each subunit within an organization has its own single function database, such as sales, shipments, and marketing in a corporation.

Data marts can be branched out top down from an enterprise data warehouse or developed from bottom up to be part of a data warehouse. An individual data mart can serve all data applications and reporting functionalities for an individual program, and it can generate quarterly reports, annual reports, and data submissions to funding agencies. This part of applications is business as usual for the individual program. A data mart can serve additional data analytic functionalities that cannot be achieved by the individual program. Common to both business as usual and add-on parts of a data mart are data operational efficiency and report replicability, which can hardly be achieved based on the traditional operational database.

However, there are some technical obstacles to having a centralized data warehouse:

1. Each program stewards its own data due to statute and funding requirements. This implies that even the director of a public health agency may not have the authority to force a fully integrated data warehouse, where all datasets could be stewarded by a crosscutting informatics program.
2. The top-down approach requires changes in the current record ID convention and authorization. For example, the Nebraska Trauma Registry uses hospital ID in combination with hospital incidence ID to create the unique registry ID that has more than 70 character spaces. Since it is a record ID, it cannot be referenced with the hospital MPI. Such an ad

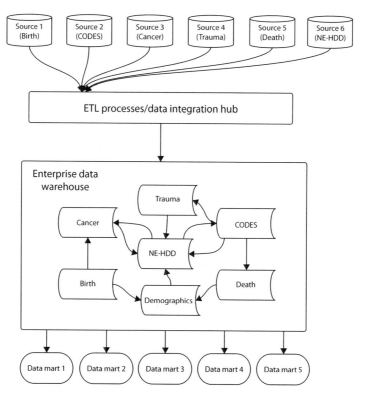

FIGURE 5.3 Top-down approach to data integration. NE-HDD, Nebraska hospital discharge data.

hoc ID convention has to be changed in the enterprise data warehouse. Even if two datasets have a common key, it may not be authorized for record linkage. For example, the SSN is available for many programs. However, an SSN from a birthing mother can only be used for social services verification. According to Nebraska statute §71-604.05(2), "social security numbers (a) shall be recorded on the birth certificate but shall not be considered part of the birth certificate and (b) shall only be used for the purpose of enforcement of child support orders in Nebraska as permitted by Title IV-D of the federal Social Security Act." Given the need for efficient data linkage, the issue of using SSNs has been raised several times, but it has not been resolved, because the statute that created the Nebraska Department of Health and Human Services (DHHS) (§81-3118) allows the chief executive officer (CEO) to adopt rules and regulations prescribing the standards and procedures for sharing confidential information between programs. However, no one has initiated this process yet. This example shows that each program may not champion change that only benefits informatics, and collectively, no one champions streamlined and efficient data integration that is required for a top-down data warehouse.

3. Changing current data systems to the enterprise system requires extra effort not defined by funding sources. Not all programs use the same database system. Variables are often coded differently in different systems. Sex can be coded either as 1 for male and 2 for female or as M and F, respectively. Even in a single program, such as EMS, different jurisdictions have different legacy systems. Although data harmonization can be broadly defined, we use the term here to describe a process of enabling consistency among tables between different data systems, so that common terminology, data format, coding, and other data features can be developed into integration tables from different data systems for the same data items. Since data harmonization takes time and effort, in many cases, it requires standardization moving forward. All of this requires extra effort from individual programs.

4. Public health practices are driven by emerging disease trends. Without knowing this demand, the data warehouse would face evolving developments that are hard to manage and maintain. Prioritizing datasets for linkage often means that linkage practice can be sustained among the most often used and interrelated programs. However, balancing future priorities and routine practice needs is an art that depends on many parameters out of IT's or informatics' control.

5. The required professional standards must be maintained to protect personal health information. If the informatics program has too much power to integrate program data, the linked information could be misused without the knowledge of data stewards. One solution to ensure privacy protection is to keep integrated datasets at a minimum level to limit data access, but this is inconsistent with the enterprise data warehouse.

Funding and program stability are additional limitations to deploying a top-down MPI design and associated data warehouse. Most public health program funding streams do not support data integration. And even though some programs do support data integration, public health programs come and go. In addition, shifting priorities from the funding agencies may curtail some data collections while expanding others. Finally, public health programs are funded with different requirements. Some require an epidemiologist or data analyst with a funding level from 0.25 to 1 full-time equivalent (FTE), while some have no funding for a data analyst. This problem is often exacerbated by the propensity for each program to manage its own data analysis personnel. Since 0.5 FTE is not sufficient, acquiring student interns or other arrangements are often made. Different levels of training and frequent turnovers in positions that are less than full-time cause a great deal of inefficiency in reporting mandatory items, leaving no time for data integration and quality improvement.

It is clear that the list of issues and limitations above prevents us from initiating the top-down approach to building an MPI within a fully integrated data warehouse. Even though the centralization scheme may be discounted under the current program-based data management and steward system, the top-down analytical approach should not necessarily be abandoned. In fact, when the time is right, the top-down analytical approach is still preferred to the bottom-up approach. However, since few people realize the value of an MPI and data integration, and because there are many statute limitations, now is not the right time to apply the top-down approach in developing an MPI within a public health agency.

5.2.3 BOTTOM-UP APPROACH TO DATA INTEGRATION

When there is a genuine need for data integration for the current public health programs in an organization, a bottom-up approach with selected programs may be a good starting point. Since data integration calls for a common set of linkage variables and a common ID within an organization, an MPI can be provisionally developed with potential for expansion. In other words, by selecting some data-intensive programs that can see the value of data integration, an MPI is a natural outcome not only for these programs, but also for all the programs with a need to share integrated data.

A bottom-up approach for data integration and data sharing has a number of advantages. First, a bottom-up approach tends to produce tangible results quickly. Deploying an enterprise data warehouse within an agency takes time. In an organization with many programs in silos, many people may be leery about providing data in the integration process. This attitude is especially true when a previous attempt at an integration process did not go well. Program people must therefore be convinced that they will be able to easily retrieve their own data and will gain information from other programs about individuals in their own dataset. Second, the lack of appreciation from the leadership for data integration means that most program managers and administrators either are new to or have not been sufficiently exposed to an integrated database concept. Program managers need to see the benefits of data integration, while upper management needs to be educated in order to appreciate the strengths of the integrated or linked data, gain additional insight, and improve operational efficiencies. Third, when an organization cannot hire consultants to design a loosely connected data warehouse via an MPI, in-house informatics specialists need to gain experience in designing and working with different and interrelated databases and database functions for variable harmonization. Finally, program users also need to gain experience and confidence with newly linked data and new tools for analyzing them. When sufficient experience is accumulated, and when other programs buy in, an organization-wide MPI for cross-dataset query and integration becomes easy. Hence, the key function for the bottom-up approach is to find and carry out demonstration projects, so that the benefit of integration and the use of an MPI can be appreciated by most, if not all, program managers.

Given that most public health programs would allow access to their data freely while fulfilling the needs of linking program data, the best option seems to be developing an agency-wide MPI through limited demographic and identification variables, so that datasets can be loosely coupled using an MPI. By loose coupling, we mean that two program datasets may not be able to be seamlessly integrated, but they can use a common ID, in this case an MPI, to link them if both programs agree to share. This development strategy fits perfectly with bottom-up piloting.

Figure 5.4 displays a number of bottom-up data integration processes. Each program is a silo, and its data source needs to be integrated with another data source either within or outside of the agency. As we went through each silo process, we developed a host of keys interlinking different program data. Even though it seems complicated, we could document and learn each data integration process one at a time. In addition, the bottom-up approach allows us to experiment with different pilot projects. In Section 5.3, we describe three pilot projects, each representing a unique feature. The first, a proof-of-technology pilot, is useful because it is able to check methodological feasibilities. However, the scope of a pilot test tends to be small, and it can only cover a small part of the collection of all the technologies. The second is a proof-of-concept pilot project, which tries

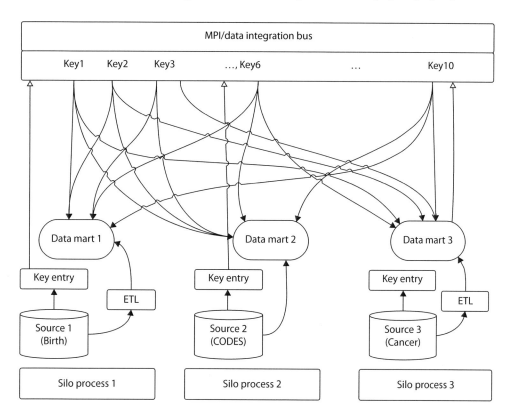

FIGURE 5.4 Bottom-up approach to data integration.

to find a viable solution for using information in order to inform public health practice and influence decision support. The third is a seed pilot project, which attempts to bring attention to the utility of an informatics initiative. The hope is to come up with an obvious program value that can be easily seen by other programs.

5.3 PILOTING BOTTOM-UP PROCESS TO GAIN EXPERIENCE

5.3.1 CVD PILOT PROJECT: PROOF OF TECHNOLOGY

This is a data linkage project not mandated by any standing programs. We wanted to consider a pilot project that meshed the need for informatics in data linkage with legitimate program needs for gaining additional information. We decided to pilot data linkage because of team members' experience and enthusiasm. Although three program specialists under informatics support had hands-on experience with data linkage, they used a free but restrictive software package that could not be shared with other programs. Hence, there was a need to test another, less expensive linkage software package. We selected Link Plus, a public domain software package, to test appropriate linkage software or linkage techniques.

The next step was to identify an enthusiastic program. We found that the cardiovascular disease (CVD) program wished to investigate EMS run times and hospital outcomes for CVD patients. The program had a report that was about 10 years old that reported CVD mortality from death records and CVD incidence from HDD, but HDD and death records were not linked. In addition, the EMS data had never been linked to HDD as a whole, even though EMS runs associated with motor vehicle crashes were routinely linked to HDD. There were three EMS systems (i.e., Omaha, Lincoln, and the rest of the state) that cannot communicate with each other, because variable names and formats among the three systems were different. Hence, there was also a need for data harmonization, at least to harmonize linkage variables such as name, date of birth, sex, residential address, and event date for ambulance service.

The Nebraska Hospital Association (NHA) provides the Nebraska DHHS HDD annually, but the data do not contain identifiable information, such as patient name. After meeting with the NHA officials, it was decided that the linkage would be performed at the NHA. A significant drawback of working at the NHA site was that the project had to rely on the NHA schedule, which meant that it had to be planned and negotiated well in advance.

There were two linkage operations: one linking EMS data to the HDD and the other linking death records to the HDD. All datasets need some cleaning and standardization, but death records and HDD have linkage variables within each database system, while statewide linkage variables from EMS have to be compiled from three EMS data systems. The latter task was assigned to the student intern, who put all linkage variables and EMS data system IDs and newly generated IDs in a file. In addition, a half-dozen important variables, such as chief complaint by patient, chief impression by EMS responders, and EMS run-time variables (call time, time to arrival at scene, and time to arrival at the hospital), were included. The rest of the variables were intended to be attached later, after linkage.

5.3.1.1 Successes

It took two experienced informatics analysts less than 3 weeks to link 2005–2009 EMS data to the HDD for 2005–2009 and to update (link) 2005–2009 HDD with 2005–2010 death record data. The linkage operation was a success. We learned the pros and cons of the Link Plus freeware, some of which were described in Chapter 4. We concluded that linkage and data integration could rely on Link Plus in combination with other data management software when there was no clear business model to sustain commercial linkage software, such as deduplication and data integration software packages that are commonly used by hospitals. We also learned that it was necessary to write our own working manual, as Link Plus has limited user guides. Another issue was that when doing data linkage, one needs to limit the number of variables to as few as possible so that Link Plus can handle the task without any errors. When a large dataset with a large number of variables is imported to Link Plus, it can cause errors that cannot be easily detected. Finally, everyone needed to be trained for using Link Plus, even the experienced people, so that everyone was on the same page when working together.

After linkage, we tried to conduct some demonstration data analyses. For instance, we showed that female stroke patients were more likely to die after 30 days than male stroke patients. Based on chief complaints and death records, we showed that about 3% of the CVD mortalities were found at the EMS scene, 7% of the CVD mortalities occurred at the hospital, and 90% of the patients who died from CVD did not receive either EMS or hospital care right before their deaths. Median EMS run times for CVD mortalities among different racial groups were all within half an hour, with a variation of less than 3 minutes. Blacks and Asians were more likely to have an inconsistent EMS impression and hospital primary diagnosis.

5.3.1.2 Lessons Learned

When we attempted to use this MPI to go back to the EMS data to attach additional variables for EMS–HDD–mortality data analysis, we ran into blocks. The main problem was that the state EMS ID was not unique, as it did not distinguish between the ambulance that delivered the patient and vehicles that arrived at the scene. An emergency service call often involves more than one ambulance. The first at the scene may treat the patient, while the better-equipped one may take the patient to the hospital. The third one could either be on standby and released or be the one transferring the patient from a smaller hospital to a larger one. Hence, it is critical to generate a person ID and link it to the many services (e.g., ambulances) when indexing an individual with one-to-many relationships in two or more data files. This index serves the function of master index within the EMS system. In this case, ambulance IDs associated with dispatches should be accompanied with the person-specific event. When the program (EMS) does not have a unique ID for a person- or dispatch-specific ambulance record, one has to create the event-based EMS record ID first, and then index it by patients, which serve the basis for the MPI. Even with regular supervision, miscommunication between the supervisor and the student, and different priorities for the student in the course of the semester, hindered the desired outcomes. The bottom line is that data harmonization or the ETL process should be handled by a professional rather than a student intern.

5.3.2 Cancer Data Linkage Project: Proof of Concept

This is the second pilot project, in which the project team was approached to carry out a proof-a-concept project for the Nebraska Cancer Registry (NCR). Traditionally, the NCR provides surveillance separately for major cancer sites by incidence and mortality. Survival surveillance had never been done. A midterm goal was set by the registry's advisory board for conducting survival surveillance. However, data quality for treatment and comorbidity, which are often used as control variables, needs to be enhanced. Since treatment was passively collected, and some cancer registers submit more complete treatment history than other cancer registers, it is necessary to update treatment information from additional data sources. In addition, comorbidity conditions are listed in the cancer registry data items, but they are seldom filled in by local cancer registers. Without controlling for comorbidity, the project might not be able to rigorously evaluate disparities in survival among cancer patients. The project team was asked if it was possible to use Nebraska HDD to regularly update treatment and comorbidity information.

HDD have more than 20 diagnostic conditions for each patient, and they are coded by International Classification of Diseases, Ninth Revision (ICD-9) codes. A number of comorbidity indices have been constructed according to hospital-diagnosed conditions by various authors; a notable one was the Charlson Index commonly used in cancer studies. However, it had never been done in Nebraska at the time of the study. In addition, there was no published report about using HDD to update treatment at the population level. However, we knew that chemotherapy (CT) and radiation therapy (RT) were often seen in outpatients. A pilot study linking NCR data to Nebraska HDD would be able to prove whether the outpatient data could provide substantial improvements in treatment completeness for cancer patients in the NCR.

From the informatics support and management points of view, the project had additional value. Both Nebraska HDD and NCR had high-quality patient IDs and record IDs. The former is indexed by patient identification variables, while the latter is indexed by incidence. If we associated the patient ID from the cancer registry with the unique patient ID in the HDD, we would have a master patient ID by combining the two IDs. If we kept doing this for each additional linkage dataset, an agency-wide MPI would become a reality.

5.3.2.1 Successes and Lessons Learned

With the above motivations in mind, we carried out the pilot to link 2005–2009 NCR data to 2005–2010 Nebraska HDD data. The linkage work and subsequent summary was contracted out. At the conclusion of the project, we learned that calculating the Charlson Index was relatively easy. We also learned that CT and RT rates could be increased by additional 15% when using outpatient data. Finally, we learned that for a project of this scope, we need about 0.10 FTE, because the contractor hired a master-level trained and experienced person part-time (20 hours a week) for about 4 months. Since it was the first time, the first month was spent identifying current procedure terminology (CPT) codes and other procedure codes that the contractor was not familiar with. In addition, some efficiency would be gained if this work could be made routine.

When designing a pilot project, the project leader should think ahead because the contractor does not see the big picture. Certain questions should be asked, such as "If the pilot project is successful, should we embark on a sustained operation for cancer–HDD data linkage?" Another question that the project leader should ask is whether the analysis program codes are reusable. If they are only for the pilot project, such as updating CT and RT, and calculating a comorbidity index, it may not be necessary to design the computer program codes to be portable and user-friendly among experienced users. However, if reuse is immediate, then the project leader should participate in the programming process. It turned out that none of the project team members, including the contractor, were familiar with CPT codes for generating CT and RT treatments, and therefore the codes for generating these treatments cannot be reused.

After the pilot, the treatment updates were included in the standard workflow of the NCR. In this regard, the proof-of-concept project supported the decision of whether a long-term project should move forward. However, we fell short in providing proof of concept for a wider range of issues. In retrospect, we should have provided a sampling of all the major issues and solutions, assuming that the decision would be made to go ahead. For instance, we did not document the importance and procedure of using event information for record linkage. When the project went ahead for production next year, we found that ignoring event information reduced the linkage rate by almost 20%, and the product cannot be used without a "redo."

5.3.3 Linking PRAMS to Birth Data Seed Planting Pilot

After successfully piloting the two data linkage projects in Sections 5.3.1 and 5.3.2, our confidence in integrating in-house grew substantially. We provided a "seed grant" to generate data integration interest among public health programs. The Pregnancy Risk Assessment Monitoring System (PRAMS) program was one of earliest to show interest. PRAMS is a surveillance system on state-specific maternal attitudes, behaviors, and experiences before, during, and after pregnancy. However, PRAMS is not designed to follow individual mothers across multiple pregnancies, hindering risk reduction efforts for subsequent births. The Nebraska PRAMS program had done a preliminary study looking into identical mothers being sampled and responding to PRAMS within four waves of surveys from 2000 to 2008. The program essentially wanted to extend the preliminary study to include the 2010 survey and to generate an internal report or external publication. The informatics team, in the meantime, wanted to test the idea of indexing mothers to the birth certificate data, so that all babies in the birth certificate data could be associated with their birthing mothers. In addition, the team had recently experienced a staff turnover and had just hired a biostatistician, who was not experienced in data linkage. The need for internal data linkage between the two programs provided a training opportunity.

Although the linkage and the approach to the linkage were straightforward, there was an unseen tension between the informatics approach to linkage and the application-oriented approach to data linkage. The former is a systematic approach to building an index of mothers for birth records of the study period (e.g., 2000–2010). Linking mothers from PRAMS is just a simple query of the index of mothers, which is essentially application independent. The latter approach is application dependent in that the linked database might be indexed by mother, but the index is only a by-product of the final data analysis project. The informatics team wanted to develop a master mother index within the birth certificate system and then use it for any projects that require mother–child paired data. However, we decided to make the informatics goal secondary, because many programs suspected that the informatics support program might take over other data operations in a big brother type of role. However, in so doing, the program PRAMS insisted on seeing tangible analysis results through data linkage, which drove the informatics specialist into an analysis mode. In the end, the pilot project became two separate projects.

The first part of the project was a linked data analysis of longitudinal outcomes for mothers who had some risky behaviors with their previous births. In this analysis project, linking mothers of multiple births was done deterministically and manually, because the project team leader from PRAMS had the most experience with this low-tech and time-consuming approach. In addition, since each wave of PRAMS had a slightly different coding for survey questions, the harmonization of survey items from multiple waves of surveys consumed additional time. In retrospect, variable harmonization should have been done by program analysts, because they have more experience with it due to the years they have been working on data analysis. Variable harmonization happened because (1) the new informatics specialist was thrown into the project without sufficient data linkage training; (2) the informatics specialist had biostatistics experiences, where variable harmonization is common; (3) the program wanted a rigorous study for free; and (4) the informatics specialist was not properly supervised by the informatics team leaders.

The second part of the project was the informatics. It followed the protocol of indexing mothers of multiple births through deduplication for mothers in all digitally available years, 1995–2012. The process of generating the master index of mothers had a number of steps. First, a sequential ID for all birth certificates was generated. Second, mothers were deduplicated using the data linkage method. This step included cleaning up mothers' names and other identifiable information and assigning a unique ID for each deduplicated mother. Third, the system-generated unique mother ID was translated to a more meaningful master index of mothers and assigned to mothers who had multiple births. Finally, a few hundred records indexed by mothers with multiple children were selected and sent to the vital statistics unit for review and verification. Since the vital statistics unit had the original paperwork and a legacy computer system that could query mothers by name and other identifiers, the verification process was not difficult. After generating the master index of mothers, the informatics project used birth certificate numbers to deterministically link the birth certificate number for PRAMS samples. The project was then able to associate each PRAMS respondent to each mother through the master index of mothers.

The master index of mothers also had the following characteristics: It identified the year that the mother first appeared in the birth certificate database, and it also had an indicator for major name or birth date corrections, such as last name, maiden name, and middle name, or transposed birth dates. Since the index can be applied to many queries and applications, its utility was demonstrated on several occasions. To query how many births each mother had had, we needed only to do a Structured Query Language (SQL), which showed that several mothers in Nebraska had 12 or even 13 children. The average number of children from a mother was 1.69.

5.3.3.1 Lessons Learned

The biggest lesson learned from this pilot project was that moving forward in a parallel manner with the dual goals wasted several resources. When planning a seed project, the informatics team should have the final say about priorities. A program in a public health agency is often lacking in resources. Once technical or statistical support becomes available, a program is going to use it to its advantage. The seed planting project should have indexed only mothers of birth certificate data and PRAMS data. The project could then have offered in-kind analytical support similar to a seed grant. The informatics team provided the index of mothers and linkage; the program could then have obtained additional funding or resources to analyze the linked dataset. The project was too much to handle for an informatics specialist who had to face two learning curves (learning linkage skills and substantive data analysis for the program). In the end, the project took more than 5 months of FTE without immediate impact, such as germinating interests from other programs. The only tangible product was the index of mothers from 18 years of data, from 1995 to 2012. These lessons taught us that we should start a simple yet valuable project to guarantee success.

To summarize the three pilot projects, we think that starting simple is the most important ingredient for success. If a program wants a complicated project, the project should be divided into pieces, and the piece that fits both program and informatics needs should be selected. When we tried to accommodate the EMS linkage in the testing technology, we made our resources thing, and failed EMS component. When we tried to accommodate the statistical analysis of PRAMS to the creation of a master mother index, we also failed. Had we separated these projects and made them simple, we would have had more success.

5.4 DEVELOPING AN AGENCY-WIDE STRATEGY FOR MPI FOR DATA INTEGRATION

5.4.1 Integrating Pilot Projects

As mentioned earlier, pilot projects serve both an exploratory and a bottom-up purpose. By bottom up, we efficiently extend the shelf life of pilot projects. This means that they not only serve their primary purposes, but also expand them, hopefully in an integrated way. Even though the proof-of-concept pilot for the cancer registry had one primary goal of proving the validity of updating treatment and comorbidity, the linked data could be used and updated operationally. Understanding the role of each pilot project within a broader data integration environment is essential for reuse and efficient use of data linkage projects. One way to expand the cancer treatment project is to add additional years, for example, from the 2005–2009 period to the 2000–2011 period. Another way is to expand the linkage operation horizontally to integrate cancer data with other program data, such as cancer screening. In most post pilot project assessments, integration, rather than a simple expansion, is the preferred way to put all experiences and lessons learned together. Common to all three projects was the data linkage element. If we could build on this element, all related programs may benefit from each other. One way to relate them is to build a DHHS systemwide MPI for multiple programs (e.g., vital statistics, WIC, child immunization, HDD, NCR, CODES, PRAMS, trauma).

5.4.2 Setting Up MPI with Loosely Connected Public Health Programs

In an enterprise environment within a closed health care system, an MPI is often used to keep track of every patient registered at a health care system, such as the Veterans Affairs (VA) system, or in a health care organization, such as Kaiser Permanente. An MPI ensures that the same set of demographic and registration information for a patient is represented only once and kept the same in all data systems within an organization. In a public health system, few programs have such a need. Perhaps the closest one is the vital statistics system that maintains birth, marriage, and death certificates. However, in the absence of a person index, even the same person in the three subsystems (birth, marriage, and death) might not have the same demographics. Within each certificate database, an index is feasible. For the death certificate database, a death has to be unique, but an index would be trivial due to a one-to-one relationship. For birth mothers, an index would be useful and possible, as shown in the PRAMS pilot. For marriage certificate data, it is not feasible to keep track of all married persons due to out-of-state marriages.

As we learned earlier, program data normally cannot be shared without data use protocols, and it is therefore impractical to index program registrants by directly accessing each program's data. Since almost all programs that have program registrants have a set of identifiers to be associated with a unique person, it is possible to build the public health MPI. The common elements for building the MPI are the person's name, date of birth, gender, race, ethnicity, period-specific address, alias or previous name, SSN-generated ID, program identification, registration date, and event date. Below, we summarize what would be a better practice in a state public health agency to build such an index:

1. Select the datasets that are most frequently used for record linkage and data integration. In Nebraska, these include (1) cancer registry; (2) trauma registry; (3) CODES; (4) cancer screening data; vital statistics data system for (5) birth, (6) death, and (7) marriage records; (8) Parkinson's disease registry; (9) National Electronic Disease Surveillance System (NEDSS) for reportable infectious diseases; and (10) PRAMS.

2. Extract the ID variable from each dataset together with identifiable information that could be used for record linkage. Normally, we include first name, middle name, last name, date of birth, sex, race, ethnicity, street address, city, and zip code. If event- or time-dependent information is available, include it. Date of birth for a newborn is also the date of giving birth for the mother, who is most likely admitted to a hospital before or on the infant's birth date. Motor vehicle crash date, EMS transport date, and date of transfer to the hospital are also likely to be close to the admission date.

3. Harmonize all linkage variables from all datasets and derive a combined set of the high-quality data linkage variables listed above. This is essentially an internal linkage or deduplication process, in which all identifiers are standardized and then linked. If a commonly used personal identification number is available, such as an SSN, a forward unique ID number generator or utility should be used among all the participating programs to generate an ID that cannot be reverse-engineered. For the event variable, we would retain only the most recent event that potentially points to hospitalization. Once we link to one hospitalization, the MPI from the HDD can be used to recover previous hospitalizations.

4. Based on step 3, create an internal, unique person-specific ID for each deduplicated record. This process is analogous to the process of generating the master index of mothers in the seed pilot. For a state like Nebraska, with less than 2 million residents, we would never expect to have more than 99 million, or eight-digit, patient records. We therefore reserved eight digits for person records, one digit for type of revision, and one digit for future use. In addition, we also generated the year and month of record entry with four digits for year and two digits for month. In reality, we also added a 4-digit field for other use indicators of MPI, which can be combined with the 10 digits or kept a separate variable. But for now, we use the 10-digit index as an example.

5. Based on step 3, create an indicator variable in addition to the index variable to indicate the existence of a record in each contributing dataset (by each program). If a record appears in all datasets of the 10 selected, a value of 1111111111 is entered in the variable field. If a record appears only in the third dataset, then the value of 0010000000 is entered. This method is not generalizable. If we had 100 datasets, we would not expect a 100-column-width indicator variable. For a small number of datasets, it served the purpose well. In addition, we also attempted to create a two-digit variable indicating the primary source for the high-quality identifiable fields. In general, we consider the cancer registry to be of

a higher quality, followed by the death and birth registries. Between the trauma registry and CODES, we consider the trauma registry to be of higher quality. Another reason for creating the primary source variable is that the race and ethnicity variables are missing in a number of datasets, including the Nebraska HDD. Hence, by keeping track of race quality, we can borrow the race variable from one dataset and compensate it by another dataset.

6. Combine the new ID field in step 4, the new index field in step 5, and the set of high-quality data linkage variables in step 3 to build the MPI data sheet or file for participating programs. This MPI dataset is returned to each program by using the original program ID. From the program's perspective, we sometimes also label the MPI program file as the program-specific master key data file. Each program's master key data file differs by the format of the program ID and the number of observations. The MPI data file has the complete records from the indexed datasets, and it points to their availability in each participating program. By looking at the agency-wide MPI data file, or the program-specific master key data file, each participating program can determine if another program has its own "patients" or program participants. In addition, if some demographics are missing, the program can use the corresponding demographic variables from another system to update them.

7. Designate a data usage protocol for each dataset. After legal review, an internal dataset with limited identifiable variables can be created for general use among participating programs, in which the common identifier is the MPI. In addition, this limited dataset is also linked to the HDD, so that all participating programs will share some information from the HDD. In the case of Nebraska, 2004 would be the earliest year we could use; before that year, data either had limited hospital coverage or poor item-specific data quality. Certainly, those contributing datasets can also be further integrated, but for now, we are leaving them intact as a single relational database for internal use. This internal linked dataset could be released to an internal program by the informatics support unit upon request.

8. For programs that only contributed program master keys, they would have to rely on data requests from each other to share program data. This type of protocol allows each program to determine when it wants to share its data with other programs through the data mart. Each data mart is like a program in a public health agency that sets its own data application policy, while informatics helps each program to develop program-specific applications or report functions.

5.4.3 Connecting Databases through Federation

We noted that HDD needed to be improved for its race and ethnicity variables, because more than 90% of them had missing values. To improve these two fields, we used race and ethnicity fields from multiple datasets, which provided a good opportunity to put all linked data together. Figure 5.5 provides a linkage scheme to update the race variable in HDD. After each dataset from the left, from cancer to driver's license, was linked to HDD, it was indexed and returned to each program. Note that the returned files are interlinked.

There is no direct reference to our loosely linked or distributed data integration method. While the MPI can be bottom up in an enterprise environment, once it is built, one entry will populate all related suborganization databases. In a public health agency, this is not possible at this time. Slightly inconsistent entries from different programs are the norm rather than the exception, and it is up to each program to decide updates or corrections to its own database. Our data integration and data sharing protocols are close to the federation approach in the warehouse design. Our data linkage and integration process is analogous to a data mart. Below we briefly describe these two dimensions.

Data federation in our public health program databases refers to managing multiple datasets within a loosely connected data system with multiple data ownerships (Haas et al., 2002).

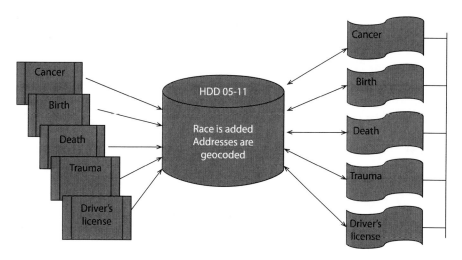

FIGURE 5.5 From individual datasets to indexed data.

The original idea for data federation was more technical, dealing primarily with using midware with some user interface to integrate different legacy data systems. Another feature of data federation is its ability to query against a single source to multiple sources. Although the public health data federation does not allow direct query over another program's database, the public health MPI provides linkage variables, and data management tools allow direct query if the program data are allowed to be accessed.

Our approach to building an agency-wide integrated data system or warehouse is similar to connecting data marts one at a time. Data marts can independently draw data directly from their operational databases in the process of ETL data. For example, breast cancer data from the NCR are used by at least four programs within the division of public health: Nebraska Department of Health and Human Services (DHHS), Every Woman Matters (for breast cancer screening); the Nebraska cancer prevention and control program; the chronic disease prevention program that includes cancer in its general reporting; and health statistics, which summarizes annual and trending data for cancer. Each program has its own data. Since all of them also use the NCR data, the best way to develop each program's data applications is through a data mart that generates uniform data with commonly used variables and time stamps. Since the ETL data are a clean and uniform dataset with all the records indexed, all programs would use the same dataset with identical outputs when tabulating the same data tables. Although independent public health data marts may sound inefficient for a public health agency, they provide a near-term solution for data integration, and this type of data warehouse may become efficient when more efficient distributed database technologies are developed.

5.5 CHAPTER SUMMARY

In this chapter, we described some inefficient data linkage methods common in state public health agencies. Data requests are redundant because program-to-program data linkage tends to be piecemeal and ad hoc. Ideally, a top-down approach of building an integrated data warehouse would meet most data linkage requests with ease. However, due to various technical obstacles and organizational limitations, this approach would not work in the current environment. We proposed a bottom-up approach to building an MPI for public health program participants, as well as for patients under public health surveillance. The primary purpose of having an MPI within a state public health agency is to efficiently deal with redundant data requests to informatics support. Although computational codes are reusable, the data analyst still has to repeat the process each time for two overlapping data requests. In addition, once an MPI is built, it tends to generate extra communication and collaboration among participating programs. Hence, the secondary purpose for having an MPI is to improve collaboration among public health programs with integrated and shared data resources. From a public health manager's perspective, it is all about performance improvement and better practice of public health.

In the process of exploring ways to integrate data, we provided three examples of how to implement a bottom-up approach to build an MPI within the state public health agency. Based on examples and other practice-based data linkage projects, we loosely connected data marts or program data. Each program is responsible for the maintenance of its data mart as if business as usual. The informatics unit maintains and updates the MPI by unifying a set of personal identifiers and providing a link to each program data mart. All data marts together are in a federation that follows the existing protocol for data sharing. This experiment is still ongoing.

REFERENCES

Devoe, J.E., R. Gold, M. Spofford, S. Chauvie, J. Muench, A. Turner, S. Likumahuwa, and C. Nelson. 2011. Developing a network of community health centers with a common electronic health record: Description of the Safety Net West Practice-Based Research Network (SNW-PBRN). *Journal of the American Board of Family Medicine* 24(5): 597–604.

Haas, L.M., E.T. Lin, and M.A. Roth. 2002. Data integration through database federation. *IBM Systems Journal* 41(4): 578–596.

Ponniah, P. 2008. *Data Warehousing Fundamentals for IT Professionals*. Hoboken, NJ: John Wiley.

Ziegler, P. and K. Dittrich. 2007. Data integration: Problems, approaches, and perspectives. In *Conceptual Modeling in Information Systems Engineering*, ed. J. Krogstie, A.L. Opdahl, and S. Brinkkemper. Berlin: Springer-Verlag, pp. 39–58.

6 Using GIS for Data Integration and Surveillance

6.1 INTRODUCTION

Traditionally, a geographic information system (GIS) is an enabling technology that includes a complete process: spatial data gathering, conversion, storage, manipulation, data analysis and visualization, finally mapping, and spatial analytical reporting. Nowadays, a GIS system can be a separate and loosely connected system. Some GIS functions can also be served by a computational module or Internet application in a distributed and loosely connected system. Regardless of centralized and distributed GISs, the core functions remain the same. In public health practices, the most direct application is during emergency, natural, and man-made disasters. Spatially coordinating critical facilities for emergency preparedness and spatially managing emergency responses to vulnerable and affected populations have been the core services provided by GIS. For instance, GIS allows a disaster response team to quickly access and display critical assets by location.

Managing location-based facility data and assets is relatively easy. After all, there are only a limited number of hospitals, nursing homes, fire stations, emergency service stations, school hazardous material storages, and so forth. While emergency preparedness remains the main GIS capacity in a local health department, the use of GIS has expanded to address many public health issues, such as social determinants in health, physical access to care, community health indicator mapping, and geospatial analysis of diseases and health conditions, as well as health outcomes. More and more geospatial analytics require that patient data to be coded at a finer geographic scale than the zip code. While we still maintain facility-based GIS infrastructure, we began to systematically expand it to program participants within the public health system using GIS.

There are many areas that can be covered by GIS applications; we narrowed our approach to building a GIS data infrastructure for all public health programs and exploratory analyses. We picked geocoding because many public health programs work in the same priority areas, and many individuals appear in multiple programs. The master patient index (MPI) framework proposed in Chapter 5 would not solve location-based individual information because people move, and the same individual may not live in the same neighborhood over a period of time, and therefore would not share the same location information. Hence, there is also a need to develop a master address index, so that those who live in the same address would share the same location information over time. To some degree, such an approach should be expanded to interagency collaboration within a state government, so that duplicated efforts can be reduced.

In this chapter, we proposed an infrastructure approach to attach geographic information to each individual public health program participant. In a typical state public health agency, geocoding is required for the cancer registry, vital records (birth and death), and a number of other administrative data. Geocoding most often refers to a process of matching residential addresses to geographically identifiable locations, using identifiers such as latitude and longitude. Geocoding belongs to record linkage because, in most applications, it is part of the process that links individual records to their contextual environment. Even though textual information other than addresses can also be geocoded, such as coding a county or a zip code to its center location, this chapter deals with geocoding address and how to attach census data to a location.

The remainder of the chapter is organized as follows. In Section 6.2, we stress the importance of geocoding by its utility from point to area-based measures. In Section 6.3, we propose a geocoding process using a master address index. In Section 6.4, we describe how to attach census tract data to each geocoded patient. Section 6.5 introduces spatial surveillance. Finally, we summarize the chapter.

6.2 GEOCODING-RELATED MEASURES IN SPATIAL ANALYSIS

The purpose of geocoding is to associate geographic information with the geocoded location. Although geocoding can be done for facilities, workplaces, and pollution point sources, it is mostly done for residential addresses. The focus on residential addresses is natural, as people spend most of their time at home. The literature dealing with residential-based segregation and environmental justice is extensive. Residential segregation deals with racial segregation and its societal consequences, such as inequalities in income, education, and occupation. More segregated communities are associated with high crime, fewer opportunities for minorities, and greater health risks. Given that racial and ethnic minority groups are more likely to live near a neighborhood with more environmental hazards, residential environmental qualities (both indoor and outdoor) have been linked to environmental justice. To achieve the public health objective that everyone should live in environmentally sound conditions, we have to link individuals to their neighborhood environments through geocoding.

As pointed out in Chapter 2, neighborhoods and social environments, such as poverty, are associated with various behavioral risk factors and health outcomes. In order to gain insight from these perspectives, it is necessary to geocode people's addresses and associate them with neighborhood conditions. In the old days, geocoding could be done by pins on a map. As addresses have become more and more digitally available, and street centerline files have become more frequently updated, computer-assisted geocoding has become the main method of deriving various geographic measures.

6.2.1 EXACT POINT MEASURE FOR INDIVIDUALS

Geocoding is important because it can associate a cluster of disease to potential environmental contamination or pollution. Before geographic information systems were introduced into public health, most disease mapping and epidemiological analyses were carried out at a given geographic unit that typically followed existing reporting units, such as county and zip code. Point-level analysis of disease data was rare. Although it is still rare, people have begun analyzing data at the point level. Methods for point cluster identification, such as spatial adaptive filters and the spatial scan test, have been developed that do not rely on any existing boundary.

6.2.2 PHYSICAL ACCESS MEASURE FOR INDIVIDUALS

Geocoding is the basis for objectively measuring geographic accessibility to health care. Access to health care facilities can be measured through questionnaire surveys, where respondents can report distance to their usual health care facilities. But more and more, distance to care tends to be determined through geocoding. Geocoding is especially useful with administrative data such as hospital discharge data, where we would have the patient address from billing information and the hospital address. Although different distance measures, such as Euclidean, Manhattan (rectangular), and network distances, can be used to measure access, the basis of these measures is geocoded locations for a patient residence and for the hospital.

6.2.3 PROXY MEASURE FOR INDIVIDUAL SES

As mentioned in Chapter 2, most administrative data in public health do not have income and education variables. In the absence of individual socioeconomic status (SES) variables, a viable option to study social determinants of health is to attach census tract (or neighborhood) information to each patient or public health registrant through geocoding. For diseases with few observations in each census tract, neighborhood SES variables can be used as proxy measures for individual SES variables.

6.2.4 MULTILEVEL MEASURES FOR INDIVIDUALS

The development of a multilevel statistical approach in the early 1990s and its later introduction to public health research have redefined the theoretical underpinning of social determinants. In addition to individual-level variables that we may have, such as income, education, and occupation, the neighborhood effect, such as where the person lives, can still exert an independent effect on a person's health. In other words, even if we are able to obtain an individual's SES variables, the contextual variables must be obtained for more rigorous analysis.

Besides geocoding individuals and their health care facilities, point sources, such as houses with lead poison, radon, or household water wells having a high concentration of toxins, are often coded for pollution studies. Recently, studies have coded other locations, such as grocery stores for nutritious food, convenience stores for tobacco points of sale, and liquor stores for behavior risks.

6.3 GEOCODING STRATEGIES: TOWARD A MASTER ADDRESS INDEX

Most geocoding is based on a street centerline file that has an address range on each side of the street. When a patient has a correct and standard address, we can normally identify the address location with acceptable accuracy. The quality of geocoding has become a concern more recently because poor geocoding can result in erroneous conclusions (Oliver et al., 2005; Zimmerman et al., 2007). For example, if multiple unmatched patient addresses are all represented by the centroid of a zip code, the census tract that is assigned to that zip code would have an artificially inflated relative disease risk (Boscoe, 2008). Although one could retain only those matched addresses by treating them as a random sample, the unmatched addresses tend to be unevenly distributed. Unmatched addresses are more likely to occur in rural areas and newly developed suburban areas, and less likely to occur in inner city areas. In this situation, disease risks would be lower in rural and suburban areas, and high in central city areas.

Although there are some new developments in geocoding, such as web services, due to confidentiality concerns and lack of certification for data quality, most state public health agencies cannot use such services as the main tool for geocoding. Address data qualities differ between states, and there are also minor differences in address style among different parts of the country. In many spatial cancer epidemiological studies, there is no other choice but to improve match rate and location specificity for unmatched records.

Borrowing the concept of MPI, geocoding should be done once at the state government level for all addresses within a state territory. Addresses can then be updated with new constructions and demolitions and be geocoded accordingly. Address recording can be standardized by subscribing to a web service and typing the zip code or city, and then the standard address can be selected by typing the first few letters. Each entered address is verified in real time within a statewide database. In this way, all the addresses can be geocoded once either by the private sector or by the public sector. Each address can then be assigned a unique ID and used to develop a master address index that includes latitude and longitude, type of properties, and so forth.

In reality, geocoding tends to be repeated many times for a single address by different agencies and organizations. For states with a population of a few million, the number of cases that are required for geocoding from a cancer registry and vital records may not sustain a high-quality and efficient business model within the state public health agency. Programs tend to contract out geocoding. However, since few people within the agency are knowledgeable about geocoding, the quality is rarely checked. For example, the quality of geocoding in a cancer registry ranges from the gold standard of global positioning system (GPS) location and street address range location to less exact centroids of geographic areas, such as zip codes, cities, and counties. But an assessment of cancer geocoding data quality in 2010 indicated that more than 30% of addresses were coded to either cities or zip codes, or could not be geocoded. Upon inquiry, we discovered that geocoded addresses for latitudes and longitudes and associated census tracts were not required to have a high quality.

One could simply put a geocoding quality variable in place indicating the quality, such as exact match, GPS location, or approximated by city, zip code, county, and so forth. Given these requirements, a contractor tends to go the easiest route by simply geocoding as much as he or she can and leaving "unable to code" with simply a zip code or city code. In a separate assessment of geocoded data from 1991 to 2005, we found that 66.4% of records had a pair of GIS coordinates coded at the street level. We also found that 82.8% of urban addresses were coded at the street level, in contrast to 51.0% of nonmetro addresses. Since African Americans in Nebraska are more likely to live in urban areas than in rural areas, we also cross-tabulated the quality variable by race. We found that addresses for African Americans were more likely to be coded at the street level (84.4%) than were addresses for Whites and other races (66%). These results suggest that georeferenced studies conducted for metro areas or for the African American population would require much less additional geocoding. They also suggest that geographic coordinates coded in the past must be updated in order to study geospatial health disparities.

For an informatics unit that provides geocoding support, a good starting point is to develop a within-agency master address database and geocode it for all. It should be mentioned that geocoding at a low-quality level could be very efficient in terms of time and labor. The concern is accuracy and consistency. The accuracy issue will be dealt with in Chapter 15 in the case study. Here we briefly describe steps to ensure consistency. In a database with a million patient records, together with addresses, the following steps should be considered:

1. Standardize addresses as much as possible while considering efficiency. This step is necessary not only for geocoding, but also for deduplication. Although there are some vendors and web services that would standardize addresses, it is not too difficult to do so in-house. In the standardization process, old addresses should be preserved, and both old and standardized addresses should have a common ID or key to go between the two. Another reason to do this in-house is that some low-quality addresses can serve as a reminder for the program to take some quality control actions in the future.

2. Develop an address index in combination with patient records, and assign a unique patient–address identifier for each combination. Let's take a hypothetical example of John and Joe, who both appeared in the database five times; John never moved, and Joe moved twice in the patient database. For John, one address would be sufficient for his five incidents in the database. For Joe, however, the patient–address combination has to be preserved. We need to ID three addresses with time stamps for him, one for his current address and two for his previous addresses. In the meantime, if another person resided at Joe's address, we would index that address to the new patient too.

3. Deduplicate address records so that each address is unique and only needs to be geocoded once. Each unique address should be indexed to the same patient for multiple visits, and multiple patients who live at the same address. In this way, each unique address can be linked back to the patient–address index in step 2. It should be pointed out that the deduplication of addresses does not have to be perfect, because the address matching process can take care of some hard to deduplicate addresses. In this regard, as long as the majority of addresses are unique, some minor differences in two addresses that refer to the same physical address can coexist in the deduplication process.

4. Geocode each unique address and link it back to the patient–address index in step 2. Interested readers can refer to Chapter 15 for detail. As a starter, one needs to find a geocoding engine from either a GIS or an address matching service. After sets of latitudes and longitudes are derived from a geocoding process, one needs to display them to make sure they are on the right spots.

5. When a new wave of data is to be integrated into the existing database, use the MPI to check whether a patient has moved.

   a. If the patient has not moved, use the most recent geocoded information for the patient to update the geocoding fields.

   b. If the patient has moved, one needs to standardize the address and geocode it.

   c. Before geocoding movers, one should check the standardized address against the master address index. If the address can be found, populate geocoding fields by using the geocoded information from the master address index; otherwise, proceed with geocoding. Note that one should not spend too much effort in standard address comparison for legacy address data because (1) early address entries into different systems could have many potential inconsistencies before address standardization was implemented, and (2) minor discrepancies due to typos or other uncertainties can be easily overcome by probability comparisons of the address fields (street, city, and zip code) in a geocoding process. When additional resources become available, the agency or program can then standardize legacy addresses that were entered into the database a long time ago.

Finally, establishments also need to be geocoded on many occasions. For example, in order to determine the nearest distance to a hospital, one must geocode hospital addresses for all hospitals accessible to a patient. Likewise, if the interest is access to liquor or tobacco, then store locations are needed. The data analyst should be reminded that most establishments have been geocoded by other entities. If it is easier to get the information through data requests between agencies, then a protocol of requesting and updating facility locations should be developed beforehand. In any case, geocoding establishments is not taxing. In the process of obtaining and geocoding establishments, one needs to add IDs that are linkable to the patient database or to a master person index. For cross-sectional analysis, an establishment ID, such as the federal income tax ID for hospitals and a liquor license ID for liquor stores, is sufficient. However, those IDs cannot be used over a longitudinal analysis, as they change even if the address is the same. In addition, one should never use automatically generated ID fields in geocoding or other geodatabases, as they tend to be just conveniently generated.

6.4 ATTACHING CENSUS TRACT DATA TO EACH PATIENT

Once each address is provided with latitude and longitude, it can be associated with any geographic unit on earth. In the United States, geographic units can be organized according to census geography, which includes census region, census division, state, county, census tract, census block group, and census block. Starting from state, the Federal Information Processing Standards (FIPS) codes are attached to all census geographic units. There are two digits for state FIPS codes, three digits for county codes, six digits for census tracts codes, and one digit for block groups. Census tract is the most commonly used geographic unit for SES measures, such as percentage of people living under the poverty level or the federal poverty line (FPL). When state and county codes are combined, FIPS codes are also used for county designations as five-digit codes. When county codes and census tract codes are combined, FIPS codes are used for census tract designations, which are called GeoID2 and have 11 digits. When the 1-digit census block group code is attached to the 11-digit census tract codes, the combined code is called GeoID and has 12 digits. FIPS codes at the county, tract, and block group levels are the building blocks of census geography.

Since latitude and longitude define a point, and a census tract defines an area or polygon, we can use a point in polygon operation in a GIS to associate any point within a census tract. In the ArcGIS environment, the following steps are in order:

1. Use geocoded longitude (X) and latitude (Y) to generate a point map, which can be a temporary event theme or a permanent shape file.

2. Load a corresponding census tract polygon map (shape file).
3. Conduct a point in polygon operation so that each point from 1 is associated with a census tract ID or GeoID2.
4. Once a census tract GeoID2 is attached to each address, add SES variables by downloading the American Community Survey (ACS) or other census tract–level data either within or outside the GIS environment.

Most recent SES data from the Census Bureau are based on the ACS. However, in order to understand ACS, we need to mention its predecessor—census long-form data from early decennial censuses, such as the 2000 census. In each decennial census, all households are required to fill out the standard short-form questionnaire asking only seven questions for each household member: name, sex, age, relationship, Hispanic origin, race, and whether the housing unit occupied is rented or owned. In addition, about one in every six households answered an expanded list of questions in the census questionnaire, called the census long form. The census long form was designed to meet various information needs from federal agencies. Examples include poverty status for welfare programs, car ownership and commuting for transportation planning, and veteran status for determining the needs of veterans and to evaluate the impact of veterans' programs.

Other census variables include marital status; place of birth, citizenship, and year of entry; school enrollment and educational attainment; ancestry; residence 5 years ago; language spoken at home and English fluency; disability; grandparents as caregivers for grandchildren in the home; labor force participation; place of work and journey to work; occupation, industry, class of worker, work status, and income in the previous year; units in structure and year when structure was built; when person moved in; and number of rooms in house and bathroom and kitchen facilities.

The above census variables and their cross-tabulations by demographics were released by the Census Bureau as summary tape file 3 (STF3) for the 1990 census and summary file 3 (SF3) for the 2000 census. For an at-risk population, census summary file 1 (SF1) is the primary source. All of the above variables are available at the census tract level, and some are also available at the census block group level. For demographic variables on the short form, data are available at the block level. The Census Bureau website of American FactFinder can lead to both 1990 and 2000 long-form data. Both the Census Bureau and the private sector offer census CDs and DVDs.

Starting with the 2010 census, the Census Bureau no longer provides the census long-form data. It instead provides similar data from the annual ACS, a 5-year pooled sample at the census tract. Since the ACS is done annually, it is very flexible. In addition to the traditional sets of questions, other timely questions, such as health insurance status, can be added and tabulated in the next year's release, at least at the state level. Furthermore, the 5-year pooled sample is on a rolling basis (e.g., 2005–2009, 2006–2010, and 2007–2011), which provides more timely monitoring of neighborhood SES changes than data from the decennial census long forms, which were only available every 10 years. When an inner city neighborhood has gone through gentrification, or a rural area is hit by a natural disaster (e.g., drought or flood), the 10-year census interval is clearly not sufficient. The ACS can monitor SES pulses in smaller areas in a timely manner.

Even though most studies would use the decennial census population as the basis for at-population calculations, sometimes people need SES as a control for an at-risk population. If one does want to include SES-based indicators, the corresponding universe should be the ACS population. For example, if a census tract has 5000 residents 25 years and older, and among them 3000 had a college degree or higher, the 3000 residents can serve as the at-risk population for college graduates who had a disease or health condition. Moreover, the ACS data are weighted by age, race, sex, and Hispanic origin according to the official Census Bureau population estimates at the county level. Since minority populations in the United States grow at much faster rates than the White population, ACS data at the county level may be more appropriate when decennial census data are more than 5 years old. In Nebraska, due to slight population growth, the ACS populations from 2006 to 2010 and from 2007 to 2011 are smaller than the census 2010 population by a fraction. However, if

TABLE 6.1

Age Distribution by Sex and Census Tract Poverty Level in Nebraska

Percent of Population under the Poverty Level

Female	<5%	5%–10%	10%–15%	15%–20%	>20%	All
Age < 35	45.59	42.85	44.76	48.29	55.87	46.50
35–44	14.48	12.35	11.36	12.26	11.63	12.45
45–54	15.64	15.22	14.43	13.66	12.44	14.53
55–64	12.65	12.84	11.71	10.87	9.16	11.75
65–74	6.18	7.84	8.10	7.78	5.24	7.12
Age ≥ 75	5.47	8.90	9.64	7.14	5.65	7.65
Female total	190,556	260,110	211,053	82,462	143,461	887,642
Male						
Age < 35	47.67	46.26	47.97	51.91	57.87	49.38
35–44	14.59	12.46	11.98	12.44	12.40	12.79
45–54	15.50	15.36	14.37	14.23	12.61	14.60
55–64	12.39	12.53	12.11	10.51	9.20	11.67
65–74	6.03	7.21	7.39	6.03	4.59	6.46
Age ≥ 75	3.82	6.19	6.18	4.89	3.34	5.09
Male total	186,582	256,370	205,165	80,482	142,464	871,063

Note: The 5-year ACS data are from 2006 to 2010.

poverty and other SES variables need to be included in an assessment, it is most straightforward to use ACS after 2005.

Table 6.1 lists the age distributions by sex and poverty level. The numbers in percent are aggregated from census tracts, and they cannot be produced by the 2010 census due to lack of census tract–specific poverty level for each age and group. The original counts from the table can be derived by multiplying each percent by the column total, which are essentially the at-risk population. We will use this table in many chapters that use at-risk population to calculate hospitalization rates and odds ratios. From the table, we would expect that females in the last two age groups are more likely to use health care resources than males, because females in the last two age groups account for 14.78% of the total females compared to 11.55% for males. Another interesting pattern is that both low-poverty neighborhoods (<5%) and high-poverty neighborhoods (>20%) had relatively low percentages for the category of age 75 and older. On the one hand, for the poor, poverty may reduce the surviving population for the last age group. On the other hand, for the well off, either the last age group is underrepresented or individual elderly persons moved out of those neighborhoods. Although we use these numbers at face value, we at least need to be cautious that ACS numbers are estimates rather than census counts, which are much more accurate at the census tract level.

A major concern with the ACS is sampling size. The ACS normally selects 5 in 55 households per year in its annual survey, which translates to 11 households over a 5-year period, compared to more than 16 households in the census long-form data. Consequently, the sampling error associated with the ACS is much higher than that of the census long-form data. For this reason, the Bureau of the Census added the margin of error (MOE) for each tabulated variable, either as a percentage or as a count. If the MOEs are too large for a particular variable, the bureau suppresses them. In the decennial census, sample size was not a problem because the complete census was also taken at the same time as the long-form sample. In other words, if a respondent did not respond by mail, the bureau would visit the respondent at least three times. With the ACS, in contrast, only one out of three was followed up for an interview if a household did not respond. Figure 6.1 presents census tract–level MOEs for Nebraska based on the 5 years of ACS data for

2010 Tract
MOE2PVT
■ 0.3990–29.10
■ 29.11–34.10
■ 34.11–38.50
□ 38.51–43.40
□ 43.41–48.80
□ 48.81–55.00
■ 55.01–65.00
■ 65.01–84.80
■ 84.81–200.0

FIGURE 6.1   (See color insert) Poverty MOE to poverty rate ratio by census tracts in Nebraska: 2007–2011. The rate is the ratio of MOE divided by the poverty rate within each census tract. If the rate is greater than 100, the MOE is greater than the estimated poverty rate. The box below the state map is Omaha and its adjacent area. (From 2007–2011 ACS 5-year data for poverty.)

2007–2011. As one can see, the MOEs are much higher adjacent to the Omaha area, where the MOEs of the poverty rates are greater than the poverty rates themselves (rate > 100). In other words, if a poverty rate is 10% in a census tract, its MOE is more than 10%. At this point, there is no acceptable way to treat large MOEs.

In addition to sample size, not all ACS variables are compatible with corresponding variables from the census long form due to the incompatibility of the 5% ACS sample versus the 5% long-form sample. The income level in a census tract for the long form is a snapshot of 1 year, while the income level in the ACS, to put it simply, is the average of 5 years of snapshots. Both the census long form and the ACS data can be retrieved from the Census Bureau's website (www.census.gov). If one wants to have census tracts for a few states, American FactFinder (http://factfinder2.census.gov) provides easy interactive access to each table: age and sex, race, Hispanic or Latino origin, educational attainment, employment status, income and earnings, industry and occupation, poverty, and work status. A nice addition to the ACS is table B19083, "Gini Index of Income Inequality." In the 5-year pooled sample, the Gini index table can be linked at the census tract level, and scholars have used it to associate low birth weight with the Gini index in South Carolina (Nkansah-Amankra et al., 2010).

6.5   SPATIAL VISUALIZATION AND DISEASE SURVEILLANCE

Spatial disparities in disease are common due to spatial gradients of SES and spatial distribution of natural and man-made hazards. There are many spatial surveillance methods, and several books are devoted to this topic. Spatial disease surveillance and assessment take mainly three approaches: detecting disease cluster, detecting potential relationship between a point source and its health impact, and assessing the environmental dose–response relationship between environmental variables and public health outcomes. In this section, we briefly summarize major approaches used by special surveillance specialists and provide some simple examples.

6.5.1 Exploratory Spatial Data Analysis

An exploratory spatial data analysis (ESDA) emphasizes data visualization. Early ESDA tool developments were mainly attributable to Getis and Ord (1992) and Anselin (1995) for spatial cluster and outlier detection. Spatial cluster detection has many methods. For continuous data, local indicators of spatial association (LISA) is popular. It is built into GeoDa, a freeware, and ArcGIS, a commercial GIS software. LISA is based on Moran's I statistic, which compares values within a defined area (usually by neighboring units vs. other areas) (Anselin, 1995). The method usually can spot either positive or negative spatial associations. The positive association includes two spatial patterns of high values near high values and low values near low values. Negative association is when high values are juxtaposed with low values, which is not common. LISA is often used for an exploratory analysis, and it usually does not consider multiple testing. Since spatial units are overlapping rather than mutually exclusive in a spatial weight matrix, a spatial unit can be tested multiple times depending on its neighboring association. When a detected spatial association has a significant p-value, it might not be significant when it accounts for multiple testing. A simple way to account for multiple testing is to divide the p-values by n, where n is the total number of spatial units in a study region. However, a more rigorous spatial statistic might be necessary to confirm the existence of a spatial cluster (Tango, 1995). LISA in GeoDa is most often used by geographers and social scientists. Below, we provide some examples.

Recall that in Chapter 2 we cross-tabulated the Gini index and poverty categories, and we showed that low-birth-weight gradients can be captured by these two dimensions. However, we do not have an idea where high-poverty or high-Gini index places are located. To demonstrate how we may use ACS data in conjunction with low-birth-weight data, we use the same geocoded birth data (2005–2011) as in Table 2.8. The process is (1) geocode the birth data for Douglas County, where Omaha is located; (2) aggregate them to each census tract by the number of newborns with low birth weight and the total number of newborns; (3) download census tract poverty data and the Gini index from the 5-year ACS data for 2006–2010; and (4) link them all by GeoID2, which is the census tract ID. We are now ready to map them. We map their distributions by quintiles that divide the census tract within the county into five equal categories (20%, mid-high 20%, mid-20%, mid-low 20%, and bottom 20%).

Figure 6.2a displays poverty rates, and it shows that census tracts that sit near the east border area had a concentration of high-poverty tracts. This area is close to downtown Omaha. There was also a spot in western Omaha where poverty rate was relatively high, which could be due to sampling error, as the area was not as populated as the eastern part of Omaha. Figure 6.2b displays local clusters of poverty rate according to LISA using the same data as in Figure 6.2a. In the calculation of LISA, we used the ArcGIS spatial statistics toolbox. The spatial weight was based on the rook rule (contiguity edges only). In a study area with variable population densities and census tract sizes, using spatial neighbors for the spatial weight matrix is better than using distance. Figure 6.2b showed that there was a significant cluster of high-poverty values close together. In addition, there were two low-value clusters in the Omaha suburbs. If we compare Figure 6.2a and b, the clusters make sense, and they simply add statistical significance to census tracts if we want to look for high-poverty areas.

Figure 6.3a displays the Gini indices. As we mentioned in Chapter 2, the Gini index is a measure of income inequality: the higher the number, the greater the inequality. The distribution of the Gini indices is broadly similar to that of poverty, but spread out more from the downtown area. Even the two most western census tracts had very high values of the Gini index. Statistically, there were still two low-value clusters in the western suburbs, but the high-value cluster or clusters tend to meander. If one knows the geography of the area, it is closely aligned with Center Street on either the north or south side. Note also that we used the rook contiguity rule, so we had a number of corner connections of high-value clusters. If we used the queen's rule or other spatial weight matrices, the results would be slightly different.

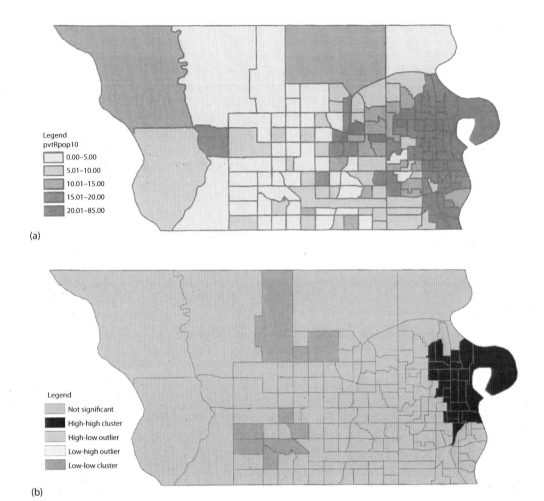

FIGURE 6.2   (See color insert) (a) Poverty rates in Douglas County (Omaha): 2006–2010. (b) Poverty rate clusters in Douglas County (LISA): 2006–2010.

Given the observed SES distributions from the above figures, we expected that low-birth-weight rates are concentrated in areas with high poverty rates and high Gini indices. Indeed, when we examined Figure 6.4a, which displays low-birth-weight rates, we found that our expectation was generally true, although low birth weight is also less concentrated than poverty concentration. In the figure, we also include a thicker outline in the north central city area indicating a cluster, which we talk about next. But now, let's look at Figure 6.4b's LISA results. First, we observed a significant high-value cluster in central city neighborhoods that corresponds to both the poverty cluster and Gini index clusters. Second, we observed a low-value cluster along the southwest that is close to the southwest cool spots from both Gini and poverty clusters and rates. Third, there were some negative local associations (high value next to low values, and vice versa), and we labeled them as outliers. The two low values next to the high value ("surrounded" by high-value census tracts) are next to the high-value cluster. The high value (high low-birth-weight rate) next to the low-value outliner is in the Ralston area.

The above results were exploratory and do not account for multiple testing problems (Tango, 2000). After an exploratory analysis, more rigorous statistics are necessary to confirm the

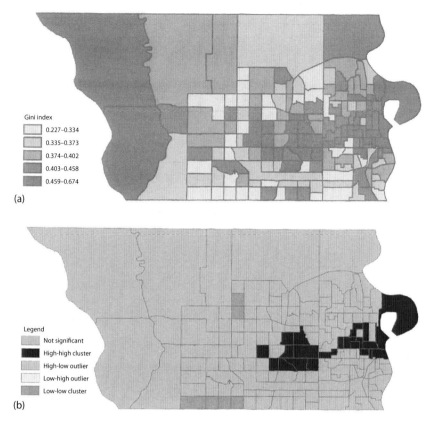

FIGURE 6.3 **(See color insert)** (a) Gini indices in Douglas County: 2006–2010. (b) Gini index clusters in Douglas County (LISA): 2006–2010.

existence of a cluster by spatial epidemiologists. Spatial epidemiological surveillance uses many spatial statistic tools, mostly for count data that are based on the Poisson assumption. The most popular one, perhaps, is SatScan, software that performs a spatial scan test for various retro-spective and prospective cluster detection and surveillance (Kulldorff, 1997). Federal, state, and city agencies, such as the National Cancer Institute, the Washington State Department of Public Health, and the New York City Department of Mental Health and Hygiene, use SatSan. The spa-tial scan statistic is shown to be the most powerful for local cluster detection (Kulldorff et al., 2003). It uses a moving circle of varying size to detect a set of clustered regions or points that are unlikely to happen by chance. Even though the moving window may circle a spatial unit multiple times, only the most likely cluster circle is used. It therefore does not have the multiple testing problem.

Based on the scan statistic from SatScan, Figure 6.4a shows a cluster based on the likelihood ratio test of one degree of freedom for 41.86 deviance. In spatial statistics, most people use relative risk, whose calculation is straightforward. It is the local rate divided by the whole study area rate. The cluster had the expected number of low-birth-weight cases (267), and there were 415 observed cases with a relative risk of 1.71. The cluster is centered at the 31055000700 census tract with a radius of 1.28 miles. Once a cluster is detected, one can then zoom in to a local area to associate the cluster with etiological factors. For instance, after identifying local breast cancer clusters by the spatial scan method, Roche et al. (2002) compared local socioeconomic conditions within and outside of the

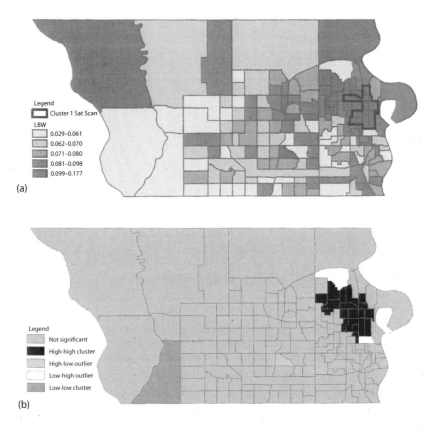

FIGURE 6.4   (See color insert) (a) Low-birth-weight rate in Douglas County (Omaha): 2005–2011. (b) Low-birth-weight rate clusters (LISA) in Douglas County: 2005–2011.

clustered areas. For instance, the Black population in the cluster area accounted for 67% of the total population in the clustered area. The Black population outside the cluster area, in contrast, accounted for 9% of the total nonarea population. Such descriptive analyses are fairly common. Recuenco et al. (2007) examined spatial and temporal patterns of enzootic raccoon rabies with multiple covariates.

SatScan is required to have a predefined scan window size by either population or some geographic units. If the window size is not clearly defined, one can end up with a cluster about half the size of the study area. How to determine to cluster location and size is one of the practical issues often encountered by spatial surveillance specialists. In the above example, we used SatScan's default population size of 50% as the window, which results in window sizes varying over the scanned area. Properly determining the window size according to either geography, such as spatial contiguity, or at-risk population size is not easy due to overdispersion and other problems associated with Poisson-based tests. The Poisson distribution assumes that its variance equals its mean, but in reality, they are often not equal. Cases are overdispersed when their variability exceeds that predicted by the Poisson distribution. The scan statistic may have an unacceptable type I error probability in the presence of overdispersion. In such a situation, the user has to decide whether an alarm or signal of a cluster is false. For example, dispersed cases were noticed in the syndromic surveillance of lower respiratory infection based on the spatial scan statistic in eastern Massachusetts (Kleinman, 2005). After the adjustment of known factors that caused overdispersion, the false alarm rate was reduced by as much as 30%. In addition to adjusting known factors, there are at least two ways to account for overdispersion statistically. One is to add a parameter to a generalized linear model that allows an extra variation from the mean, and the other is to model spatial random effects. Both require a model-based approach that can incorporate an additional parameter into the spatial scan statistic. In addition, the traditional scan statistic was unable to incorporate ecological covariates, but this problem has been resolved by and large (Zhang and Lin, 2009, 2013).

The spatial scan test was unable to infer point source, which can be achieved by focused tests. Both Getis–Ord G (Getis and Ord, 1992) and Tango C_F (Tango, 1995) are focused tests, but they are not designed primarily for area unit–based counts or rates. The Diggle's parameterized test was designed to work with a known point source, such as a municipal solid waste incinerator, power plant, or nuclear power station (Diggle, 1990). The original method was based on a spatial point process for disease incidents and used a circle to define and test a cluster existence. The point process framework was later extended to generalized linear models for individual and aggregated data (Diggle and Elliott, 1997). Depending on types of data, one can run either a logistic, logit model, or lograte model around the point source. Both cases and noncases (or exposures) would be aligned according to distance from the point source. In the parameterized model, one would use a distance decay function, such as an exponential function. If the expected rate follows the distance decay function, then the point source has a significant impact on disease incidence. Certainly, one can also use a series of dummy variables according to the distance from the point source. If the parameter estimate for a zone closer to the point source has a more positive and significant value than the one farther away, then the point source would significantly elevate the disease incidence. Since point sources are determined by public health concerns, and diseases or health conditions can be rare or common, such a test often requires some kind of study design that is hard to include in a surveillance system.

Besides point source and area SES, there is area-based pollution exposure. Obvious population exposures include air, water, and agricultural chemicals. If we know the geographic distribution of hazards and some dose–response relationship, we could use them to explain spatial patterns. Most environmental health tracking systems tend to track known pollutants and environment hazards as environmental exposures on the one hand and potential health outcomes on the other. Interested readers can refer to the Centers for Disease Control and Prevention (CDC) environmental tracking system website (http://www.cdc.gov/nceh/tracking/). Essentially, using area-based environmental variables to correlate with spatial distribution of disease is analogous to including ecological covariates in cluster detection. If environmental hazard A causes disease B, then the disease cluster would not be evident if we use environmental hazard A as the explanatory variable for disease B's distribution.

In many cases, environmental exposures are known, but their relationship to a particular disease incidence is unknown. In this case, some surveillance practices might be warranted. For example, Paraquat is known to be associated with Parkinson's disease, so we map its distribution and then relate it to the distribution of Parkinson's disease. Figure 6.5 shows Paraquat usage by census tract, which was derived from 1×1 kilometer grid exposure estimates from 1 of 19 agricultural chemical maps in Nebraska (Wan, 2015). Although there are many methods to assign a value to a census tract from a set of values within the census tract, we simply extracted the exposure map value according to the centroid of each census tract. Based on values at the census tract level, one can then attempt to associate spatially observed disease patterns with spatially estimated exposure patterns. An example will be presented in Chapter 20.

Finally, it is necessary to point out how to protect confidential data from visualization. Sometimes, sensitive data cannot be displayed by geographic units due to potential Health Insurance Portability and Accountability Act (HIPAA) concerns. To protect the confidentiality of public health program participants, a smoothed map without a clear boundary or an accurate rate can be shown. A case in point is HIV positive data. Figure 6.5 shows HIV positive cases in a 27-year period. The original data were geocoded at the address locations and then aggregated to census tracts. Since exact rates

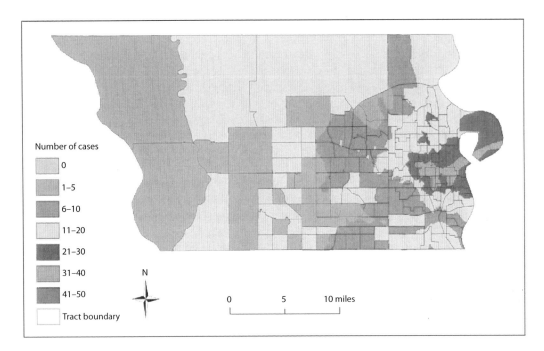

FIGURE 6.5   **(See color insert)** Smoothed HIV cases in Douglas County: 1983–2010.

cannot be displayed, a spatial smoothing method (e.g., Kriging) was used. In this case, it is still obvious that the areas are close to central city or low-income neighborhoods.

6.6   CHAPTER SUMMARY

In this chapter, we highlighted the importance of geocoding and recognized that there are going to be some quality issues associated with geocoding. We proposed a master address index approach to sustain geocoding operations. Theoretically, each standardized address should be coded only once to a very high quality. After a while, there will be very moderate effort in geocoding each year within the public health agency. In reality, low-quality geocoding is easy and cheap; people might just take an easy route to geocode. The drawback is low and inconsistent geocoding. Since a major application of geocoding is to associate patient residential location with neighborhood SES variables, we also described how to attach census tract data to geocoded coordinates, and issues that people should watch for. Specifically, SES variables are now most commonly from ACS 5-year data, which also include the MOE for each variable. The MOE concern is more pronounced in rural and sparsely populated areas due to small sample size.

In addition, we conducted some exploratory spatial data analyses by visualizing poverty, Gini index and low birth weight, and LISA cluster maps. One thing about ESDA is that the results are preliminary and need to be confirmed by more rigorous spatial statisticst. To that end, we described the spatial scan statistic and demonstrated its application. Finally, we briefly commented on other spatial surveillance concerning environmental exposure and displaying potentially protected data.

REFERENCES

Anselin, L. 1995. Local indicators of spatial association-LISA. *Geographic Analysis* 27: 93–115.
Boscoe, F.P. 2008. The science and art of geocoding: Tips for improving match rates and handling unmatched cases in analysis. In *Geocoding Health Data: The Use of Geographic Codes in Cancer Prevention and Control, Research and Practice*, ed. G. Rushton, M.P. Armstrong, J. Gittler, B.R. Greene, C.E. Pavlik, M.M. West, and D.L. Zimmerman. Boca Raton, FL: CRC Press, pp. 95–110.

Diggle, P. and P. Elliott. 1997. Regression modelling of disease risk in relation to point sources. *Journal of Royal Statistical Society A* 160: 491–505.

Diggle, P.J. 1990. A point process modelling approach to raised incidence of a rare phenomenon in the vicinity of a prespecified point. *Journal of Royal Statistical Society A* 153: 349–362.

Getis, A. and J.K. Ord. 1992. The analysis of spatial association by use of distance statistics. *Geographical Analysis* 24: 189–206.

Kleinman, K. 2005. Generalized linear models and generalized linear mixed models for small-area surveillance. In *Spatial and Syndromic Surveillance for Public Health*, ed. A.B. Lawson and K. Kleinman. London: Wiley, 2005, pp. 77–93.

Kulldorff, M. 1997. A spatial scan statistic. *Communications in Statistics, Theory and Methods* 26: 1481–1496.

Kulldorff, M., T. Tango, and P.J. Park. 2003. Power comparisons for disease clustering tests. *Computational Statistics and Data Analysis* 42: 665–684.

Nkansah-Amankra, S., K.J. Luchok, J.R. Hussey, K. Watkins, and X. Liu. 2010. Effects of maternal stress on low birth weight and preterm birth outcomes across neighborhoods of South Carolina, 2000–2003. *Maternal and Child Health Journal* 14(2): 215–226.

Oliver, M.N., K.A. Matthews, M. Siadaty, F.R. Hauck, and L.W. Pickle. 2005. Geographic bias related to geocoding in epidemiologic studies. *International Journal of Health Geographics* 4: 29.

Recuenco, S., M. Eidson, M. Kulldorff, G. Johnson, and B. Cherry. 2007. Spatial and temporal patterns of enzootic raccoon rabies adjusted for multiple covariates. *International Journal of Health Geographics* 6: 14.

Roche, L.S., R. Skinner, and R.B. Weinstein. 2002. Use of a geographic information system to identify and characterize areas with high proportions of distant stage breast cancer. *Journal of Public Health Management Practice* 8: 26–32.

Tango, T. 1995. A class of tests for detecting general and focused clustering of rare diseases. *Statistics in Medicine* 14: 2323–2334.

Tango, T. 2000. A test for spatial disease clustering adjusted for multiple testing. *Statistics in Medicine* 19: 191–204.

Wan, N. 2015. Pesticides exposure modeling based on GIS and remote sensing land use data. *Applied Geography* 56: 99–106.

Zhang, T. and G. Lin. 2009. Cluster detection based on spatial associations and iterated residuals in generalized linear mixed models. *Biometrics* 65: 353–360.

Zhang, T. and G. Lin. 2013. On the limiting distribution of the spatial scan statistic. *Journal of Multivariate Analysis* 122: 215–225.

Zimmerman, D.L., X. Fang, S. Mazumdar, and G. Rushton. 2007. Modeling the probability distribution of positional errors incurred by residential address geocoding. *International Journal of Health Geographics* 6: 1.

7 Methodological Preparation for Health Disparity Assessment

7.1 INTRODUCTION

In preparation for health disparity assessment, we have introduced conceptual and measurement issues of health disparities, record linkage, data integration, geocoding, and exploratory spatial data analysis in the previous chapters. This chapter covers the remaining methods in study design and analysis for disparity surveillance or assessment. We define health disparity surveillance as using existing administrative or observational data to systematically detect health disparities along major disparity dimensions, such as race and ethnicity, gender, socioeconomic status (SES), and rurality. The materials in this chapter are mostly covered by an intro-level epidemiological course.

Although we use the term *surveillance*, we want to make some distinctions between *disease surveillance* and *disparity surveillance*. First, disease surveillance most commonly refers to timely monitoring a set of reportable disease occurrences so that a disease outbreak can be captured at an early stage. Although health disparity surveillance can be based on a set of diseases or conditions, the list is much longer than that for reportable diseases. As mentioned in Chapter 2, health disparities include not only disease, but also health behaviors and access to care. In addition, due to the dynamics of disease-specific importance in public health and health policy, the list of diseases often changes in disparity surveillance. At a minimum, as one disease disparity is eliminated, one moves on to another. Second, disease surveillance happens in real time or near real time. Disparity surveillance, on the other hand, is not critical to conduct in real time, because most factors contributing to health disparity are chronical rather than instantaneous or time sensitive. Third, in disease surveillance, the next disease outbreak is unknown. In health disparity surveillance, it is already known that a certain degree of disparity exists, and the task is to measure disparities as a baseline and monitor changes in disparities while controlling for known risks. Finally, once a disease outbreak is detected and confirmed, it triggers an immediate public health action. Once an important health disparity is confirmed, it might lead to further disparity assessments or studies to identify modifiable risk factors.

It also worth noting that even though both disease surveillance and disparity surveillance use the same methodology and statistical tools, disparity surveillance represents much wider interests from multiple public health programs. It implies that there might not be a standard protocol for conducting disparity surveillance, as different programs have different requirements. Given the above cautionary note, we tried to cover study design and statistical measures for cross section and time series disparity surveillance.

7.2 SETTING THE SURVEILLANCE SCOPE

We started with setting the surveillance scope, which is different from most epidemiological studies that start with setting the hypothesis. Typically, when initiating a research project, we phrase a specific research question as the hypothesis; for example, there is no treatment difference between Blacks and Whites for small-cell lung cancer patients. In disparity surveillance, we normally do not have a specific research question; hence, determining the scope, such as which diseases or health conditions should be included in the surveillance, becomes a nontrivial question. Normally, defining the scope is an iteratively narrow-down process. Factors that influence the scope of disparity surveillance include a list of needs, existing data and auxiliary data sources, and a list of diseases or conditions.

Needs. A list of needs often comes from programs and the literature. A rural health program would want to know access to care disparity, a stroke program would want to know stroke rehabilitation, a health service–related program would want to know readmission, and a minority program would want to know morbidity and mortality by race and ethnicity. Many state public health programs have an advisory council that can provide some input. In the preparation of this book, we interacted with minority health, cardiovascular disease (CVD), vital statistics, cancer prevention, cancer screening, and Pregnancy Risk Assessment Monitoring System (PRAMS) programs. The consensus was that many programs need to know morbidity, mortality, and birth outcomes. Since the latter two have been documented extensively in the literature and internal reports, the largest information gap existed in morbidity.

Data. Some morbidity information can come from survey data, such as Behavioral Risk Factor Surveillance System (BRFSS) and PRAMS. However, these surveys provide a very limited number of diseases or conditions, and most on prevalence. We decided to use hospital discharge data (HDD) because it has never been used systematically for disparity surveillance. In the absence of data from physician clinics, the HDD represent the most comprehensive data for calculating population-based morbidity for the state. However, a major limitation of the HDD was that they do not have race and SES information. With race information from other program data and with geocoding for SES data, we could add this auxiliary information through data linkage.

Depth and width of scope. Most surveillances use descriptive statistics, so the focus should be on the width of major disease categories. A good starting point for disease selection is the list of top 10 diseases with largest disparities or significance. For example, most state and local health departments regularly report the top 10 causes of death, with the top 5 normally including heart disease, cancer, chronic obstructive pulmonary disease, stroke, and unintentional injury. One consideration for diseases to be included in surveillance is statistical power, or the number of observations needed for testing statistical significance among different combinations of groups and categories. Since diseases and events can be common or rare, it is difficult to use the power to decide the width of surveillance. If we want to examine health disparities by race or SES for the top 20 diseases or the principal diagnosis for hospital admission, we will likely have a sufficient number of cases in Nebraska, which has a little more than 200,000 admissions annually. However, if we want to examine quality of care by using the indicator of readmission, then the top 20 diseases for admission are unlikely to yield sufficient readmissions for disparity surveillance.

Even though we describe data and the width of scope separately, the two in general often go together: we need to identify an appropriate dataset and assess its feasibility. For example, if we want to assess racial disparities in low birth weight by neighborhood poverty, we would want to see how racial categories are distributed by neighborhood poverty categories. If a neighborhood category had few observations of a particular race, then we would have to think about combining race or neighborhood categories. Furthermore, if the study is expanded to consider residential mobility during the pregnancy, we have to know if both the race and neighborhood categories have a sufficient number of movers. In other words, in expanding the scope of study, it is necessary to assess existing datasets for their potentials and limitations.

7.3 STUDY DESIGN

7.3.1 CROSS-SECTIONAL AND CASE CONTROL

Most surveillance studies are cross-sectional with a predefined study population, which can be divided into having and not having the disease. One can then examine the two groups by a disparity dimension, such as race or SES. Here, we use a disparity dimension to mimic the degree of exposure. For example, the Nebraska BRFSS is an annual survey of adults age 18+ with many disease indicators, such as asthma. It also has standard income, education, occupation, and race and ethnicity groups, which could be used as disparity dimensions. In such a cross-sectional study, we can use the term *case control* loosely, where we could start by dividing the sample into having asthma or not as cases versus controls, and then use farmer versus not as exposure, assuming farmers would be exposed to a lot more agricultural dusts than nonfarmers.

Since most administrative data are collected for a number of years, they can easily be adopted for a case-control design. A case-control study is retrospective, and it starts with defining cases and controls and then associates them with a past exposure. For instance, when we examine asthma admissions, we treat having the asthma as cases and not having it as controls. We can then correlate them with neighborhood exposure to air particulates prior to admissions.

7.3.2 TIME SERIES OR PRE- AND POSTSTUDY DESIGN

The time series is a research design with repeat measures taken at different points in time. Time series data do not require following the same individuals over time. If two cross-sectional designs use the same set of variables, then they become a time series design. A commonly used time series design in public health is BRFSS, which has a set of the same questions and variables each year. Its annual report can track changes of many behavior risk factors over time. When time series data are divided by an event, intervention, or something that is expected to have some impact on key variables of interest, it becomes a pre- and poststudy design. The smoking rate is perhaps one of the most monitored variables. When assessing the impact of a citywide smoking ban in public places, one can use this design to collect data on smoking prevalence before and after the smoking ban. We will use an example based on BRFSS data to elaborate this design in Section 7.5.3. In many surveillance studies, it is necessary to compare temporal trends with an implicit pre- and postdesign. For example, after the Affordable Care Act, we would expect a decline in the use of emergency departments (EDs) for ambulatory care sensitive conditions (ACSCs). Since the use of EDs for ACSCs is disproportionate for the poor and racial minorities, we would also expect a reduced disparity of using EDs for ACSCs along SES after the Affordable Care Act.

Time series designs are also widely used in environmental exposure studies. Increased modeling sophistication, such as generalized additive models, can smooth seasonal and other time-dependent trends while teasing out the relationship between environmental exposure and disease. A classic example is the study that associated air population to mortality in Philadelphia in 1974–1988 (Kelsall et al., 1997). Time series designs can also incorporate multiple locations. In a study by Lin and Zhang (2012), birth weight and extreme weather were associated at the county level in the United States using the National Natality Index from 1969 to 1988.

7.3.3 LONGITUDINAL

The longitudinal study design requires study of the same individuals over time. In an administrative database, it can be achieved by indexing the same individuals and following them over time either prospectively or retrospectively. The retrospective design can be incorporated by epidemiologists for a cohort study, in which individuals normally divide into exposed and not exposed groups to see whether they develop a disease or not within the dataset. With few exceptions (cancer registry,

HDD), most datasets within a state public health agency are not indexed at the individual level and cannot meet the standard of the perspective cohort design. However, most data can be indexed to build a retrospective cohort. For example, annual motor vehicle crash injury data are cross-sectional. However, the police-reported data include a driver's license number together with the name. We can easily index drivers by their license numbers and study their behavior change over time retrospectively. For example, drunken drivers who crashed and caused injuries are often required by the court to take alcohol education classes as well as an evaluation. We can then study which drivers are more likely to be repeat offenders in terms of crash injuries due to driving under the influence of alcohol. Since most statistical measurements in cohort studies are similar to those in cross-sectional and time series studies, we will not separately cover them in this chapter.

7.4 CROSS-SECTIONAL MEASUREMENTS

Regardless of the study design, most surveillance studies use the same set of statistical measures. We first briefly describe some of the common measures of health disparities, and then proceed with examples on how to guide disparities assessment. Although mean is used for a number of health indicators, such as birth weight and body mass index, most comparisons of health disparities are based on rates. We will describe primarily the rate and its various measures for group comparison.

7.4.1 Rate

Let's assume that Table 7.1 shows deaths due to gun violence in a section of a city in a given year, where the total population is 11,620 and the average death rate due to gun violence is 0.0182, or 18 deaths per 1,000. Note that there are sharp racial differences, where the death rates for Blacks and American Indians and Alaska Natives (AI/ANs) are 10 and 5 times, respectively, that for Whites. However, when comparing different racial groups, some consideration must be given to which reference group to use.

Without a preoccupied position, a good starting reference group would be the one with the largest number of observations, because the largest group has the most stable rate. For a health disparity assessment project, one could also select the group with the lowest or best rate; if one group can achieve this rate, all other groups should also be able to achieve it too. Numerically, if all other groups are compared with the best group, the directional effects will be the same. In Table 7.1, Asian is the best group, and the death rates for the White, Black, AI/AN, and Native Hawaiian and Pacific Islander (NHPI) groups are 0.006, 0.096, 0.036, and 0.046, respectively, higher than the rate for Asians.

Now, let's look at different ways of using a reference group based on data from Table 7.1. We started with a simple log-rate model:

$$\log(\text{Death/Total}) = \text{Race} \tag{7.1}$$

TABLE 7.1

Hypothetical Data on Gun Violence by Race

Race	Alive	Dead	Total	Death Rate
White	9,900	100	10,000	0.01
Black	900	100	1,000	0.1
AI/AN	96	4	100	0.04
Asian	199	1	200	0.005
NHPI	19	1	20	0.05
Column total	11,114	206	11,320	0.01819788

TABLE 7.2

Log-Rate Model Parameter Estimates Based on Different Reference Groups

	Estimate	Standard Error	95% Confidence Limits		Chi-Square	*p*-Value
Model I						
C (ref = White)	−4.6052	0.1	−4.8012	−4.4092	2120.76	<0.0001
AI/AN	1.3863	0.5099	0.3869	2.3857	7.39	0.0066
Asian	−0.6931	1.005	−2.6629	1.2766	0.48	0.4904
Black	2.3026	0.1414	2.0254	2.5798	265.09	<0.0001
NHPI	1.6094	1.005	−0.3603	3.5792	2.56	0.1093
Model II						
C (ref = NHPI)	−2.9957	1	−4.9557	−1.0358	8.97	0.0027
AI/AN	−0.2231	1.118	−2.4144	1.9682	0.04	0.8418
Asian	−2.3026	1.4142	−5.0744	0.4692	2.65	0.1035
Black	0.6931	1.005	−1.2766	2.6629	0.48	0.4904
White	−1.6094	1.005	−3.5792	0.3603	2.56	0.1093
Model III						
C (ref = Asian)	−5.2983	1	−7.2583	−3.3384	28.07	<0.0001
AI/AN	2.0794	1.118	−0.1119	4.2707	3.46	0.0629
Black	2.9957	1.005	1.026	4.9655	8.89	0.0029
NHPI	2.3026	1.4142	−0.4692	5.0744	2.65	0.1035
White	0.6931	1.005	−1.2766	2.6629	0.48	0.4904

This model can be estimated by moving the total to the right-hand side of the equation as the offset. Most statistical software packages have a Poisson regression module that can estimate this type of model (Powers and Xie, 2000). The result from Model I (Table 7.2, upper panel) is based on outputs from SAS 9.3 Proc Genmod, with Whites being the referent:

```
proc genmod data=fakedata;
class Race (ref="White" param=ref);
model dead=race/dist=poisson offset=LogTotal;
```

Note that LogTotal is the natural logarithm of the *total* population (LogTotal = log(total)) in SAS. Since we have five cell frequencies of dead and five parameter estimates, it is a saturated model that fits each cell exactly. We observe that the log rates for Blacks and American Indians are significantly higher than those for Whites. However, if we use NHPI as the reference group in Model II (Table 7.2, middle panel), none of the results of racial group comparisons are significant. This clearly shows that one should avoid using a small group as the reference, especially when calculating confidence intervals (CIs). In this case, using the best outcome group (Asian; Model III) is fine. As expected, all coefficients have the same directional effects, even though the effects for Whites and NHPIs are not significant. Although which reference group to use is statistically trivial, one would not want to use NHPI in Model II due to the small cell size. It is better to use Model I due to its group importance in race recognition.

As shown in Table 7.2 under "Estimate," log rates are not intuitive. Sometimes, it is necessary to convert log rates to rates by taking the exponential of the log estimate. Let's use Model I as an example. The exp(−4.6052) = 0.01 is also called the grand mean. In the saturated model, it is identical to the rate for Whites in Table 7.1. The confidence limits can be calculated in rate using exp(estimate ± 1.96*Std Err), in this case exp(−4.6052 − 1.96*0.1) = 0.008219877 for the lower limit and exp(4.6052 + 1.96*0.1) = 0.012164906 for the upper limit. Certainly, one can also use the

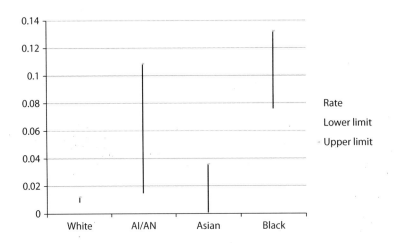

FIGURE 7.1   Hypothetical rate and confidence limits by race.

exponential upper and lower confidence limits directly, when they are parts of the output. Note that calculating the rate for a nonreference group needs to take the constant C or the estimate for the reference group into consideration, because coefficients for nonreference groups are relative to the coefficient for the reference group. For example, in order to calculate the rate for Blacks, we need to add the coefficient for Blacks (2.3026) to the constant C (−4.6052) in the following steps:

1. (−4.6052 + 2.3026) = −2.3026.
2. exp(−2.3026) = 0.1.
3. If the confidence interval is needed, then take exp(−2.3026 ± 1.96*0.1414), which equals 0.075793405 and 0.131933668, respectively, for upper and lower confidence limits.

Once all the confidence limits are calculated, they can be placed together with each expected rate, or the observed rate in this case, to make a graph. Figure 7.1 presents the results for four races. We dropped the results for NHPI because the confidence interval between the lower and upper limits was too wide to be reasonably displayed. Rates displayed by a bar chart implicitly suggest an absolute measure of disparity, because a difference between a group rate and a specified reference rate can be inspected graphically or measured arithmetically. As we already knew, the rate difference between Blacks and Whites is 0.1 − 0.01 = 0.09. In other words, the magnitude of difference is 90 per 1000 gun violence patients.

7.4.2   Relative Rate

Disparities can be measured by either a rate ratio for a particular reference group or the mean rate of the total population. The rate for the total population is a weighted average of the group rates. The total population rate is more stable than any reference group rates. Basically, any race-specific rate in Table 7.1 divided by 0.0182 is a ratio to the mean rate. A rate ratio close to 1 suggests that a race-specific rate is close to the mean or average rate. However, it is not common to use the rate ratio in reference to the mean, because of complications when a rate is controlled by another category, such as age group. In addition, different areas may have different population means, and it is not straightforward to compare rate ratios across regions or over time periods. For this reason, a particular group rate is usually used as the reference.

    Although the confidence limit can also be calculated by hand, the log-rate model provides a consistent method, especially when data are presented in a multiway table. Table 7.3 lists the same data as Table 7.1, but adds sex. Upon inspection of the table, we find males have proportionally more

TABLE 7.3

Hypothetical Data on Gun Violence by Race and Sex

Race	Sex	Alive	Dead	Total	Death Rate
White	M	4,940	60	5,000	0.012
Black	M	440	60	500	0.12
AI/AN	M	47	3	50	0.06
Asian	M	99	1	100	0.01
NHPI	M	9	1	10	0.1
White	F	4,960	40	5,000	0.008
Black	F	460	40	500	0.08
AI/AN	F	49	1	50	0.02
Asian	F	100	0	100	0
NHPI	F	10	0	10	0
Column total		11,114	206	11,320	0.018198

deaths than females across all racial groups. If we assume that the effects of sex by race are similar, and their average effect can fit the model sufficiently, we can then specify the independent log-rate model of race and sex:

$$\log(\text{Death} / \text{Total}) = \text{Race} + \text{Sex} \tag{7.2}$$

Table 7.4 presents the results for the log-rate model in Equation 7.2. The last column shows the relative rate result, which is another way of presenting results. In fact, when data are presented in a multiway table, it is more common to present the result as a relative rate. The grand mean or $\exp(-4.4116) = 0.0121$ is the omitted category in calculating the relative rate, as its relative is 1. All other relative rates are listed by taking their exponential terms from their logarithm estimates. The confidence limits can then be calculated directly from their logarithms. For instance, the lower and upper limits for Asians are 0.069746 and 3.5844, respectively. They are not significant because the interval contains 1. For females, the relative rate is 0.648, or about two-thirds the male rate. This average effect applies to all races.

One caveat of relative rate is that it removes the sense of magnitude in the expected rate. If we take $\exp(-4.4116)$, we would have the expected rate of 0.0121 for White males. However, when we say that the rate for Blacks is 10 times the rate for Whites, we still lose sight of the absolute measure for rate. When two rates are very small in comparison, 10 times may represent a huge gap, such as in the case of measles incidence between regions or between arbitrary racial groups. When two rates in references are high in magnitude, two times higher can have quite different disease prevention implications between regions or any other categories.

TABLE 7.4

Log-Rate Model Parameter Estimates by Race and Sex

Model I	Estimate	Standard Error	95% Confidence Limits		p-Value	RR
C (ref = White)	−4.4116	0.1147	−4.6363	−4.1869	<0.0001	1
AI/AN	1.3863	0.5099	0.3869	2.3857	0.0066	4
Asian	−0.6931	1.005	−2.6629	1.2766	0.4904	0.5
Black	2.3026	0.1414	2.0254	2.5798	<0.0001	10
NHPI	1.6094	1.005	−0.3603	3.5792	0.1093	5
Female (ref = male)	−0.4339	0.1426	−0.7134	−0.1543	0.0024	0.648

7.4.3 Odds and Odds Ratio

Odds and odds ratio are most commonly used for group comparison. When calculating the odds ratio, the numerator is the odds in the intervention group, while the denominator is the odds in the control or placebo group. The ratio equals 1 if the outcome between the two is the same. Odds ratio is used frequently in exposure studies, where the odds of the exposed group are often used as the numerator. Take the first two rows of Table 7.1, for example. We can assume that Blacks had the exposure of a poor neighborhood, while Whites did not. The odds of living or dying for Whites are $100/9900 = 0.0101$, and the odds for Blacks are 0.11111. To measure the relative effect of Blacks versus Whites, we would use an odds ratio. Since we assume that Blacks had exposure, the odds ratio would be derived from the odds of Blacks divided by the odds of Whites, which equals 11. It should be noted that the odds ratio can be calculated directly by the cross product ratio of $(9900*100)/(900*100)$. Compared to Whites, the likelihood of dead versus alive is 11 times higher for Blacks. Based on Table 7.1, Table 7.5 lists the results from a logit model. Odds ratios are listed in the last column relative to Whites.

In public health practice, the use and interpretation of exposure are not straightforward, and they often depend on data availability and communication strategy. The use of disparity exposure is essentially to devise a disparity scale or level for comparison. When multiple disparity measures are available, one needs to choose the one based on the scope of study and ease of interpretation. The scope of study dictates the intended intervention mechanism; the ease of interpretation helps to deliver the message. For instance, race and education are easier to explain than the Gini index. However, if the target is to reduce income inequality, one would prefer a Gini index. Table 7.6 shows the results from four logistic regressions. It uses low birth weight as the outcome and compares odds ratios for four disparity measures: race and ethnicity and education of the mother, percent of people living under the federal poverty line, and the Gini index. The latter two are based on the census tract of the mother. The birth data were restricted to mothers with Nebraska residence from 2005 to 2011 birth certificates. Both poverty and Gini index are census tract–based measures, while other measures in the table were individual based. In all models, the odds ratios were age-controlled for mothers and sex-controlled for newborns. The results from the educational level and the Gini index were the most consistent; the lower the educational level, the greater the chance of having a low-birth-weight baby; the greater the concentration of income, the higher the odds ratios. For race and ethnicity, neither the race of others nor Hispanic origin was significant. By dropping the Hispanic origin variable, the result for race does not change. Poverty variables were also consistent in terms of point estimates, but the odds ratios for the two neighborhood groups with 15% of people below the poverty level were not significant.

7.4.4 Standard Mortality Rate

Age is a known risk factor. When comparing rates among different groups at the population level, age structure differences between two groups can affect the conclusion. For this reason, most people use the age standard rate. There are two ways to adjust age: direct standardization and indirect

TABLE 7.5
Logit Model Parameter Estimates

Model I	Estimate	Standard Error	95% Confidence Limits		p-Value	Odds Ratio
C (ref = White)	−4.5951	0.1005	−4.7921	−4.3981	<0.0001	
AI/AN	1.4171	0.5201	0.3977	2.4365	0.0064	4.1251
Asian	−0.6982	1.0075	−2.6729	1.2765	0.4883	0.4975
Black	2.3979	0.1456	2.1124	2.6834	<0.0001	11.0001
NHPI	1.6507	1.0309	−0.3698	3.6712	0.1093	5.2106

TABLE 7.6
Odds Ratios of Low Birth Weight by Four Disparity Measures

	Odds Ratio	Confidence Limits	
		Low Limit	Upper Limit
Race/Ethnicity			
Black vs. White	2.266	2.139	2.4
AI vs. White	1.279	1.084	1.508
Asian vs. White	1.242	1.111	1.387
Other vs. White	1.012	0.914	1.121
Hispanic vs. non-Hispanic	1.031	0.945	1.126
Education			
Associate's degree vs. college	1.138	1.061	1.221
High school vs. college	1.376	1.312	1.443
Less than high school vs. college	1.596	1.504	1.694
Poverty			
5%–10% vs. <5%	1.037	0.981	1.097
10%–15% vs. <5%	1.047	0.987	1.111
15%–20% vs. <5%	1.149	1.07	1.234
20%+ vs. <5%	1.394	1.317	1.477
Gini Index			
0.347–0.385 vs. <0.346	1.073	1.019	1.131
0.385–0.419 vs. <0.346	1.06	1.006	1.117
0.420 vs. <0.346	1.133	1.076	1.193

Note: Race and ethnicity are self-reported by mothers. Both poverty and Gini index were from census tract–level information extracted from 2007 to 2011 ACS data. The four categories in the Gini index are based on quantiles (top 25%, upper middle 25%, lower middle 25, and bottom 20% census tracts).

standardization. The former requires standard population, while the latter requires age-specific death rate among the standard population. Since we normally have access to the standard population, the direct method is commonly used.

The directional effects between standardized and crude mortality rates are often different, and it is important to adjust the age effect when making comparisons. The heart disease mortality rate has been in decline for many years nationwide, and we thought it would be interesting to compare trends for Nebraska by race and ethnicity. Table 7.6 displays crude rates and age-adjusted rates for comparing 2004–2008 (time 1) and 2009–2013 (time 2) heart disease mortality trends in Nebraska. When calculating mortality by race, one needs to pay attention to race-specific at-risk populations. The population numbers in the table are based on single-race populations, where the multiple races are ignored as judged by the health statistic of the Nebraska Public Health Division. The ignored multiple-race group had a population of about 60,000 in 2008, which would affect the results differently among different race groups. First, it might assume that the proportion of multiple-race persons within each race population is about the same. This assumption would affect all racial groups in the same way. However, if most multiracial persons self-identified with a minority group, then the equal share assumption would not hold, as there would be <50% of multiracial persons identified with Whites. Hence, if more than 50% of multiracial persons were to allocated among the minority population, their standard mortality rates would be much lower for minority racial groups. Hence,

TABLE 7.7

Changes in Heart Disease Mortality between 2004–2009 and 2009–2013 Periods

	Total	White	Black	Native	Asian	Hispanic
			Male			
Number of deaths 2004–2008	8,638	8,318	236	46	21	109
Number of deaths 2009–2013	8,290	7,904	285	34	23	107
Population 2004–2008	878,006	800,512	39,848	10,255	14,881	73,942
Population 2009–2013	914,690	822,883	44,821	12,302	17,922	91,899
Crude rate 2004–2008	196.8	207.8	118.5	89.7	28.2	29.5
Crude rate 2009–2011	181.3	192.1	127.2	55.3	25.7	23.3
Standard mortality 2004–2008	218.4	217.6	249.5	261.3	119.9	118
Standard mortality 2009–2013	188.3	186.7	256.4	155.6	88	90.5
Changes in standard mortality	–13.8	–14.2	2.8	–40.5	–26.6	–23.3
			Female			
Number of deaths 2004–2008	9184	8866	251	33	18	79
Number of deaths 2009–2013	8282	7976	222	33	21	67
Population 2004–2008	894,687	816,422	39,341	10,064	15,879	64,539
Population 2009–2013	927,059	835,551	43,000	12,122	19,325	82,798
Crude rate 2004–2008	205.3	217.2	127.6	65.6	22.7	24.5
Crude rate 2009–2011	178.7	190.9	103.3	54.4	21.7	16.2
Standard mortality 2004–2008	138.8	137	200.8	169.3	71.1	95.1
Standard mortality 2009–2013	119.9	118.7	158.5	124.7	59.6	54
Changes in standard mortality	–13.6	–13.4	–21.1	–26.3	–16.2	–43.2

Source: Nebraska health statistics. Rates are per 100,000, annualized by dividing by 5 for each period.

omitting multiracial persons from the at-risk population would likely affect different racial groups differently in terms of calculating standard mortality rates. It would affect the White group little due its large population size.

As shown in Table 7.7, the crude rates for White males and females in time 1 were 196.8 and 205.3, respectively. The corresponding rates for Blacks, in contrast, were 119 and 127, respectively. With a more than 40% racial difference, one could blindly conclude that directional effects by race would not change when age is introduced. However, the race-specific population age structure here is critical, because the standard rate based on the standard 11 age groups changed directional effects. Compared to Whites, the standard mortality rates for Blacks were about 15% higher for males and 46% higher for females for the time 1 period. In addition, we found that changes in mortality between time 2 and time 1 were much larger among minority groups than those for Whites. One important fact was that the rate for Black males increased slightly between the two periods. Here, the decline was not always in favor of minority groups.

In addition to racial grouping of the at-risk population, age grouping can also change standard rates. Here, we use stroke data from Nebraska between 2005 and 2009 to assess the influence of the last age group. An extremely high rate for the last age group can swing the results from one direction to another. In Table 7.7, we list the results for each step and demonstrate how the last age group may influence the results. Data are arranged by age and sex, because sex is an important predictor of stroke mortality. The mortality rates in the table are presented per 100,000, which can be calculated by dividing the number of age-specific deaths and the corresponding population. The corresponding population is based on the 2007 population (midyear of 2005 and 2009). Since deaths are for the 5-year period, we use the annualized population that is five times that of the at-risk population.

Standard population weights are in rate, and their sum should equal 1. The standard mortality rate is weighted by population, which can be derived by calculating (1) age-sex specific rates, (2) age-specific weights in rate, and (3) the sum of multiplications of 1 and 2.

The upper panel of Table 7.8 lists seven age groups, while the lower panel list six groups. The difference is the last age group in the six-age-group panel combines the last two age groups (75–84 and 85 and over) into a single age group, 75+. In so doing, a sex difference in favor of females (males 44.505 vs. females 40.685) in the seven-age-group calculation becomes a difference in favor of males (males 44.074 vs. females 45.579) in the six-age-group calculation. The reason is that age distribution versus mortality distribution of the last two groups differs sufficiently to swing the results. Even though we used stroke, the influence of the last age group is quite common in other diseases, such as in cancer.

In addition to the standard mortality, age standardization can be used for incidence and health behavior too. We present a surveillance application in which standard cancer incidence rates are calculated by census tract poverty level. In the past, standard incidence rates for major cancer sites were often included in the annual Nebraska cancer registry reports. Age-standardized rates are often presented by county, race, and other dimensions, but SES-based incidence has never been presented. With geocoded census tract information, it is very easy to present small-area data. One of most commonly used methods is to aggregate census tract–level data according to some neighborhood characteristics and present them in tabulated forms. Table 7.9 presents overall and late-stage incidence rates for males and females by three poverty levels: low-poverty census tracts with <5% people under the poverty level, midpoverty census tracts with between 5% and 15% people under the poverty level, and high-poverty census tracts with >15% people under the poverty level. In the table, we used the 2007–2011 American Community Survey (ACS) population at the census tract level by age and sex as the at-risk population and 2005–2011 cancer data. We chose this period because later on we will use this period extensively in hospital-based disease surveillance. As one can see, lung cancer and colorectal cancer had an SES gradient: the higher the poverty level, the greater the incidence rate.

Finally, the concepts of sample and sample population are important in determining standard mortality. In general, standard mortality and incidence rates are all based on population rather than a sample. In the above stroke case, stroke deaths were for all residents of Nebraska in the 5-year period. An age-specific number of deaths divided by the age-specific at-risk population becomes an age-specific death rate. We can compare and describe straightforwardly any pairs of age-specific rates for their differences; it is not necessary to test them for statistical difference. However, if we consider time series data, for example, 1960–2010, and consider Nebraska as a sample of 50 states, then the 5-year data from the time series can be considered a sample, for which statistical tests would be fine.

7.5 INTERTEMPORAL MEASUREMENTS

Temporal patterns are an important aspect of disparity surveillance. As shown in Section 7.4.4, standard mortality and standard incidence are measures for comparing intertemporal trends. For long-term trends, such as over a 15- or 20-year period, intertemporal comparisons with few discrete time intervals are not effective. For the population change governed by fertility, mortality, and in- and out-migration, the standard mortality is a summary measure that masks age- and time-dependent effects. Warning labels on all cigarette packages started in 1970, and waves of "stop smoking" campaigns have reduced men's lung cancer incidence rate for a particular age segment, while population aging increases lung cancer incidence, which can be adjusted by the standard incidence rate.

7.5.1 SEASONAL TRENDS

Another time-dependent effect is seasonal in nature. Influenza and asthma are two examples. The former tends to have an annual peak season around winter months, while the latter tends to have

TABLE 7.8
Calculating Standard Stroke Mortality by Sex (2005–2009)

Age Group	Males			Females			U.S. 2000 Standard Population	Weighted Rate	Standard Rate	
	Population	Deaths	Mortality	Population	Deaths	Mortality			Males	Females
0–14	189,055	5	0.53	180,429	3	0.33	214,700	0.215	0.114	0.071
15–29	200,003	7	0.70	189,662	5	0.53	203,176	0.203	0.142	0.107
30–44	169,438	30	3.54	163,512	24	2.94	233,657	0.234	0.827	0.686
45–59	179,511	165	18.38	181,849	106	11.66	183,288	0.183	3.369	2.137
60–74	92,416	353	76.39	101,637	286	56.28	104,830	0.105	8.008	5.900
75–84	34,948	636	363.97	48,966	775	316.55	44841	0.045	16.321	14.194
85+	11,402	578	1,013.86	27,084	1,536	1,134.25	15,508	0.016	15.723	17.590
Total	876,773	1,774	40.47	893,139	2,735	61.24	1,000,000	1.000	44.505	40.685
				Using 6 Age Groups						
0–14	189,055	5	0.53	180,429	3	0.33	214,700	0.215	0.114	0.071
15–29	200,003	7	0.70	189,662	5	0.53	203,176	0.203	0.142	0.107
30–44	169,438	30	3.54	163,512	24	2.94	233,657	0.234	0.827	0.686
45–59	179,511	165	18.38	181,849	106	11.66	183,288	0.183	3.369	2.137
60–74	92,416	353	76.39	101,637	286	56.28	104,830	0.105	8.008	5.900
75+	46,350	1,214	523.84	76,050	2,311	607.76	60,349	0.060	31.613	36.678
Total	876,773	1,774	40.46657	893,139	2,735	61.24467	1,000,000	1.000	44.074	45.579

Note: The at-risk population is based on the 2007 population. Mortality is annualized per 100,000.

TABLE 7.9

Annualized Cancer Incidence for Top 10 Cancer Sites by Sex and Poverty: 2005–2011

	Males			Females		
	Low Poverty	Midpoverty	High Poverty	Low Poverty	Midpoverty	High Poverty
Overall						
Breast				147.6	119.5	116.5
Prostate	165.5	137.6	143.9			
Lung	69.9	71.2	98.2	55.3	46.1	63.5
Colorectal	60.5	56.7	66.2	48.0	45.4	51.8
Skin	30.4	20.8	17.8	23.6	14.9	12.6
Non-Hodgkins	28.6	23.8	23.8	20.7	17.6	19.0
Urinary bladder	21.0	16.7	17.6	5.2	4.3	5.0
Kidney	25.3	20.4	23.3	12.0	11.5	13.5
Brain	11.0	7.9	7.7	7.4	7.1	5.6
Leukemia	21.7	16.9	19.4	12.8	11.2	11.8
Late Stage						
Breast				54.6	45.2	45.1
Prostate	34.5	31.4	33.8			
Lung	58.2	59.6	83.3	41.9	36.9	50.7
Colorectal	36.9	35.4	42.1	28.3	28.3	32.3
Skin	6.4	5.5	5.6	4.7	3.2	3.1
Non-Hodgkins	20.2	18.8	17.9	15.5	13.1	15.6
Urinary bladder	5.7	4.4	5.4	1.4	1.4	1.9
Kidney	8.2	8.3	9.4	3.5	4.0	4.4
Brain	2.4	2.0	2.0	1.9	2.0	1.9
Leukemia	21.7	16.8	19.3	12.8	11.1	11.8

Source: Data are from Nebraska Cancer Registry; only invasive cancers are included.

Note: Cancer sites are based on incidence from 10 most commonly diagnosed.

biannual peaks in spring and fall. The recurrence of influenza has been modeled at the national or city level, and excessive morbidity can be assessed by Serfling et al. (1967) or more complicated Fourier regressions and nonparametric regression methods. More recent developments tend to decompose aggregated models into multiregional models (Viboud et al., 2006) or different age groups (Olson et al., 2007), so that the interactions among different regions or different age groups can be assessed and linked to influenza diffusion. Viboud et al. (2006) studied the spatial hierarchies in the spread of influenza in the United States. They found that spatially lagged influenza mortality at the state level is highly correlated with air passage flows based on a gravity or spatial interaction model. Olson et al. (2007) estimated age-specific models using Serfling regression, and they found that influenza epidemic period peaks occurred earliest among school-aged children each season between 2001 and 2006.

A Serfling regression can be defined according to ordinary least-squares (OLS), log-rate, and logit regressions. Conceptually, log-rate regression and logit regression are more appealing, because both use count data. When data are complete and the sample is large, OLS regression using count data would be fine. When data are incomplete either by week or by submitting hospital, modeling expected rate is more appropriate. Let $\omega = \pi/52.18$, or annual cycle, and t be the index weeks from the beginning period to the end period; then a simple OLS Serfling regression model can be set up as

$$M_t = \alpha_0 + \beta_1 \cos(2\omega t) + \beta_1 \sin(2\omega t) + \gamma_2 \cos(4\omega t) + \gamma_2 \sin(4\omega t) + \varepsilon_\tau \qquad (7.3)$$

where M_t is the expected ED visits in week t, which could expressed in rate or count. β_1 and γ_2 are annual sinusoidal and semiannual terms, respectively. For convenience, the SAS codes for OLS and log-rate models are included:

```
omega= (2* CONSTANT ('PI')) /52.18; omega2= (4* CONSTANT ('PI')) /52.18;
cos1=cos (omega*t); sin1=sin (omega*t);
cos2=cos (omega2*t); sin2=sin (omega2*t);

**Staring an OLS Serfling model**;

proc reg data=c;
/*C is a SAS dataset containing ILI rate, the numbers ILI & total ED
visits by week*/
  model ILI_Rate=cos1 sin1 cos2 sin2 ;
  output out=c1 p=Pre_ILI_rate ; /*C1 is an output dataset containing the
expected ILI rate as Pre_ILI */

** log-rate Serfling mode**l;
proc genmod data=c;
  model ILI_count=cos1 sin1 cos2 sin2 /dist=poisson offset=logtotal;
output out=c1 pred=pre_ILI_count;
```

In general, an OLS regression is easier for model interpretation than a log-rate model, while the latter has a more accurately predicted power. When sample size is large, the expected values from both models tend to be very close. As an example, we used data from January 1, 2005 to June 27, 2010, weekly ED visits for influenza-like illness and modeled the expected rate. The observed and expected weekly ED visit curves are shown in Figure 7.2. Note that the expected visits were calculated from the expected rate in Equation 7.3. The highest observed rate happened in the third and fourth weeks of October 2009, which was the peak period of the H1N1 pandemic that year. If we want to calculate excessive morbidity for the 2-week period, we simply use the difference between the two curves, which was 3116.

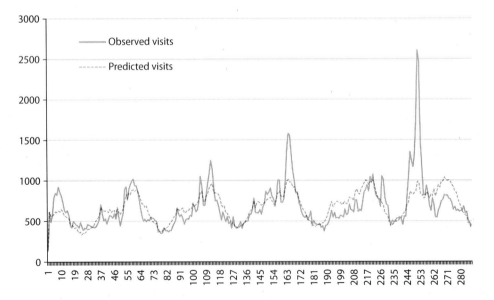

FIGURE 7.2 Observed and predicted influenza-like illness (ILI) ED visits in Nebraska from January 2005 to June 2010.

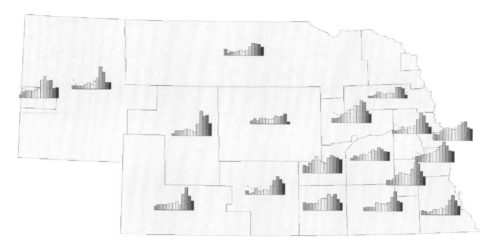

FIGURE 7.3 Weekly H1N1 flu ED visit rate by local health department in Nebraska (August 10, 2009–November 7, 2009).

It is also possible to model local Serfling curves and identify regions that started epidemics earlier (Schanzer et al., 2011). Before modeling, one might want to conduct some exploratory analyses. Figure 7.3 displays the H1N1 flu season by local health departments. There are 20 local health departments in Nebraska. Two local health departments in the northeast corner had too few cases, as the area is close to Sioux City, Iowa, and those affected tended to visit EDs there. From the figure, the taller the bar is in the beginning period, the higher was the early incidence for the H1N1 influenza. It is evident that high rates started from Omaha, although the Central District Health Department, where Grand Island is located, also had a high rate in the beginning.

7.5.2 Intertemporal Trends

In Section 7.5.1, we presented a method for seasonal trends based on a continued timescale. There are many other disease trends that may not cyclic in nature. Keeping track of cancer incidence is an important responsibility of a state cancer registry, but it is not always straightforward to compare rates over time, because screening program effects sometimes interact with secular trends, such as population aging. For example, we know that when a low-cost screening test is available, such as the prostrate-specific antigen (PSA) for prostate cancer, cancer incidence tends to increase and then gradually declines to a normal range. The increase is due to the introduction of the test, which captures some cancer at an early stage. Assuming that a cancer incidence rate is stable, when relatively more cases are captured in the early stage upon introduction of the test, there are fewer late-stage cases in the years after the introduction. In between, there should be a peak time for positively screened cases or incidence. One way to conduct surveillance among different racial groups is to have pre- and postdesigning of the introduction of the PSA screening and compare peak times using standardized incidence rates. In Nebraska, it was reported that the age-adjusted rate for Black men peaked a few years behind that for White men for prostate cancer, suggesting that although the PSA test might be available for all in the introduction period, Black men tended to use the test slightly later than White men did (NE-DHHS, 2011).

To calculate age-adjusted incidence, one would need to use an age-specific population. However, if we want to examine trends over 15+ years, adjusting each year for an age-specific population is taxing from the surveillance point of view. This task is also complicated by an SES-specific population. If we use poverty as a measure of neighborhood SES, then we need to obtain an age-specific population in each poverty category. Such data may not be available each year. In addition, when cancer cases are further divided by poverty level and age group, there might not be a sufficient number of cases in each cell for calculating standard incidence rates. Furthermore, for researchers engaging in cancer screening surveillance or exploratory data analysis, data are often in an aggregated form, and it is not easy to go back and forth with several aggregated data requests. One simple adjustment is to use all other cancer cases to derive the normalized incidence ratio, where a trend for a particular cancer site is relative to the trend from all other cancer sites, which is equal to 1. Suppose we are looking for a trend of 20 years of data for prostate cancer, during which the PSA test was introduced. Instead of using an age-specific at-risk population for incidence rate calculation, we can use all other cancer cases as a reference. If prostate cancer cases increase at the same rate as all other cancer sites, then the introduction of the PSA has little impact. In this case, the incidence ratio between prostate cancer and other cancers would remain constant. Since the incidence ratio is dependent on the magnitude of incidence, it is not easy for comparisons when a number of incidences are plotted. For this reason, we can normalize the incidence ratio by adjusting the two incidence distributions to a standard z-scale (with a mean set to zero).

1. Calculate the frequencies for prostate cancer and all other cancers by years (e.g., 20 years).
2. Calculate the standard z-score for each incidence (dividing each year's incidence by the mean incidence over the study period). If a rate in a particular year is >1, then the incidence in that year is greater than the average. If, on the other hand, the rate is <1, then the incidence in that year is below the average.
3. Divide standard the z-scores for the prostate and reference cancers. This is the z-score ratio.
4. Plot the z-scale incidence ratios by year. If the ratio is above 1, then it is above the average trend of all other sites. If it is below 1, then the incidence of the interest, or prostate cancer, has relatively declined compared to the reference group.

The z-scale ratio approach does not need the age adjusted by the reference group, as it assumes that age-specific cancer incidence rates for selected cancer sites follow a similar trend. Since the reference group is generally large enough, the ratio based on it generally reflects the trend for the cancer site of interest.

Figure 7.4 is an example of z-scale incidence ratios for breast cancer, colorectal cancer, prostate cancer, and thyroid cancer relative to all other cancers (see Appendix 7A for the original data). In the other cancer group, we excluded skin cancer, as it not only increased rapidly since 1990, but also had sizable cases that may unduly cause an increased trend for all other cancers. Let's focus on prostate cancer first. The PSA test was approved by the Food and Drug Administration (FDA) for the purpose of monitoring disease status in prostate cancer patients in 1986 and for aiding in the detection of prostate cancer in men 50 years and older in 1994. The use of the PSA test for prostate cancer screening started in 1988, causing a dramatic increase in prostate cancer incidence between 1989 and 1992, which was captured by Surveillance, Epidemiology, and End Results (SEER) data (Legler et al., 1998). As one can see, the peak in Nebraska was in 1992, and the relative incidence (to other cancer sites) declined. There was a small increase in the relative incidence of prostate cancer in 1999 and 2000, which could be attributed to the 1994 FDA decision, because it takes time to trickle down from the approval to the screening process. From the figure, we also observed a general decline in colorectal cancer relative incidence, an oscillating pattern in breast cancer, and a steady increase in thyroid cancer. We want to point out that if we had used an age-adjusted incidence rate instead, we would not have detected the above trends. Prior to 1995, especially in 1990–1992, the Nebraska Cancer Registry missed a hospital or two in each year of reporting. Consequently, the overall cancer incidence was relatively low for all cancer sites. However, since missing cases tend to be systematic, if the missing hospital did not report breast cancer cases, it also would not report other cancer cases. Hence, the z-scale incidence ratio is a quick way to conduct exploratory surveillance.

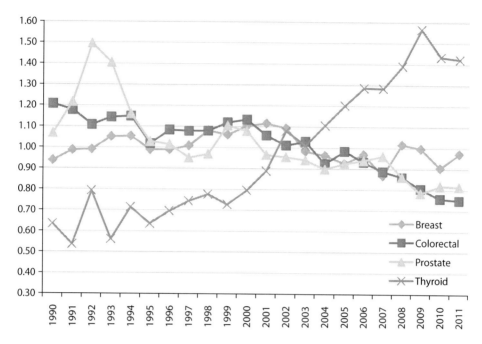

FIGURE 7.4 Relative incidence by four cancer sites: 1990–2011.

7.5.3 PRE- AND POST-ASSESSMENT

Pre- and postevent assessment is often used in the evaluation. Below is a simple example about the smoking ban effect in Omaha. On October 2, 2006, the city of Omaha passed smoking-free legislation that banned indoor smoking in businesses and government agencies citywide. To measure its effect, we compiled 6 years of data from BRFSS 2004–2009, with 3 years prelegislation (2004–2006) and 3 years postlegislation (2007–2009). The total unweighted sample was 17,672. We divided the sample into Omaha, or Douglas County, as the case, and counties adjacent to Omaha as the control. The adjacent counties include Cass, Dodge, Sarpy, and Washington. This design is necessary because Omaha respondents were exposed to the smoking ban after 2006, while respondents from the adjacent counties were not directly exposed throughout the 6-year period. Since these counties are all part of Omaha metropolitan counties, respondents in these counties could be considered the same as those in Omaha in responding to other tobacco control campaigns.

Table 7.10 lists by geographic location the number of weighted observations for current smokers (with corresponding percentages in parentheses), the total numbers of respondents before and after the smoking ban, and the percent of change from pre- to postsmoking ban. A simple conclusion can be made from the upper panel: 17.49% of survey respondents were current smokers before the smoking ban, and the percentage dropped to 14.28 after the smoking ban. Although evidence is strong, we may not be able to attribute the 18.35% drop in current smoking rate to the smoking ban because other tobacco control campaigns and unknown reasons may have contributed to the decline. From the lower panel of the table, we found a 15.92% drop in smoking rate after their neighbor, Douglas, banned smoking. It could be speculated that without the smoking ban, we could expect a 15.92% drop in the smoking rate in Omaha instead of the observed 18.35%. The difference between the two percentages of 2.43% (18.35 − 15.92) could be potentially attributed to the smoking ban. This method is often labeled the difference in differences in quasi-experimental designs.

We could test its significance by using the chi-square test, but a simple logit model can also be set up, with the dependent variable being current smokers or not, and explanatory variables being adjacent county status, post- versus preban time periods. The Omaha-specific effect contrasts Omaha's post- and preban periods while controlling for the two effects above. The results from

TABLE 7.10

Impact of Omaha Smoking Ban in 2006

	Current Smoker	% Change	Total N
Omaha			
Presmoking ban	117,417 (17.49)		671,393
Postsmoking ban	104,260 (14.28)	18.35	730,182
Total	221,677 (16.36)		1,401,579
Counties Adjacent to Omaha			
Presmoking ban	19,558 (16.14)		101,603
Postsmoking ban	15,823 (13.57)	15.92	116,642
Total	35,381 (15.3)		202,422

Source: Data were based on 2004–2009 BRFSS.

Note: Presmoking ban, weighted BRFSS 2004–2006; postsmoking ban, weighted BRFSS 2007–2009. Counties adjacent to Omaha include Sarpy, Cass, Dodge, and Washington.

TABLE 7.11

Odds Ratio Estimates for Omaha Smoking Ban in 2006

	OR	95% CI: Low	95% CI: Up
Adjacent counties vs. Omaha	0.908	0.898	0.919
Post- vs. presmoking ban	0.815	0.799	0.832
Omaha postban effect	0.964	0.943	0.985

Source: Data were based on 2004–2009 BRFSS in Nebraska.

Table 7.11 showed that the odds ratio after the smoking ban for Omaha was 0.964 (95% CI = 0.943–0.985). The smoking ban contributed to a 3.6% reduction in current smokers, given an overall trend of the decline of current smokers in the study area due to other reasons (OR = 0.815, 95% CI = 0.799–0.832).

7.6   CHAPTER SUMMARY

In this chapter, we introduced steps, measurements, and examples of how to conduct health disparity surveillance for cross-sectional and time series data. We stressed that disparity surveillance is observational, and it normally does not need to meet the standards of the experimental and cohort study designs. Different from a research study, disparity surveillance may not need a research question to drive the study, but it does need a scope to frame surveillance questions.

We introduced a list of measurements by examples. They included rate, relative rate, odds ratio, and rate ratio. We also introduced a log-rate model in the context of Poisson regression, Surfling regression for cyclic effects, and a quasi-experimental design for pre-and post-Omaha smoking ban effects. Although we tried to cover as much as possible, we could not cover all statistical and practical measures for health disparity surveillance, and our hopes were to introduce other necessary concepts in later chapters as we go through each health disparity surveillance dimension and case study.

APPENDIX 7A: CANCER CASES FOR SELECTED CANCER SITES IN FIGURE 7.4

Year	Breast	Colorectal	Prostate	Thyroid	Skin	Other Sites
1990	1118	1084	1156	79	185	3866
1991	1198	1078	1341	68	191	3936
1992	1208	1018	1657	101	185	3955
1993	1254	1029	1522	70	177	3870
1994	1285	1056	1285	91	192	3956
1995	1296	1005	1223	87	307	4246
1996	1305	1077	1216	96	303	4277
1997	1388	1119	1189	107	362	4464
1998	1506	1133	1226	113	321	4517
1999	1478	1176	1398	106	363	4519
2000	1552	1202	1378	117	377	4561
2001	1610	1151	1266	134	397	4682
2002	1609	1122	1277	166	460	4773
2003	1452	1147	1263	156	409	4791
2004	1499	1084	1269	180	460	5047
2005	1466	1175	1332	199	475	5138
2006	1550	1123	1365	215	495	5186
2007	1449	1116	1456	224	504	5411
2008	1656	1057	1274	237	515	5286
2009	1590	967	1135	261	544	5171
2010	1509	952	1239	250	569	5409
2011	1589	923	1209	243	645	5298

Note: Cancer cases were from Nebraska residents. Other sites exclude skin cancer.

REFERENCES

Kelsall, J.E., J.M. Samet, S.L. Zeger, and J. Xu. 1997. Air pollution and mortality in Philadelphia, 1974–1988. *American Journal of Epidemiology* 146(9): 750–762.

Legler, J.M., E.J. Feuer, A.L. Potosky, R.M. Merrill, and B.S. Kramer. 1998. The role of prostate-specific antigen (PSA) testing patterns in the recent prostate cancer incidence decline in the United States. *Cancer Causes & Control* 9: 519–527.

Lin, G. and T. Zhang. 2012. Examining extreme weather effects on birth weight from the individual effect to spatiotemperal aggregation effects. *Journal of Agricultural, Biological, and Environmental Statistics* 17: 490–507.

NE-DHHS (Nebraska Department of Health and Human Services). 2011. Cancer disparities by race and ethnicity in Nebraska: 1991–2005. Lincoln, NE: NE-DHHS. http://dhhs.ne.gov/Documents/CancerDisparitiesReport19912005.pdf.

Olson, D.R., R.T. Heffernan, M. Paladini, K. Konty, D. Weiss, and F. Mostashari. 2007. Monitoring the impact of influenza by age: Emergency department fever and respiratory complaint surveillance in New York City. *PLoS Medicine* 4(8): e247.

Powers, D.A. and Y. Xie. 2000. *Statistical Methods for Categorical Data Analysis.* New York: Academic Press.

Schanzer, D.L., J.M. Langley, T. Dummer, and S. Aziz. 2011. The geographic synchrony of seasonal influenza: A waves across Canada and the United States. *PLoS ONE* 6(6): e21471.

Serfling, R.E., I.L. Sherman, and W.J. Houseworth. 1967. Excess pneumonia influenza mortality by age and sex in three major influenza A2 epidemics, United States, 1957–58, 1960 and 1963. *American Journal of Epidemiology* 86: 433–441.

Viboud, C., O.N. Bjørnstad, D.L. Smith, L. Simonsen, M.A. Miller, and B.T. Grenfell. 2006. Synchrony, waves, and spatial hierarchies in the spread of influenza. *Science* 312(5772): 447–451.

Section II

Health Disparity Surveillance
Based on Hospital and
Emergency Department Data

8 SES Disparities in Hospitalization

8.1 INTRODUCTION

Most administrative data, such as registry and hospital discharge data (HDD), do not have income or education variables. Some datasets have socioeconomic status (SES) (e.g., occupation) and race variables, but they are often not usable because of poor data qualities, such as a large number of missing values. A case in point is the Nebraska HDD, which lack income variables and have unusable race and ethnicity variables due to missing values. Using the Nebraska HDD to conduct disparity assessments means adding SES variables through data linkage. In this chapter, we use different measures—such as incidence of major diseases, procedures, readmission, and 1-year out-of-hospital mortality—to conduct SES-based surveillance. We emphasize practical steps and considerations in constructing and presenting SES gradients around those measures. Our assessments are based on surveillance principles for chronic diseases, in which time was not as sensitive as it is for communicable diseases. Major steps of the surveillance include data collection and continued data updates, data improvements, data analysis of selected diseases, and in our case relating to SES, disparity detection and result dissemination. Normally, after disseminating results, we expect some public health actions to reduce disparities in in-depth data analysis, controlling for further disparity assessments.

We intend to conduct hospitalization disparity assessments in Chapters 9 through 12 based on neighborhood SES variables and hospital claims data. A number of studies using neighborhood SES gradients and hospital claims data to assess hospitalization disparities have been conducted. One of the earliest studies in the United States compared hospitalizations due to ASSCs between high- and low-SES neighborhoods in New York City (Billings et al., 1993). If a zip code had <17.5% of households with an income below $15,000, then it was in a high-income area. If a zip code had more than 60% of households with an income below $15,000, then it was in a low-income area. The study found higher standardized hospitalization rates in low-income areas than in the comparison group. Although the zip code area unit has been used quite often in both the United States and Canada (Lin et al., 1999; Trachtenberg et al., 2014), it suffers some criticism of not representing neighborhood SESs (Krieger et al., 2002). Studies using census enumeration areas to present neighborhood SESs have been done mostly in Canada. An early study using the Ontario Diabetes Database (ODD) found that patients in the lowest-income quintile were 43% more likely to have a hospital admission than those in the highest quintile, and they also had a greater likelihood of readmissions (Booth and Hux, 2003). The income quintiles in the study were based on Canadian census enumeration areas, which had an average population of 700, and hence are close to the U.S. census blocks in population size. A more comprehensive study (Roos et al., 2005) was conducted in Manitoba that examined association between neighborhood income and visits to physician offices and hospitals due to ambulatory care sensitive conditions (ACSCs). The study also used the quintile classification of the enumeration area, and the at-risk population is based on the registered residence in the provincial health plan. The study found that both visits had an SES gradient, with residents of the lowest-income neighborhoods having the highest rates. Many other studies have individually associated acute myocardial infarction (AMI), asthma, cystic fibrosis, influenza, and kidney disease with neighborhood SES gradients (Alter et al., 1999; Lin et al., 1999; Stephenson et al., 2011; Yousey-Hindes and Hadler, 2011). Our surveillance approach broadens neighborhood SES analysis by (1) including a set of diseases or conditions rather than a single disease or condition and (2) assessing a host of hospitalization outcomes.

In the United States, there has been no study at the state level using census tract SES to examine hospitalization disparity. For this reason, this chapter serves as a methodological introduction to set out an analytical framework using census tract SES variables and baseline estimates. In this way, we do not have to elaborate the procedure of acquiring census tract–level information. In the remaining parts of the chapter, we first introduce our approach to conducting disparity surveillance using the census tract poverty variable—the percent of people living under the federal poverty line. We then separately assess the differences in hospital-based comorbidity, procedure utilization, hospital readmission, and 1-year mortality. Finally, we provide some concluding remarks.

8.2　ANALYTICAL APPROACH TO NEIGHBORHOOD SES DISPARITY ASSESSMENTS

It is known that income is positively related to health at the population level. However, at a certain point, even with rising income, population health may not be further improved markedly without tackling health disparities. To some degree, the income level in the United States is already very high compared to that of most countries. It might be that further improvement in population health would be more effective if we could identify *soft* spots of health disparities, by which we mean diseases or health conditions that have a wide disparity gap with a relatively great number of cases. In addition, even with rising income, it would be extremely hard to remove income disparities in medium terms, such as 10–20 years. Monitoring SES-based health disparities would be helpful to establish baseline indicators for eliminating health disparities.

In the absence of income-based SES measures, most previous health disparity studies using HDD have used insurance information as a proxy measure of SES, because the variable is readily available. It is very taxing to derive SES variables at the neighborhood level because of the amount of geocoding. We geocoded patient addresses and assigned them to a census tract. Afterwards, we linked census tract poverty status to each patient. Our work was made a little easier because we built a master address index during the process of geocoding (see Chapter 6). We first standardized each patient's address according to the U.S. post office standard. We then deduplicated addresses for the 2005–2008 HDD, resulting in about 561,000 unique addresses from the patient records. Since each unique address is indexed to a patient, and patients are also indexed, we can find patient mobility information easily. If a patient moved once, he or she would have two unique addresses. If the patient moved three times, and the last address is the same as the first, he or should would still have two unique addresses. If a patient is repeatedly hospitalized from one address, that address would be the unique address. Hence, once patients' addresses are geocoded, they can be used for more purposes than just getting neighborhood SES variables.

Hospital-based health care utilization is likely to follow an SES gradient. Poorer individuals have poorer health status, which in turn leads to greater need for health care. This relationship is exacerbated in hospital-based care due to limited access to primary care resources. In other words, in spite of their greater need for it, poor individuals employ less ambulatory care than well-off individuals due to the lack of good health insurance, time, and purchasing power. It is therefore necessary to control for differences in need when measuring disparities in health care utilization. Health economists often use horizontal equity to describe medical ethics: equal health conditions (from different individuals) should lead to equal treatment (van Doorslaer et al., 2000; Wagstaff et al., 1991). However, other than age and sex, administrative data rarely have an indicator of medical care need. We therefore cannot really evaluate horizontal equity. We will revisit how to control for medical care need in later chapters. For the same reason (lack of need indicators), we cannot evaluate vertical equity—appropriate (different) treatments of different needs. In this regard, disparity in health care utilization at the population level should not be interpreted as "the more utilization, the better off," because no one would like to be hospitalized if he or she had a better choice. Rather, we view health care utilization in general as a burden on states, cities, and neighborhoods.

In the context of SES disparity surveillance, we did not aim at any specific disease. We used the poverty status as the SES indicator to survey diseases based on their prevalence. Since most commonly used diagnoses for admissions represent more prevalent diseases, we relied on the National Health Statistics for 2010 to provide a ranking by principal diagnosis category for number of discharges (Pfuntner et al., 2013a). Appendix 8A lists the top 50 discharges by the principal diagnoses, and these International Classification of Diseases, Ninth (ICD-9) diagnosis codes can be extracted from the Clinical Classification Software (CCS). We selected about 40 diagnoses by removing some diagnoses that either were related to giving birth or had too few cases in Nebraska.

Once a set of diseases was selected, the next step was choosing a proper method. There were a number of issues, some conceptual and some practical. First, should we treat 7 years of statewide data from 2005 to 2013 as a sample or as a population? The Nebraska HDD cover all nonfederal hospitals in Nebraska; only veteran hospitals are not included. We treat the HDD as a sample because we think 7 years of data were selected from a sample of more than years, and 40 comorbidities were selected from more than a hundred of them. Second, we had to consider how to compare health outcomes by a number of poverty groups when facing 40 diagnosis groups. To compare hospitalization disparities by neighborhood poverty levels, hospitalization data alone are not sufficient, because a high frequency of total hospitalization within an SES category could be due to variation in the at-risk population. Assuming that the total hospitalization follows the population distribution of poverty status, we could compare a particular disease as the proportion of total hospitalization. However, if the total hospitalization had an SES gradient, we would not be able to detect disparities in many disease incidences. For this reason, we added the at-risk population, and we treated hospitalized patients as incidence, with the at-risk population being the total civilian population.

For the numerators, we wanted to survey comorbidities to gauge the frequency of a particular disease or condition according to neighborhood poverty. Ideally, we should include both physician visits and hospitalizations to represent neighborhood disease incidences, similar to Roos et al. (2005). However, we only had access to hospital claims data, so the rate calculated would represent more severe conditions and be restrictive to hospitalized patients. Since it's difficult for humans to inspect or detect a large number of rates, odds ratio provided a uniform framework for comparison. A natural choice was the log-rate model or Poisson regression, with the offset being the log of the neighborhood population by age and sex. Another distinction is that we were interested in incidence of hospital stays rather than the reason for hospitalization. We screened comorbidities based on nine diagnosis codes for each patient because it is the common denominator for all the available years in terms of number of diagnosis codes listed for each patient. In our preliminary analysis of cardiovascular disease (CVD) data, we found that 9 diagnosis codes would cover 98% of CVD patients, assuming that 25 diagnosis codes cover 100%. Finally, calculation of disease-specific odds ratios includes all hospitalizations without distinguishing if they are multiple or not. This is common in many hospital-based studies (Goodman et al., 1997). Although the primary reason for record-based study was the lack of patient-level identification, a major policy implication is that each hospitalization places an additional burden on society, and we wanted to know how hospitalizations are distributed in addition to how hospital patients are distributed. The latter will be discussed in Chapter 9.

For the at-risk population, we downloaded the American Community Survey (ACS) data for 2007–2011 (ACS summary table B17001), which has the poverty rate per capita for each census tract, together with the at-risk population. For surveillance, this table provides one-stop downloading that can determine both poverty and the at-risk population. Two caveats exist, however: (1) the top age group is 75+, which may not be sufficient for some diseases, such as stroke, and (2) ACS data are not census data, so the age- and sex-specific population at the census tract level may not be as certain as that of the decennial census. However, since our interest is a group of census tracts sharing a similar poverty status, rather than a particular census tract, the uncertainty is almost negligible after aggregation. With the focus of chronic diseases, it is necessary to group some younger age groups together. A preliminary analysis suggested that some diseases had few cases in young age groups, such as ages 15–24 or 25–34; we therefore grouped all those aged 34 and below as a single group. Therefore, the six age groups are <35, 35–44, 45–54, 55–64, 65–74, and 75+. These six age groups together with two sex groups serve as controls, and they are cross-tabulated with five poverty categories (<5%, 5%–10%, 10%–15%, 15%–20%, 20+%) in a three-way table. Our main effect log-rate model can then be specified as

$$\log(\text{Disease}(x)) = \text{Age} + \text{Sex} + \text{Poverty} + \text{Offset}(\text{log of population}) \qquad (8.1)$$

where disease x represents a selected disease, such as diabetes, by age–sex and poverty in a $6 \times 2 \times 5$ three-way format. x's are from 1 to 40, and we used D1 to D40 to represent disease names, so that computer codes could be programed more efficiently.

To summarize, we assessed the age- and sex-adjusted odds ratio by poverty levels with the following steps:

1. Geocoding hospital data and assigning census tract codes to each patient, which we named TractFips.
2. Downloading ASC poverty together with population data from American FactFinder, keeping the GeoID2 variable and renaming it TractFips, and then checking to make sure the summary level was 140 (census tract level).
3. Linking patient data to population data to obtain poverty-level information. (The link ID is TractFips.)
4. Aggregating ASC data by age groups, sex, and poverty level. (It is a three-way cross-tabulation with six age groups, two sex categories, and five poverty levels in percent, representing the at-risk population file.)
5. Selecting a disease or a set of diseases and aggregating them to frequencies according to age–sex and poverty level. This is the case file. (In our dataset, it was based on ICD-9 codes from diagnoses 1–9.)
6. Merging the at-risk population file from step 4 to the case file from step 5.
7. Performing a log-rate model analysis for each selected disease or condition to detect significant odds ratios, using a low-poverty neighborhood as the reference group.
8. Organizing results for reporting. (The interpretation of an odds ratio for comorbidity has a number of implications, and we used, for example, poverty level 2 vs. poverty level 1 for diabetes.)

8.3   SURVEILLANCE RESULTS

8.3.1   SES DISPARITIES IN COMORBIDITIES

Table 8.1 presents age- and sex-adjusted odds ratios from each output of the log-rate model shown in each row. Only poverty contrasts are shown, because age groups and sex were used as controls. Note that our aim was to survey comorbidities from each hospitalization, and select 39 of the 50 most frequent principal diagnoses for inpatient admission. We used the term *comorbidity* to distinguish it from the principal diagnosis, because we searched all nine diagnoses to find a particular condition. In addition, we included overall hospitalization (the first) and ACSCs (AHRQ, 2007) that include many diseases on the list, such as diabetes, hypertension, congestive, heart failure, angina without procedure, asthma, bacterial pneumonia, and urinary infections. There were a total of 164 odds ratios in reference to 41 diseases from low-poverty neighborhoods. If an odds ratio is >1, then it is more likely to happen. For example, let's examine the odds ratio of 1.53 of congestive heart failure (CHF) from poverty group 5 (>20%). Patients from the highest-poverty neighborhood group were 53 times more likely to have CHF admissions than those from low-poverty neighborhoods

TABLE 8.1

Age- and Sex-Adjusted Odds Ratios for Each Selected Comorbidity Model

Disease ID	Diseases and Conditions	Pvt2 vs. 1	Pvt3 vs. 1	Pvt4 vs. 1	Pvt5 vs. 1
0	Overall hospitalization	1.11	1.14	1.25	1.37
1	Pneumonia	1.21	1.34	1.52	1.55
2	Osteoarthritis	1.04	1.03	1.09	—
3	CHF	1.11	1.18	1.40	1.53
4	Septicemia (except in labor)	—	—	1.25	1.48
5	Mood disorders	1.20	1.28	1.55	1.92
6	Cardiac dysrhythmias	1.01	1.02	1.10	1.16
7	COPD and bronchiectasis	1.19	1.34	1.62	1.86
8	Back pain	1.06	1.14	1.16	1.28
9	Skin and subcutaneous tissue infections	1.14	1.23	1.44	1.77
10	Coronary atherosclerosis	1.15	1.15	1.28	1.31
11	Urinary tract infections	1.08	1.12	1.35	1.51
12	Nonspecific chest pain	1.33	1.47	1.49	2.01
13	AMI	1.09	1.13	1.22	1.37
14	Acute cerebrovascular disease	1.05	1.06	1.19	1.46
15	Diabetes mellitus with complications	1.17	1.29	1.57	2.30
16	Biliary tract disease	1.14	1.21	1.26	1.32
17	Fluid and electrolyte disorders	1.05	1.10	1.31	1.51
18	Asthma	1.13	1.27	1.53	2.10
19	Schizophrenia and other psychotic disorders	1.55	2.09	3.07	5.57
20	Acute and unspecified renal failure	0.94	0.90	1.21	1.65
21	Respiratory failure, insufficiency	—	1.09	1.39	1.80
22	Gastrointestinal hemorrhage	1.11	1.18	1.43	1.53
23	Intestinal obstruction without hernia	1.07	1.08	1.25	1.22
24	Pancreatic disorders	1.18	1.27	1.49	1.94
25	Diverticulosis and diverticulitis	0.93	—	1.10	—
26	Hip fracture	1.07	1.10	1.12	1.11
27	Epilepsy, convulsions	1.32	1.50	1.82	2.24
28	Appendicitis and appendiceal conditions	1.11	1.14	—	1.11
29	Alcohol-related disorders	1.36	1.53	1.95	3.28
30	Fracture of lower limb	1.21	1.18	1.32	1.52
31	Other nervous system disorders	1.05	1.12	1.22	1.47
32	Syncope	—	—	1.14	1.20
33	Substance-related disorders	1.31	1.47	2.09	3.74
34	Intracranial injury	1.13	1.12	1.37	1.61
35	Hypertension with complications	—	—	1.27	1.84
36	Other fractures	1.09	1.10	1.10	1.17
37	Intestinal infection	1.07	1.07	1.22	1.27
38	Secondary malignancies	—	—	1.05	1.14
39	Anemia	1.06	1.08	1.29	1.55
40	ACSCs	1.15	1.23	1.44	1.69

Note: —, not significant; all others are significant. Except ACSCs, all selected diseases are based on the top 50 discharges, excluding labor and related complications. Pvt, poverty level, all in odds ratios. Pvt2 vs. 1, census tracts with >5% and ≤10% people under vs. <5% under the poverty level; pvt3 vs. 1, census tracts with >10% and ≤15% people under vs. <5% under the poverty level; pvt4 vs. 1, census tracts with >15% and ≤20% people under vs. <5% under the poverty level; pvt5 vs. 1, census tracts with >20% vs. <5% under the poverty level.

(controlling for age and sex). Even though it is the most straightforward interpretation, we need to point out that the unit of analysis is incidence rather than admitted patients. Sometimes, an odds ratio of >1 does not necessarily mean people from a high-poverty group are more likely to have heart failure, because a patient can be hospitalized multiple times. If the high-poverty group had a lot more repeated hospital visits than the low-poverty group, then the odds ratio would still be >1, even though each group had an identical number of patients with CHF. In general, we assume that patients from neighborhoods with different poverty levels had similar numbers of repeated hospitalizations, and that the above interpretation is not that far off.

Among the 164 odds ratios estimated from 41 models, 151 were significant. In general, they showed a general gradient. The poorer the neighborhood the patients were from, the more common was each health condition. The 12 diagnoses with the steepest gradients were schizophrenia and other psychotic disorders, substance-related disorders, alcohol-related disorders, diabetes mellitus with complications, epilepsy, convulsions, asthma, nonspecific chest pain, pancreatic disorders, mood disorders, chronic obstructive pulmonary disease (COPD), and bronchiectasis. Upon further inspection, we found mental health–related diagnoses dominant among the 10. In a real surveillance system, these results could either be presented in graphics or ranked by the magnitude of odds ratios. If we did the ranking, we would pick schizophrenia and other psychotic disorders (no. 19), substance-related disorders (no. 33), and alcohol-related disorders (no. 29). We could then identify some high-incidence neighborhoods and refer them to intervention specialists for public health interventions that may reduce these hospitalizations. Finally, conditions contributing to ACSCs had a strong relationship with neighborhood SES. The odds were 1.69 times higher for those from high-poverty neighborhoods than for those from low-poverty neighborhoods.

Two diseases or conditions, acute and unspecified renal failure (no. 20) and diverticulosis (no. 25), had some inconsistent odds ratios. For renal failure, patients from poverty levels 2 and 3 were about 6% and 10%, respectively, less likely to be seen in hospital admissions, whereas patients from poverty groups 4 and 5 were 21% and 65%, respectively, more likely to be seen. Diverticulosis is a condition when pouches form in the wall of the colon, and it was often diagnosed while performing a colonoscopy. Those in poverty group 2 were 7% less likely than those in poverty group 1 to have diverticulosis, while those from poverty group 3 were 10% more likely. This could simply reflect a temporary trend that relates to the popularity of colonoscopy among different neighborhoods. Finally, compared to low-poverty neighborhoods, people from poor neighborhoods are injured more often, perhaps due to violence and accidents.

Although surveillance is not intended to offer explanations, some insights can be gained. It seems that chronic conditions (e.g., back pain, diabetes, heart disease, and COPD) affecting physical well-being along the poverty strata are related to commonly reported psychological symptoms. Common sense suggests that back pain (often caused by arthritis, osteoporosis, or other chronic pain–related illnesses) would deprive sleep, alter mood, and increase anxiety, which in turn can cause mental illness. Several studies have linked the experience of chronic pain and the degree of anxiety in pain (Hadjistavropoulos et al., 2002). However, the link along the SES is not clearly understood. Our findings are cross-sectional: psychological or mental health conditions had greater SES gradients than the traditional chronic diseases as far as hospitalization was concerned.

8.3.2 SES Disparities in Hospital Procedure Utilization

Theoretically, a hospital should provide appropriate treatment to all patients regardless of ability to pay. Physicians are trained to provide the most appropriate care to everyone and are normally shielded from insurance and other information when treating patients. However, due to various reasons, the same diseases and conditions may result in different treatments for those with relatively high and those with relatively low SES. Public health practitioners concerned about health care outcomes need to assess potential disparities in treatment procedures as they bridge disease conditions and health care outcomes such as readmissions and mortality.

In the "Unequal Treatment" report, the Institute of Medicine (2003) highlighted substantial treatment disparities and recommended better data collections by race and ethnicity and SES. In this subsection, we tried to use the enhanced HDD to assess major hospital procedures along the SES dimension.

We realize that treatment is highly individualized for both physicians and patients, and hospital procedure disparities do necessarily mean treatment disparities. Two patients with an identical principal diagnosis for hospital admission may end up with different treatments, which cannot be answered without a chart review or real-time peer-physician diagnosis. In addition, the lack of health insurance inhibits access to preventive care, timely diagnosis, and treatment, which may result in different treatment options later on. The hospital data for the current study were collected prior to the Affordable Care Act, which provides insurance for almost all Americans. Hence, by looking at HDD alone, it is difficult to say if a treatment is appropriate or not. Nevertheless, when a sample is large enough to represent almost the entire hospital population of a state, some procedure disparity detections are possible. For example, if osteoporosis is more common in high-poverty neighborhoods than in low-poverty neighborhoods, then the need for hip and knee replacement operations should be higher in high-poverty areas. However, if lower rates are observed in high-need areas, then patients in high-poverty areas are disadvantaged. The Dartmouth Atlas has provided many excellent examples of hospital-based treatment variation by health regions in the United States. Since different regions present different levels of SES and health care resources, SES disparities in hospital procedures are implicit in many Dartmouth studies. In the United Kingdom, health equity audits prove effective for local health authorities (DH, 2008). In the absence of established indicators, we opted to use procedure volume as the criterion for procedure selection and selected the top 20 procedures (Pfuntner et al., 2013b).

All the steps in disease surveillance above should be applicable to procedure surveillance. We only need to replace "disease" in step 5 with "procedure." Table 8.2 presents results in odds ratios for procedure surveillance. With some exceptions, the clear SES gradients evident in comorbidity (Table 8.1) are not obvious in procedures. For example, prophylactic vaccination and inoculation are often needed against influenza, and such a procedure could be administered in an ambulatory setting. However, it was much more common for patients in poorer neighborhoods to get this procedure than for patients in neighborhoods with <5% of its population under the poverty level. COPD is another condition poor patients suffered disproportionally more from it, and the procedure often needed in emergency situations is respiratory intubation and mechanical ventilation. The SES gradients for this procedure mirrored the COPD gradients in Table 8.1.

However, there are also some reversals. Hip and knee replacements are more likely for patients from well-off neighborhoods than those from poor neighborhoods. Some procedures have to be interpreted with caution, though. For instance, colonoscopies and biopsies that detect colorectal cancer had a positive relationship (odds ratios >1) in the poorest census tracts (>20%), which was inconsistent with findings from a national study, which showed an inverse relationship between the use of colonoscopy and poverty (Steele et al., 2013). The inconsistency might be due to other reasons, such as missing freestanding ambulatory surgery centers from the calculation.

Without further controls, it is hard to interpret SES differences in the procedures performed. Nevertheless, the wide range of odds ratios highlight the importance of documenting procedures along the SES dimension.

8.3.3 SES Disparities in Hospital Readmission Surveillance

Hospital readmissions cost billions each year to the U.S. economy, and reducing their number is a key strategy for improving the quality of health care and lowering associated costs. The Patient Protection and Affordable Care Act outlines the hospital readmission reduction program that limits payments to hospitals with excessive Medicare readmissions. The National Quality Strategy and the Partnership for Patients initiative includes reduction in readmissions as a national goal. To develop

TABLE 8.2

Odds Ratios of Most Frequently Performed Procedures by Neighborhood Poverty Status

	Pvt2 vs. 1	Pvt3 vs. 1	Pvt4 vs. 1	Pvt5 vs. 1
Blood transfusion	1.08*	1.23*	1.37*	1.62*
Prophylactic vaccinations and inoculations	1.24*	1.46*	1.58*	3.33*
Respiratory intubation and mechanical ventilation	1.05*	1.11*	1.39*	2.12*
Colorectal resection	0.95	0.94*	0.98	0.94
Open reduction of fracture	1.13*	1.1*	1.24*	1.29*
Cardiac catheterization and coronary arteriography	1.21*	1.21*	1.26*	1.59*
Upper gastrointestinal endoscopy, biopsy	0.99	1.04	1.24*	1.47*
Thyroid or parathyroid surgery or both	1.25*	1.44*	1.29*	1.19
Abdominal paracentesis	1*	0.92*	1.2	1.66
Hemodialysis	1.09	0.97	1.63*	3.22*
Echocardiogram	0.99	1.05	1.36*	1.49*
CABG	1.06	1.07	1.06	0.95
Arthroplasty knee	1.06*	0.97	0.99	0.86*
Enteral and parenteral nutrition	0.78*	0.8*	1.14*	1.46*
Percutaneous transluminal coronary angioplasty (PTCA)	1.08*	1	1.1*	1.08*
Colonoscopy and biopsy	0.89*	0.85*	1.04	1.16*
Laminectomy, excision intervertebral	0.94*	0.84*	0.84*	0.81*
Spinal fusion	0.94*	0.99	0.95	0.82*
Incision of pleura, thoracentesis, chest drainage	0.95*	0.94*	1.11*	1.16*
Hip replacement, total and partial	1.01	0.95*	0.97	0.82*

Note: Pvt, poverty level, all in odds ratios. All selected procedures are based on volumes for hospital admissions by excluding those due to labor. Pvt2 vs. 1, odds ratios for neighborhood poverty rate in 5%–10% group vs. <5% group; pvt3 vs. 1, odds ratios for neighborhood poverty rate in 10%–15% group vs. <5% group; pvt4 vs. 1, odds ratios for neighborhood poverty rate in 15%–20% group vs. <5% group; pvt5 vs. 1, odds ratios for neighborhood poverty rate in 20+% group vs. <5% group.

*Significant at $p < 0.05$.

effective readmission prevention strategies, it is necessary to establish some baseline rates and identify gaps by SES, among risk factors.

Hospital readmissions are defined as multiple inpatient stays by the same patient within a time period, most often in 30 days. In this definition, the readmission can be for any reason—not necessarily for the same diagnosis. Readmission surveillance by hospital is routinely conducted by the Centers for Medicare and Medicaid Services (CMS), which is tasked to reduce payments to hospitals with excessive readmissions of Medicare patients. The number of hospital readmissions has always been an important quality indicator of hospital care. The National Quality Strategy and the Partnership for Patients initiative actively tracks readmissions at a national level. The Patient Protection and Affordable Care Act also stipulates a hospital readmission reduction program that cuts payments to hospitals with excessive Medicare readmissions. Although state public health programs do not conduct readmission surveillance, it does belong to the realm of public health service surveillance.

Recent debate about the influence of SES in hospital readmissions highlights the importance of the care continuum from hospital-based care to community-based care. At its core is whether the SES of an individual patient and the community has an impact on readmission beyond the control of hospital care providers. Currently, CMS surveillance on hospital readmission does not consider SES. It is argued that SES differences in readmission could be eliminated if the quality of care by SES is the same. If hospitals have different standards of care according to patient SES (e.g., lower standards for socioeconomically disadvantaged populations), then it should be prevented and corrected.

One correction method is hospital payment reforms. However, an expert panel convened by the National Quality Forum (NQF) recommended that sociodemographic factors be included in risk adjustment of the hospital performance score. This view is based on empirical evidence that hospital care is only one component of the hospital treatment continuum. SES factors predisposed at the hospitalization also determine hospital outcomes, such as readmission. In other words, the same quality of care may produce different readmission outcomes if a patient experiences a lack of social support at home or in the community. Given that hospital treatment is comparable, it might be even more important for community-based care, such as from a spouse and family members who may have different resources and levels of health literacy. Since community intervention is traditionally in the public health domain, detecting which diseases have high readmissions along SES can be rightly justified under public health practice, so that practitioners can assist community-based social support programs for after-hospital care.

However, there is no uniform way of calculating readmissions, especially for surveillance purposes. According to the Healthcare Cost and Utilization Project (HCUP) report (Barrett et al., 2012), there are 12 existing measures of readmission, and we list those that use administrative inpatient data. The Agency for Healthcare Research and Quality (AHRQ) provides a general 30-day readmission measure, while the CMS provides four all-cause readmissions following hospitalizations for (1) AMI, (2) heart failure, (3) pneumonia, and (4) hospital-wide readmissions, and some of these measures also use outpatient data. There are also readmissions following elective total hip (THA) or total knee (TKA) arthroplasty and percutaneous coronary intervention (PCI). In addition, the 3M Health Information System provides the 3M™ potentially preventable readmission measure (part of proprietary software to calculate hospital readmission rates), the National Cancer Institute (NCI) has a readmission measure for end-of-life cancer patients, and the National Committee on Quality Assurance (NCQA) provides a plan-level readmission measure.

Based on a review of the above measures, we developed an operational definition for readmission. The analysis is at the individual level in that a readmission is only captured once for each patient who only seeks the indexed hospital care for one reason. However, since every hospital stay is counted as a separate index admission, if a patient was treated for diabetes for one hospital stay and heart failure for another stay 6 months later, the patient can be counted multiple times for readmission. (One is indexed by diabetes, and the other by heart failure.) The index admission is the basis for the denominator. We used inpatient admissions for patients discharged alive. However, since our data covered 7 years, there could be multiple discharges, and we needed to decide which hospitalization event to index or be the reference. We selected the most recent hospital admission and searched for the most recent discharge within 30 days of the admission as the index admission for the same patient. This practice follows the HCUP measure, allowing a hospitalization to be counted as both an index and a readmission. If there was no discharge within 30 days, then the last admission was the index admission. In this way, we captured the last readmission from the duration of 2005–2011.

In addition, most readmission measures have some risk adjustments. A common adjustment includes disease-specific control for clinical condition (e.g., comorbidities and severity of illness), and it is quite complicated without clinician-level knowledge. The CMS uses outpatient data to index admission for some heart disease procedures, but outpatient data are often not available for public health practitioners. We opted not to use outpatient data, even though they have already been integrated with inpatient data. Common to all is the adjustment by age and sex, which is adopted in the following analysis. In some academic studies, SES is adjusted or controlled. Since our surveillance is along the SES, we use poverty level 5 explicitly. Similar to HCUP, no attempt was made to remove from the surveillance analysis some readmissions that may have been planned or unavoidable.

Finally, we also needed to select a set of diseases for readmission surveillance. We primarily used the list of diseases from Table 8.1 based on the primary and secondary diagnoses. In addition, we added two conditions from the HCUP readmission report that are not in the table: complication of device and complications of surgical procedures. We listed the diseases the same way as in Table 8.1, and used the following procedure to generate readmission indicator:

1. Select diseases based on the primary diagnosis from the HDD file and create an indicator variable (1 = yes, 2 = no); repeat this step until finished with all the selected diseases, and save the file as File A.
2. Sort File A by master patient ID and admission date.
3. Create an admission index (ad) for each patient if it is his or her first admission; then ad = 1 for the first admission, ad = 2 for the second admission, and so on.
4. Generate a diagnosis-specific readmission indicator.
   a. Select a diagnosis indicator of 1 and search for the second admission. If the patient was only admitted once, select the record as the index event; otherwise, select the first diagnosis-specific admission record as the indexed event.
   b. Create an admission duration variable from the index event to the ad + 1 event.
   c. If the duration between the index event and index + 1 event is ≤30 days, then 30-day admission = 1; else, 30-day admission = 0.
   d. Delete all ad+ events.
   e. Name and save the diagnosis-specific readmission file.
   f. Repeat the above steps for all selected diseases.
5. Generate model-based readmission estimates using a logistic regression. The dependent variable is whether the patient was readmitted; independent variables include age, sex, and poverty level. Model outputs can be restricted to poverty-based contrasts, because age and sex effects are controls and can be ignored in model-based outputs.

Note that the readmission analysis differs from comorbidity and procedure surveillance in two ways: the data file is patient based rather than record based, and the odds ratios are from a logistic regression rather than the log-rate model. In addition, if we want to obtain disease-specific readmission, an additional condition should be added: the ad+ event has to be the same in either primary or secondary diagnoses.

Table 8.3 lists the significant results. Overall, 22 out of 42 selected diseases and conditions had at least one significant odds ratio. In most cases, the directional effects are as expected. Patients residing in high-poverty neighborhoods had a greater chance of readmission than those in low-poverty neighborhoods. For AMI, patients from neighborhoods with 20% or greater poverty were 37% more likely to be readmitted to a hospital within 30 days. The corresponding number for osteoarthritis is 170%. Also, as we move from left to the right in the table, the contrast with the reference poverty level (5%) increases. As expected, the odds ratios in the far right column had the highest number of significant values. For patients living in the poorest neighborhood category, 16 out of 22 odds ratios indicated elevated readmission rates, whereas in the poverty 2 category (5%–10%), only 6 of them were significant. The results also show that those living in poorer neighborhoods tend to have repeated ACSCs.

Besides the above general trends, there are several ways to inspect the results. First, the lack of significant results from the 42 conditions is due to small cells or frequencies. If we were to combine the last two poverty groups into a single group, we would show many more diseases as significant by neighborhood SES. Second, the readmission window of 30 days may affect some results. For example, we included two procedure-related diagnoses according to the HCUP report (Elixhauser and Steiner, 2013): (1) complications of device, implant, or graft and (2) complications of surgical procedures. Only the latter was significant for the high-poverty group. Suspecting that the device, implant, or graft may last longer than 30 days, we ran a model for 60-day readmission to check.

TABLE 8.3

Readmission of Frequent Diagnoses by Neighborhood Poverty Status

First or Second Diagnosis	Pvt2 vs. 1	Pvt3 vs. 1	Pvt4 vs. 1	Pvt5 vs. 1
Osteoarthritis	2.133*	2.592*	1.939	2.703*
CHF	0.961	0.884*	0.965	1.116*
Mood disorders	1.052	1.033	1.191*	1.292*
Cardiac dysrhythmias	1.211*	1.238*	1.154	1.192*
COPD and bronchiectasis	1.058	1.066	1.102	1.311*
Back problems[a]	1.212	1.418*	1.067	1.202
Skin and subcutaneous tissue infections	1.04	1.062	1.271	1.344*
Nonspecific chest pain	1.552*	1.36	1.403	2.197*
AMI	1.203	1.347*	1.367	1.370*
Acute cerebrovascular disease	0.861	0.871	0.666*	0.727*
Diabetes mellitus with complications	1.284	1.424*	1.468*	1.900*
Asthma	1.312	1.533*	1.076	1.536*
Schizophrenia and other psychotic disorders	1.03	1.072	1.291	1.420*
Acute and unspecified renal failure	1.111	0.986	1.05	1.239*
Respiratory failure, insufficiency, arrest	1.054	1.019	1.109	1.523*
Gastrointestinal hemorrhage	1.380*	1.187	1.379	1.22
Epilepsy, convulsions	1.304	1.401*	1.394	1.477*
Fracture of lower limb	0.724*	0.574*	0.637*	0.675*
Hypertension (HP) with complications and secondary HP	1.433*	1.325	1.393	1.999*
Anemia	1.478*	1.262	0.978	1.031
ACSCs	1.110*	1.129*	1.250*	1.311*
Complications of surgical procedures	0.981	1.065	1.228*	1.225*

Note: Readmission is based on 30 days, all causes. Pvt, poverty level, all in odds ratios. Pvt2 vs. 1, census tracts with >5% and ≤10% people under vs. <5% under the poverty level; pvt3 vs. 1, ensus tracts with >10% and ≤15% people under vs. <5% under the poverty level; pvt4 vs. 1, census tracts with >10% and ≤15% people under vs. <5% under the poverty level; pvt5 vs. 1, census tracts with >20% people under vs. <5% under the poverty level.

[a] Spondylosis, intervertebral disk disorders, and other back problems.

*Significant at $p < 0.05$.

Indeed, the odds ratio of 1.348 (p-value = 0.00257) for the poverty level 5 versus 1 was significant. Third, we can associate findings from this table to Table 8.1, as some disease-specific significant results from Table 8.3 tend to be associated with high SES gradients. Examples include schizophrenia and psychotic disorders, mood disorders, and seizures and epilepsy. Even though mental health conditions are excluded for readmission surveillance by many programs, their SES gradients should not be ignored. Finally, we also included cancer, as comorbidities are often developed after cancer treatment. It seems that cancer patients from poorer neighborhoods ended up being more likely to be readmitted to a hospital than those from low-poverty neighborhoods.

Some readmissions were more likely for patients from low-poverty neighborhoods. Fracture of lower limb and acute cerebrovascular disease were both included in this category. For acute cerebrovascular disease and fractures of lower limbs, the effects may be genuine. Acute cerebrovascular disease (mostly stroke patients from poor neighborhoods) patients may be more likely to die and less likely to engage in rehabilitation. The same can be said for patients with lower-limb fractures. We note that patients with CHF in poverty level 3 were 12% less likely to be readmitted for no apparent reason. On the one hand, the risk factor may be more prone to the well-off neighborhoods (<5%) than the middle ones. On the other hand, one unexpected effect could be due to

TABLE 8.4
Readmission by the Same Primary Diagnoses Neighborhood SES

First or Second Diagnosis	Pvt2 vs. 1	Pvt3 vs. 1	Pvt4 vs. 1	Pvt5 vs. 1
Osteoarthritis	1.997	2.663*	2.082	3.786*
CHF	0.988	0.874*	1.053	1.128*
Mood disorders	1.077	1.041	1.282*	1.458*
Cardiac dysrhythmias	1.215*	1.272*	1.358*	1.362*
COPD and bronchiectasis	0.978	1.015	1.083	1.277*
Back problems[a]	1.569*	1.709*	1.353	1.336
Skin and subcutaneous tissue infections	1.063	1.150	1.325	1.620*
Nonspecific chest pain	1.308	1.003	1.024	1.759*
AMI	1.225	1.421	1.467	2.030*
Acute cerebrovascular disease	0.682*	0.714	0.431*	0.620*
Diabetes mellitus with complications	1.175	1.245	1.409	1.657*
Schizophrenia and psychotic disorders	1.095	1.081	1.451*	1.475*
Respiratory failure, insufficiency, arrest	1.002	0.879	1.013	1.501*
Intestinal obstruction without hernia	1.215	1.149	1.340*	1.092
Epilepsy, convulsions	1.280	1.242	1.399	1.456*
Fracture of lower limb	0.708	0.356*	0.648	0.555*
Hypertension (HP) with complications and secondary HP	1.515*	1.255	1.802*	2.149*
Anemia	1.810*	1.494*	1.099	1.119
Avoidable hospitalization	1.055	1.061	1.196*	1.291*
Complications of surgical procedures	0.998	1.121	1.276*	1.314*

Note: 30-day readmission conditioned on the second admission corresponds to the first one. Pvt, poverty level, all in odds ratios. Pvt2 vs. 1, census tracts with >5% and ≤10% people under vs. <5% under the poverty level; pvt3 vs. 1, census tracts with >10% and ≤15% people under vs. <5% under the poverty level; pvt4 vs. 1, census tracts with >10% and ≤15% people under vs. <5% under the poverty level; pvt5 vs. 1, census tracts with >20% people under vs. <5% under the poverty level.

[a] Spondylosis and intervertebral disk disorders and other back problems.
*Significant at $p < 0.05$.

some unknown temporal reason or by chance, because the odds ratio from the poorest neighborhoods (poverty level 5) did show 11.6% more readmissions for heart failure patients. Since generating hypotheses is one of the functions of surveillance, one could use these results for in-depth analyses.

In the above analysis, we followed the traditional definition of readmission. The initial index admission or hospital stay was disease specific, and readmission could be for other causes. However, such a definition did get into the question of repeated hospital treatments for the same disease. In the following, we adhered to the restriction that the index diagnosis must be the same diagnosis for readmission within 30 days. The significant results are in Table 8.4. One apparent finding is that those results significant in Table 8.4 are also significant in Table 8.3. In addition, all the directional effects remain the same. Six conditions became nonsignificant: (1) CHF COPD and bronchiectasis, (2) nonspecific chest pain, (3) asthma, (4) acute and unspecified renal failure, (5) gastrointestinal hemorrhage, and (6) anemia.

8.3.4 OUT-OF-HOSPITAL MORTALITY SURVEILLANCE BY SES

Previous surveillance efforts have been on very near-term hospital mortality, such as in-hospital mortality or 30-day mortality after hospital discharge, because those hospital-based surveillances intend to improve patient care. After 30 days of the discharge, mortality variation may be related to (1) the quality of hospital care, (2) family support, and (3) community resources. Since public health focuses on community-based resources for post-hospital care, out-of-hospital mortality surveillance would be a natural extension of hospital care–based surveillance. Given that few studies deal with out-of-hospital mortality assessments, we use in-hospital mortality surveillance literature to guide us in the following analysis.

Variation in hospital mortality has been a subject of statistical inquiries since at least 1861, when the *Journal of the Statistical Society of London* started publishing a series of articles describing hospital mortality rates (Spiegelhalter, 1999). It was advocated at the time that mortality rates should be compared at both the hospital and ward levels to tease out (1) the quality of care within hospitals and (2) the case mix by neighborhoods or wards. While both effects were claimed by Guy (1867), he seemed more concerned with variation due to case mix, presumably at the ward level, and urged caution about not adjusting for potential risk factors. Even though risk adjustment from case mix and other comorbidities is an important issue, more than 100 years after Guy's study, early studies rarely take it into consideration. Fink et al. (1989) reviewed 21 studies with hospital-specific mortality data, and only 3 had adjusted demographic and comorbidities. They pointed out the importance of tracking hospital mortality, as some patients were discharged to nursing homes and hospices.

Literature on in-hospital mortality by SES has accumulated rapidly in recent years (Zarzaur et al., 2010). Colvin et al. (2013) identified an inverse relationship between zip code–based SES and in-hospital pediatric mortality. Reames et al. (2014) attributed higher in-hospital mortality of cancer patients to failure to rescue in some hospitals near low-SES neighborhoods. A study of coronary artery bypass graft (CABG) patients in California (Kim et al., 2007) showed that zip code SES is inversely related to in-hospital mortality, but the study also attributed this to high-volume hospitals. While a study in Shelby County, Tennessee, found an inverse relationship between census tract–based SES and injury admission, it failed to detect any in-hospital mortality. Vrbova et al. (2005) did not find association between neighborhood SES and in-hospital mortality among community-acquired pneumonia patients. Although most studies tend to find an inverse relationship between area SES and in-hospital mortality (Gerward et al., 2006), some fail to find any relationship, and some attribute it to hospital locations and volumes.

In recent years, there has been an increased interest in post-hospitalization care. While health care service researchers are interested in measuring many effects of hospital care that become evident only after patients leave hospitals, public health practitioners are interested in social environments that influence (1) preexisting conditions before hospitalization and (2) postsurgery or other complications after hospitalization. In addition, after quality improvement indicators are implemented, hospitals might be tempted to discharge patients with poor prognoses to minimize in-hospital mortality, which would be captured by out-of-hospital mortality indicators. Early studies on out-of-hospital mortality are primarily for 30-day mortality after discharge. Chassin et al. (1989) and Rosenthal et al. (2000) showed that 30-day postdischarge mortality was similar to in-hospital mortality. While studies of this nature required some kind of linkage between hospital data and death records (Krakauer et al., 1992), most HDD today includes 30-day mortality information, which is required for reporting purposes.

Besides 30-day mortality, some use wider time windows, such as 60 days, 180 days, and more than a year. Lubitz et al. (1985) used six different time windows following surgery, from 15 days to 1-year postdischarge mortality, and suggested that "anytime windows" have their uses. Fleming et al. (1991) developed risk-adjusted mortality indices using time-varying windows tailored to clinical conditions. Garnick et al. (1995) suggested that 30-day mortality is the best time window measurement for three types of cardiac hospitalization: (1) CABG surgery, (2) treatment for an AMI, and (3) CHF. Working with a more controlled registry dataset, Hannan et al. (2000) extended the study time window to 3 years for AMI patients who had primary angioplasty. In general, however, hospital treatment effects normally should not extend to more than 1 year, and we therefore use 1-year mortality as an indicator to gauge community effects. The 1-year post-hospital mortality was used in a recent CVD-SES relationship study.

Our surveillance approach differs from more elaborate research designs in important ways. First, it encompasses multiple diseases and conditions, and a comorbidity-based risk adjustment method for one disease does not necessarily apply to another. For this reason, we just used age group and sex as control variables. Likewise, we used the same set of diseases and conditions listed in Table 8.1, but we only used the first two diagnoses to code for principal diagnoses for admission. Second, we used the same logistic regression approach used in readmission tables, which suffer from a small number of issues in younger age groups that we cannot individually examine due to the surveillance practice: only the poverty-related odds ratios were extracted from the outputs. Third, there are some uncertainties regarding mortality by neighborhoods. We used death records provided by the Nebraska Vital Records Office. The records are very good, but not perfect. The office has data exchange agreements with many states (adjacent states and retirement destination states), but some deaths occurring outside Nebraska may not have been recorded. We relied on full name, date of birth, sex, and residential location as matching variables, but the linkage method is probabilistic, and some mismatches were deemed acceptable. Fourth, there is no acceptable way to determine discharge date for 1-year mortality surveillance, and we made our own rule. If a patient is hospitalized for CHF and a readmission occurs for some other illness, we would ignore the readmission and use the discharge date of CHF as the index date for 1-year mortality. Moreover, mortality tends to have multiple causes, and we included all-cause mortality, other than cause-specific death. Finally, different ways of coding the same diseases by name may have different outcomes. For this reason, we also included three diseases from the Charlson Index that coded comorbidity conditions in ICD-9. In particular, we included Charlson Index MI (410 and 412 in three digits), CHF ([428] or [4254, 4255, 4257–4259] or [39891, 40201, 40211, 40291, 40401, 40403, 40411, 40413, 40491, 40493]), and chronic pulmonary disease ([490–496, 500–505] or [4168, 4169, 5064, 5081, 5088]), and compared the results with corresponding diagnoses according to the HCUP definitions that we have used throughout all the tables.

Table 8.5 lists diseases or conditions with at least one significant odds ratio. Most odds ratios favor the well-off neighborhoods. Although some diagnosis-specific admissions had SES gradients, such as coronary atherosclerosis, nonspecific chest pain, mood disorders, and appendicitis and other appendiceal conditions, most listed diagnoses could only be contrasted between poverty group 1 and poverty groups 2–5. It is also evident that a cluster of CVD-related diseases had an inverse relationship between area SES and 1-year mortality (e.g., AMI, MI, CHF, peripheral vascular disease, cerebrovascular disease, and cardiac dysrhythmias). If we include chest pain, almost a quarter of the listed diseases relate to CVD, in which poor neighborhoods had higher odds of 1-year mortality. In addition, kidney to biliary tract and urinary tract diseases, pulmonary disease, and mood and alcohol-related disorders form some small clusters.

The exception to the above SES relationship is pneumonia, whose odds ratios suggest that patients from middle and upper-middle SES groups were less likely to die within 1 year of discharge than those in the highest SES group. This result, to some degree, is consistent with a provincial-wide study in Canada on in-hospital mortality among community-acquired pneumonia patients.

Finally, our sensitivity analyses showed that the results from MI in the Charlson Index and AMI were similar, but MI had more significant results than AMI. Likewise, the results from COPD and bronchiectasis were not significant according to the HCUP definition ([490, 494, 496] or [4910, 4911, 4912, 4918, 4919, 4920, 4928, 4940, 4941] or [49120, 49121, 49122]). However, the results from chronic pulmonary disease in the Charlson Index were significant.

TABLE 8.5
Odds Ratios of 1-Year Mortality by Neighborhood Poverty Status

First or Second Diagnosis	Pvt2 vs. 1	Pvt3 vs. 1	Pvt4 vs. 1	Pvt5 vs. 1
Myocardial infarction (CI)	1.092	1.151*	1.185*	1.327*
AMI	1.036	1.084	1.155	1.284*
CHF (CI)	1.057	1.04	1.134*	1.068
Coronary atherosclerosis	1.175*	1.252*	1.254*	1.517*
Peripheral vascular disease	1.154	1.201*	1.276*	1.233*
Cerebrovascular disease	1.122*	1.11*	1.108	1.233*
Cardiac dysrhythmias	1.125*	1.143*	1.171*	1.335*
Nonspecific chest pain	1.137	1.033	1.096	1.766*
Diabetes without complications	1.145	1.419*	1.278*	1.391*
Diabetes with complications	1.371*	1.256	1.328	1.277
HP with complications and secondary HP	1.043	1.197*	1.241*	1.073
Chronic pulmonary disease (CI)	1.079	1.11*	1.064	1.157*
Pneumonia	0.934*	0.898*	0.919	1.016
Asthma	1.156	1.151	1.053	1.452*
Mood disorders	1.313*	1.388	1.173	1.555*
Alcohol-related disorders	1.119	1.279	1.478*	1.073
Secondary malignancies	1.082	1.099	1.161*	1.267
Osteoarthritis	0.992	1.083	1.301	1.618*
Back problems[a]	1.195	1.317*	1.407*	1.313*
Skin and subcutaneous tissue infections	1.167*	1.153	1.235*	1.413*
Anemia	1.198*	1.117	1.119	1.083
Septicemia (except in labor)	0.959	1.021	1.121	1.157*
Urinary tract infections	1.107*	1.091	1.114	1.26*
Biliary tract disease	1.2*	1.286*	1.173	1.339*
Acute and unspecified renal failure	1.086*	1.111*	1.117*	1.028
Fluid and electrolyte disorders	1.187*	1.184*	1.161*	1.301*
Intestinal obstruction without hernia	1.191*	1.168*	1.224	1.25*
Pancreatic disorders (not diabetes)	1.102	1.219	1.188	1.257*
Diverticulosis and diverticulitis	1.195	1.05	1.272	1.481*
Appendicitis and other appendiceal conditions	1.315	1.634	2.441*	2.412*
Other fractures	1.02	1.003	0.888	1.231*
Fracture of lower limb	1.317*	1.512*	1.184	1.336
ACSCs	1.036	1.049	1.05	1.11*

Note: CI, diagnosis codes were from the Charlson Index. Pvt, poverty level, all in odds ratios. Pvt2 vs. 1, census tracts with >5% and ≤10% people under vs. <5% under the poverty level; pvt3 vs. 1, census tracts with >10% and ≤15% people under vs. <5% under the poverty level; pvt4 vs. 1, census tracts with >10% and ≤15% people under vs. <5% under the poverty level; pvt5 vs. 1, census tracts with >20% people under vs. <5% under the poverty level.

[a] Spondylosis, intervertebral disk disorders, and other back problems.

*Significant at $p < 0.05$.

8.4 CONCLUDING REMARKS

In this chapter, we systematically examined four indicators—hospitalization incidence, treatment procedure, readmission, and mortality—that intersect with public health and health care delivery. Unlike program-based practice, we did not report frequencies, incidence, and trends, which are often included in public health reports. Unlike research studies, we made no attempt to answer a research question. Instead, we attempted to answer a crosscutting question: To what degree does neighborhood poverty status affect the four indicators for a set of frequently diagnosed diseases and conditions? The answer to this question was that the impact of neighborhood poverty status trickles down from incidence to readmission and mortality for the most selected diseases and conditions.

In the incidence surveillance, we found that overwhelmingly, majority odds ratios for neighborhood effects were significant. In most disease categories, patients from poor neighborhoods were more likely to have a disease or condition than patients in neighborhoods with <5% of the residents under the poverty line. In addition, the SES gradients tend to get stronger going from midpoor to poor neighborhoods, and the effects are often more than doubled, or a disease is twice as likely to be seen in patients in the poorest neighborhoods than in the reference category. An important set of diseases with relatively high SES gradients include schizophrenia, alcohol-related disorders, and mood disorders. Since the mental health burden can be greatly reduced for hospitals if family and community support is adequate, public health can play a major role there.

In contrast to incidence, fewer diseases from readmission and mortality surveillances would be significant along SES, and fewer odds ratios within each disease would be significant. These results suggest that many disparities at the disease incidence level may not permeate through the health care system to reflect in readmission and mortality. On the other hand, we found that some, such as mental illness (mood disorder, alcohol-related disorder, etc.), were persistent through incidence, readmission, and mortality. The same goes with AMI and a few other heart disease categories. These findings are consistent with studies in both the United States and Canada (Villanueva and Aggarwal, 2013; Kapral et al., 2002).

While we believe that incidence, readmission, and mortality should be included in public health surveillance, we are less certain about procedure disparity surveillance, primarily because proper procedures depend on clinical conditions that diagnosis codes alone cannot tell us much about. Nevertheless, we had some interesting findings. For example, prophylactic vaccination and inoculation are often needed against influenza, and these procedures were much more common for patients in poor neighborhoods than for patients in neighborhoods with a poverty rate of <5%. Likewise, respiratory intubation and mechanical ventilation are often used for COPD, and they are more likely to be used for those living in poor neighborhoods.

Finally, we want to comment on ACSCs that could have been treated effectively in a primary care setting (Billings et al., 1996). In the United States, billions are spent each year on ACSCs (Kruzikas, 2004). It seems that the SES gradients in ACSC hospitalization had more to do with accessibility than availability. The inverse relationship between neighborhood SES and ACSC hospitalization has also been shown in countries with universal health care coverage, such as Italy (Agabiti et al., 2009), Canada (Agha et al., 2007; Booth and Hux, 2003), and the United States for the Medicare population (Blustein et al., 1998). Even though we used ACSCs as a broad category, many conditions within the categories also had significant effects. The effects of COPD, asthma, and diabetes were clear, and their contributions to ACSCs were also highlighted by Booth and Hux (2003) and Trachtenberg et al. (2014). Our main contribution to the ACSC literature, perhaps, is the downstream surveillance of readmission and mortality. In both cases, ACSCs had an SES gradient favoring the well-off neighborhoods. Poor neighborhoods tended to be at a double disadvantage, being more likely to have both ACSCs and repeated preventable hospitalizations.

APPENDIX 8A: RANK ORDER OF CCS PRINCIPAL DIAGNOSIS CATEGORY BY NUMBER OF DISCHARGES (2010 NATIONAL STATISTICS—PRINCIPAL DIAGNOSIS ONLY)

Rank		CCS Principal Diagnosis Category and Name	Total Number of Discharges	Standard Error of Total Number of Discharges
1	218	Live-born	3,905,550	134,912
2	122	Pneumonia (except that caused by tuberculosis and sexually transmitted diseases)	1,103,370	21,279
3	203	Osteoarthritis	974,410	45,544
4	108	CHF, nonhypertensive	966,595	22,399
5	2	Septicemia (except in labor)	934,094	25,664
6	657	Mood disorders	886,704	55,047
7	106	Cardiac dysrhythmias	763,502	20,853
8	127	COPD and bronchiectasis	702,715	17,558
9	237	Complication of device, implant, or graft	683,787	23,130
10	193	Trauma to perineum and vulva	674,191	28,140
11	205	Spondylosis, intervertebral disk disorders, other back problems	670,555	29,284
12	197	Skin and subcutaneous tissue infections	655,600	14,590
13	195	Other complications of birth, puerperium affecting management of the mother	654,243	25,533
14	101	Coronary atherosclerosis	646,064	25,277
15	159	Urinary tract infections	613,710	13,391
16	102	Nonspecific chest pain	607,388	23,463
17	100	AMI	604,784	21,675
18	109	Acute cerebrovascular disease	588,898	16,092
19	50	Diabetes mellitus with complications	547,652	12,849
20	181	Other complications of pregnancy	533,506	18,555
21	238	Complications of surgical procedures or medical care	522,128	13,464
22	189	Previous C-section	502,796	18,935
23	149	Biliary tract disease	470,329	11,082
24	55	Fluid and electrolyte disorders	466,195	10,018
25	128	Asthma	418,990	14,686
26	254	Rehabilitation care, fitting of prostheses, and adjustment of devices	411,309	29,267
27	659	Schizophrenia and other psychotic disorders	406,313	36,254
28	157	Acute and unspecified renal failure	403,772	9,882
29	131	Respiratory failure, insufficiency, arrest (adult)	362,681	10,230
30	153	Gastrointestinal hemorrhage	362,467	7,679
31	145	Intestinal obstruction without hernia	354,484	7,342
32	152	Pancreatic disorders (not diabetes)	322,115	7,570
33	146	Diverticulosis and diverticulitis	308,813	7,601
34	226	Fracture of neck of femur (hip)	304,241	7,938
35	83	Epilepsy, convulsions	298,842	11,453
36	142	Appendicitis and other appendiceal conditions	295,589	10,302
37	660	Alcohol-related disorders	294,576	20,315
38	230	Fracture of lower limb	281,337	10,160
39	185	Prolonged pregnancy	278,316	13,162
40	95	Other nervous system disorders	266,100	7,926
41	245	Syncope	255,536	8,709
42	661	Substance-related disorders	250,901	24,288
43	183	Hypertension complicating pregnancy, childbirth, and the puerperium	247,018	9,634
44	233	Intracranial injury	245,560	13,455

(Continued)

(Continued)

Rank		CCS Principal Diagnosis Category and Name	Total Number of Discharges	Standard Error of Total Number of Discharges
45	99	Hypertension with complications and secondary hypertension	240,804	8,242
46	231	Other fractures	240,438	9,434
47	135	Intestinal infection	237,497	6,574
48	196	Normal pregnancy or delivery or both	235,429	10,701
49	42	Secondary malignancies	233,930	18,412
50	59	Anemia	230,203	5,458

Source: Weighted national estimates are from the 2010 HCUP Nationwide Inpatient Sample (NIS), Agency for Healthcare Research and Quality (AHRQ), based on data collected by individual states and provided to AHRQ by the states.

Note: The total number of weighted discharges in the United States based on HCUP NIS is 39,008,298. Statistics based on estimates with a relative standard error (standard error/weighted estimate) of >0.30 or with a standard error of 0 in the nationwide statistics (NIS and KID) are not reliable. These statistics were suppressed and are designated with an asterisk. The estimates of standard errors in HCUPnet were calculated using SUDAAN software. These estimates may differ slightly if other software packages are used to calculate variances.

REFERENCES

Agabiti, N., M. Pirani, P. Schifano, G. Cesaroni, M. Davoli, L. Bisanti, N. Caranci, et al. 2009. Income level and chronic ambulatory care sensitive conditions in adults: A multicity population-based study in Italy. *BMC Public Health* 9(1): 457.

Agha, M.M., R.H. Glazier, and A. Guttmann. 2007. Relationship between social inequalities and ambulatory care–sensitive hospitalizations persists for up to 9 years among children born in a major Canadian urban center. *Ambulatory Pediatrics* 7(3): 258–262.

AHRQ (Agency for Healthcare Research and Quality). 2007. Prevention quality indicators: Technical specifications. Version 3.1. Rockville, MD: AHRQ.

Alter, D.A., C.D. Naylor, P. Austin, and J.V. Tu. 1999. Effects of socioeconomic status on access to invasive cardiac procedures and on mortality after acute myocardial infarction. *New England Journal of Medicine* 341(18): 1359–1367.

Barrett, M., S. Raetzman, and R. Andrews. 2012. Overview of key readmission measures and methods. HCUP Methods Series Report No. 2012-04 [online]. Rockville, MD: Agency for Healthcare Research and Quality, December 20.

Billings, J., G.M. Anderson, and L.S. Newman. 1996. Recent findings on preventable hospitalizations. *Health Affairs* 15(3): 239–249.

Billings, J., L. Zeitel, J. Lukomnik, T.S. Carey, A.E. Blank, and L. Newman. 1993. Impact of socioeconomic status on hospital use in New York City. *Health Affairs* 12(1): 162–173.

Blustein, J., K. Hanson, and S. Shea. 1998. Preventable hospitalizations and socioeconomic status. *Health Affairs* 17(2): 177–189.

Booth, G.L. and J.E. Hux. 2003. Relationship between avoidable hospitalizations for diabetes mellitus and income level. *Archives of Internal Medicine* 163(1): 101–106.

Chassin, M.R., R.E. Park, K.N. Lohr, J. Keesey, and R.H. Brook. 1989. Differences among hospitals in Medicare patient mortality. *Health Services Research* 24(1): 1–31.

Colvin, J.D., I. Zaniletti, E.S. Fieldston, L.M. Gottlieb, J.L. Raphael, M. Hall, J.D. Cowden, and S.S. Shah. 2013. Socioeconomic status and in-hospital pediatric mortality. *Pediatrics* 131(1): e182–e190. doi: 10.1542/peds.2012-1215.

DH (Department of Health). 2008. Tackling health inequalities: Status report on the Programme for Action—2007 update on headline indicators. London: DH.

Elixhauser, A. and Steiner, C. 2013. Readmissions to U.S. hospitals by diagnosis, 2010. HCUP Statistical Brief No. 153. Rockville, MD: Agency for Healthcare Research and Quality, April.

Garnick, D.W., E.R. DeLong, and H.S. Luft. 1995. Measuring hospital mortality rates: Are 30-day data enough? *Health Services Research* 29(6): 679–695.

Gerward, S., P. Tydén, O. Hansen, G. Engström, L. Janzon, and B. Hedblad. 2006. Survival rate 28 days after hospital admission with first myocardial infarction. Inverse relationship with socio-economic circumstances. *Journal of Internal Medicine* 259: 164–172.

Goodman, D.C., E. Fisher, T.A. Stukel, and C. Chang. 1997. The distance to community medical care and the likelihood of hospitalization: Is closer always better? *American Journal of Public Health* 87(7): 1144–1150.

Guy, W.A. 1867. On the mortality of London hospitals: And incidentally on the deaths in the prisons and public institutions of the metropolis. *Journal of the Statistical Society of London* 30(2): 293–322.

Fink, A., E.M. Yano, and R.H. Brook. 1989. The condition of the literature on differences in hospital mortality. *Medical Care* 27(4): 315–336.

Fleming, S.T., L.F. McMahon Jr., S.I. DesHarnais, J.D. Chesney, and R.T. Wroblewski. 1991. The measurement of mortality: A risk-adjusted variable time window approach. *Medical Care* 29(9): 815–828.

Hadjistavropoulos, G.J.G., D.L. Asmundson, and A.Q. LaChapelle. 2002. The role of health anxiety among patients with chronic pain in determining response to therapy. *Pain Research and Management* 7(3): 127–133.

Hannan, E.L., M.J. Racz, D.T. Arani, T.J. Ryan, G. Walford, and B.D. McCallister. 2000. Short- and long-term mortality for patients undergoing primary angioplasty for acute myocardial infarction. *Journal of the American College of Cardiology* 36: 1194–1201.

Institute of Medicine. 2003. *Unequal Treatment: Confronting Racial and Ethnic Disparities in Health Care.* Washington, DC: National Academy of Sciences Press.

Kapral, M.K., H. Wang, M. Mamdani, and J.V. Tu. 2002. Effect of socioeconomic status on treatment and mortality after stroke. *Stroke* 33: 268–275.

Kim, C., A.V. Diez Roux, T.P. Hofer, B.K. Nallamothu, S.J. Bernstein, and M.A. Rogers. 2007. Area socioeconomic status and mortality after coronary artery bypass graft surgery: The role of hospital volume. *American Heart Journal* 154: 385–390. Back to cited text no. 22.

Krakauer, H., R.C. Bailey, K.J. Skellan, J.D. Stewart, A.J. Hartz, E.M. Kuhn, and A.A. Rimm. 1992. Evaluation of the HCFA model for the analysis of mortality following hospitalisation. *Health Services Research* 27(3): 317–319.

Krieger, N., P. Waterman, J.T. Chen, M.-J. Soobader, S.V. Subramanian, and R. Carson. 2002. Zip code caveat: Bias due to spatiotemporal mismatches between zip codes and US census–defined geographic areas— The Public Health Disparities Geocoding Project. *American Journal of Public Health* 92(7): 1100–1102.

Kruzikas, D.T. 2004. Preventable hospitalizations: A window into primary and preventive care, 2000 (no. 5). Rockville, MD: Agency for Healthcare Research and Quality. http://archive.ahrq.gov/data/hcup/factbk5/factbk5.pdf (accessed November 9, 2014).

Lin, S., E. Fitzgerald, S.A. Hwang, J.P. Munsie, and A. Stark. 1999. Asthma hospitalization rates and socioeconomic status in New York State (1987–1993). *Journal of Asthma* 36: 239–251.

Lubitz, J., G. Riley, and M. Newton. 1985. Outcomes of surgery in the Medicare aged population: Mortality after surgery. *Health Care Financing Review* 6: 103.

Pfuntner, A., L.M. Wier, and C. Stocks. 2013a. Most frequent conditions in U.S. hospitals, 2010. HCUP Statistical Brief No. 148. Rockville, MD: Agency for Healthcare Research and Quality, January.

Pfuntner, A., L.M. Wier, and C. Stocks. 2013b. Most frequent procedures performed in U.S. hospitals, 2011. HCUP Statistical Brief No. 165. Rockville, MD: Agency for Healthcare Research and Quality, October.

Reames, B.N., N.J. Birkmeyer, J.B. Dimick, and A.A. Ghaferi. 2014. Socioeconomic disparities in mortality after cancer surgery: Failure to rescue. *JAMA Surgery* 149(5): 475–481.

Roos, L.L., R. Walld, J. Uhanova, and R. Bond. 2005. Physician visits, hospitalizations, and socioeconomic status: Ambulatory care sensitive conditions in a Canadian setting. *Health Services Research* 40(4): 1167–1185.

Rosenthal, G.E., D.W. Baker, D.G. Norris, L.E. Way, D.L. Harper, and R.J. Snow. 2000. Relationships between in-hospital and 30-day standardized hospital mortality: Implications for profiling hospitals. *Health Services Research* 34(7): 1449–1468.

Spiegelhalter, D.J. 1999. Surgical audit: Statistical lessons from Nightingale and Codman. *Journal of the Royal Statistical Society: Series A (Statistics in Society)* 162(1): 45–58.

Steele, C.B., S.H. Rim, D.A. Joseph, J.B. King, and L.C. Seeff. 2013. Colorectal cancer incidence and screening—United States, 2008 and 2010. *Morbidity and Mortality Weekly Report Surveillance Summaries* 62(Suppl. 3): 53–60.

Stephenson, A., J. Hux, E. Tullis, P.C. Austin, M. Corey, and J. Ray. 2011. Socioeconomic status and risk of hospitalization among individuals with cystic fibrosis in Ontario, Canada. *Pediatric Pulmonology* 46(4): 376–384.

Trachtenberg, A.J., N. Dik, D. Chateau, and A. Katz. 2014. Inequities in ambulatory care and relationship between socioeconomic status and respiratory hospitalizations: A population-based study of a Canadian city. *Annals of Family Medicine* 12(5): 402–407.

van Doorslaer, E., A. Wagstaff, H. van der Burg, T. Christiansen, D. De Graeve, U.-G. Gerdtham, M. Gerfin, et al. 2000. Equity in the delivery of health care in Europe and the U.S. *Journal of Health Economics* 19(5): 553–584.

Villanueva, C. and B. Aggarwal. 2013. The association between neighborhood socioeconomic status and clinical outcomes among patients 1 year after hospitalization for cardiovascular disease. *Journal of Community Health* 38(4): 690–697.

Vrbova, L., M. Mamdani, R. Moineddin, L. Jaakimainen, and R.E.G. Upshur. 2005. Does socioeconomic status affect mortality subsequent to hospital admission for community acquired pneumonia among older persons? *Journal of Negative Results in Biomedicine* 4: 4. doi: 10.1186/1477-5751-4-4.

Wagstaff, A., E. van Doorslaer, and P. Paci. 1991. On the measurement of horizontal inequity in the delivery of health care. *Journal of Health Economics* 10(2): 169–205.

Yousey-Hindes, K.M. and J.L. Hadler. 2011. Neighborhood socioeconomic status and influenza hospitalizations among children: New Haven County, Connecticut, 2003–2010. *American Journal of Public Health* 101(9): 1785–1789.

Zarzaur, B.L., M.A. Croce, T.C. Fabian, P. Fischer, and L.J. Magnotti. 2010. A population-based analysis of neighborhood socioeconomic status and injury admission rates and in-hospital mortality. *Journal of the American College of Surgeons* 211(2): 216–223.

9 Sex Disparities in Hospitalization

9.1 INTRODUCTION

Sex is biologically based, while gender is socially constructed. However, for simplicity, we use *gender* and *sex* interchangeably, by ignoring socially constructed connotations. Sex is an important dimension for basic and clinical medical research that evaluates underlying causes of sex differences in disease prevalence, age at onset, severity of progression, and symptom presentation. Although progress is being made on understanding sex differences in biological, clinical, pharmaceutical, and public health sciences, there is a lack of interaction among the four fields. This chapter addresses the public health or health disparity perspective. Although sex disparities in health are found for both males and females, they are most often related to disadvantaged socioeconomic status (SES) and related health status or to conditions among women. We believe that greater awareness of sex-specific disease differences will increase clinicians' awareness of diagnosis and treatment differences and improve health outcomes.

The documentation of gender differences in health is most often related to women. Paradoxically, it has long been known that women live longer than men, yet women live with a greater number of comorbidities, and have worse self-reported health than men. This phenomenon is known as the morbidity–mortality paradox (Verbrugge and Wingard, 1987). Survey data show that women report more functional limitations and more comorbidities. Women also use more health care services (Redondo-Sendino et al., 2006). Since age effects are often controlled, gender differences in health reflect biomedical, SES, and physiological differences between men and women. When biological factors are known, people tend to use sex as a control variable while assessing the influence of other risk factors. Although health disparities by sex may be biologically based, it is commonly viewed that they are mainly driven by social and economic inequalities in favor of men. However, most empirical studies are based on a few selected health outcomes, while few studies have examined a range of health outcomes (Hunt, 2002). Some studies suggest that female morbidity is overstated. Using primarily self-reports and diagnoses from primary care settings, Kulminski et al. (2008) developed four indices based on the Framingham offspring study. They showed that female morbidity is overstated in that only the oldest age group (age 75–78 years) had fewer comorbidities. In addition, even though the sex dimension is almost always included in health reports, it is not often linked to different disparity contexts, such as socioeconomic, occupation, and domestic violence.

In this chapter, we present some gender differences in disease and how gender may interact with SES and other factors. We use a demography method to compare gender difference. We treated the 7 years of data as the hospitalized patient population, so that any descriptive differences can be reported as real differences during the study period. This approach is different from that in Chapter 8, where we treated the same data as a sample from a long period of time. Although both approaches have been used by demographers and epidemiologists, demographers tend to use age standardization, while epidemiologists tend to use odds ratios. We used the direct standardization method to calculate an age-standardized hospitalization rate per 1000 population, where age was stratified into eight age groups (0–14, 15–24, 25–34, 35–44, 45–54, 55–64, 65–74, and 75+). It is often that the top age group of 75+ may not be sufficient in age standardization. However, we wanted to use an age- and poverty-specific population from the American Community Survey at the census tract level, which only had the top age group 75+. Recall from

Chapter 8 that we had five poverty levels. However, it is not easy to visually compare five pairs of standardized rates. We, therefore, used three poverty levels: low-poverty census tracts with <5% of the people under the poverty level, midpoverty census tracts with >5% and <15% of the people under the poverty level, and high-poverty census tracts with >15% of the people under the poverty level. The 2000 standard population was then used for calculating the standard hospitalization rate.

In Chapter 8, we used record-based measures or incidence. While we still use incidence whenever necessary, we also tried to use patient-based measures, because most gender difference studies are person based. For instance, the number of comorbidities by sex should not be record based, as the record with the most comorbidities should be assigned to a patient. After demographic analysis, we also assess the differences between record-based and patient-based measures. Finally, we present a complete case study of gender difference in stroke.

9.2 USING HOSPITAL INCIDENCE AND PREVALENCE DATA TO REVISIT THE MORBIDITY–MORTALITY PARADOX

The morbidity–mortality paradox suggests that females have more comorbidities. We begin the analysis by counting the number of comorbidities by sex. However, when a patient had a number of discharges, we needed to decide how to summarize the number of comorbidities. Our options were to (1) summarize unique comorbidities by comparing multiple discharges for each patient, (2) use the mean, (3) use the record with the maximum number of comorbidities, or (4) use the record with the least number of comorbidities and over the entire study period. Which option we used depended on the research questions. In the context of measuring the morbidity–mortality paradox, we chose the discharge record with the maximum number of comorbidities. In other words, if a patient had four hospitalizations, we retained only the hospitalization with the largest number of comorbidities. The algorithm works as follows:

1. Generate a comorbidity count for each record by searching nine diagnosis codes.
2. Sort all records by patient ID, comorbidity count, and discharge date.
3. Select the record from each patient that has the largest number of comorbidities.
4. If there is a tie, select the record with the most recent discharge date.
5. Perform comorbidity count by sex and other dimensions.

Table 9.1 lists the number of comorbidities according to the maximum number of diagnosis codes listed for each patient during the 7-year study period by age, sex, and poverty. Note that the maximum number of comorbidities cannot exceed 9, as we were only provided with up to nine diagnosis codes for each patient. It is apparent that males and females have about the same number of comorbidities in each age group and each poverty category. If we ignore minor differences of <0.2 in this table, males in the age groups of <15 years, 25–34 years, and 75–84 years had a higher number of comorbidities than females. As far as the number of comorbidities is concerned, we found no evidence to support that females have a greater number of comorbidities from any age group between 2005 and 2011.

The mortality–morbidity paradox also suggests that females are more likely to use health care services. Hospitalization represents the most costly health services, and we can compare gender differences in disease-specific hospitalization. In a health interview survey, the respondent would recall how many times he or she was admitted to a hospital. Likewise, in the Nebraska hospital discharge data (HDD), we count the number of disease-specific hospitalizations. In all, there were about 1.375 million hospitalizations during the 7-year period among Nebraska residents. Females accounted for about 60%, compared to 40% for males, with some of this difference attributable to obstetric hospitalizations. To be more specific, we used the 10 leading

TABLE 9.1

Number of Comorbidities by Age, Sex, and Poverty: 2005–2011

	Male			Female		
	Low Pvt	Mid-Pvt	High Pvt	Low Pvt	Mid-Pvt	High Pvt
<15	**2.7**	**2.6**	**2.7**	2.4	2.4	2.5
15–24	4.5	4.5	4.7	4.5	4.5	4.6
25–34	**5**	**5.1**	**5.2**	4.6	4.7	4.7
35–44	5.5	5.6	6	5.5	5.5	5.8
45–54	6.3	6.3	6.7	6.2	6.2	6.6
55–64	7.1	7.1	7.3	6.9	7	7.2
65–74	7.8	7.7	7.8	7.7	7.6	7.7
75–84	**8.4**	**8.2**	8.2	8.2	8	8.1
Age ≥ 85	8.6	8.4	8.4	8.5	8.3	8.3
Total	6.2	6.2	6.3	6.1	6	6.2

Note: The number of comorbidities is person based. If a patient has three hospitalizations, then during the 7-year period, the one with the largest number of comorbidities is selected. Pvt, poverty level. Low pvt, census tracts with <5% people under the poverty level; mid-pvt, census tracts with >5% and <15% people under the poverty level; high pvt, census tracts with >5% people under the poverty level.

TABLE 9.2

Primary Diagnosis for Hospitalization by Sex in Nebraska: 2005–2011

	Standard Rates		Total *N*	
10 Diseases	**Males**	**Females**	**Males**	**Females**
Heart disease	1110.9	718.1	9446	7626
Cancer	353.4	298.1	3447	4467
CLRDs	228.8	**241.0**	1925	2424
Stroke	239.4	202.3	1987	2145
Unintentional injuries	1646.8	**1658.7**	11643	14487
Alzheimer's disease	13.8	12.6	96	124
Diabetes	119.5	95.4	927	786
Kidney diseases	101.6	80.8	658	641
Influenza and pneumonia	464.0	392.7	3425	3595
Suicide	52.4	**84.7**	442	703

Note: Rates and counts are annualized. Standard rates are calculated using eight age groups, starting with 0–15 and incremented by 10 until age 75+.

causes of death as if they were leading causes of hospital admissions. Different from Chapter 8, we used causes rather than comorbidity, and the coding was from the primary diagnosis rather than from a set of diagnoses.

Table 9.2 lists annualized standard hospitalization rates (first two columns) and frequency counts according to the 10 causes of death. The 10 disease-specific hospitalizations totaled 497,000 over the 7 years, which accounted for 36% of all hospitalizations. While overall hospitalizations by sex for the 10 diseases were fairly even (48% for males vs. 52% for females), males had higher standard hospitalization rates for 7 out of 10 diseases. Females had higher standard hospitalization rates for chronic lower respiratory disease (CLRD), unintentional injury, and suicide. Together, males had 603 more hospitalizations per 100,000 than females among the seven diseases. In particular, males had 393 more standardized heart disease hospitalizations than females (1111 vs. 718). Males also had 71 more standardized hospitalizations than females (464 vs. 393). Females, on the other hand, had 32 more standardized suicide hospitalizations than males. However, since the total standardized suicide rates were not that high, the impact was at best moderate upon the overall hospitalization gap between males and females. It is clear that males used more hospital resources than females, as far as the 10 leading causes of death are concerned. The morbidity paradox is not supported by evidence from the Nebraska HDD from 2005 to 2011.

Studies of the mortality–morbidity paradox often compare disease-specific prevalence by sex, which shows no difference when disease prevalence among populations is monitored. One common way to measure disease prevalence is to use survey data, in which sample respondents are given a list of medical conditions and asked which apply to them. Based on the number of respondents having or not having a condition, disease prevalence can be derived. The advantages of using survey data include (1) a well-defined prevalence by sample is obtained that can represent the whole population, (2) prevalence can be cross-tabulated with demographic and socioeconomic variables that are often not available from administrative data, and (3) research questions can be flexibly incorporated into a survey design. A major weakness of using surveys is their reliance on self-reported diseases or conditions. Another weakness is that survey data can only accommodate a limited number of diseases or conditions due to the concerns of cost and response rate. HDD, in contrast, tend to have many diagnosis conditions for each patient, and they can overcome some of these weaknesses. However, HDD only represent those having a hospital stay, and they are not representative of the whole population.

We were interested in comparing the sex difference in the prevalence of 10 specific diseases and their potential gradients along neighborhood SES. To calculate prevalence for each selected disease, we used the follow algorithm:

1. Generate 10 comorbidity indicator variables by searching nine diagnosis codes.
2. Selected 1 comorbidity = true and sort all records by patient ID.
3. Deduplicate records by patient ID.
4. Run frequency by poverty, sex, and age for the selected comorbidity.
5. Use standardized population to calculate standard prevalence.
6. Divide the standard rate in step 5 by that in step 6 so that the rate is annualized.
7. Repeat steps 2–6 until the 10th disease.

The results are shown in Table 9.3. First, similar to the results in Table 9.2, the prevalence of 7 out of 10 diseases tended to show male disadvantages during the study period. Diseases such as heart disease, cancer, and kidney disease were at least 30% higher for males than for females. Three diseases in which females had higher rates than males are identical to those identified in Table 9.2 (CLRD, Alzheimer's disease, and suicide). These results suggest that prevalence based on 10 comorbidities tends to have similar directional effects as the rates based on the primary diagnosis. As far as hospital-detected prevalence is concerned, there was no obvious evidence suggesting that females had higher rates than males. Second, there were some interactions between sex and poverty status at the neighborhood level, but they influenced only the magnitude of the difference, not the directional effects. However, there were a few exceptions. CLRDs were more prevalent among females, and this disparity was more pronounced for those living in neighborhoods with poverty rates higher than 15%. In fact, for those living in midpoverty neighborhoods (or census tracts), there was no sex difference. For both unintentional injuries and diabetes, females had a higher rate in poor neighborhoods (poverty rate > 15%) than males, even though males had an overall higher rate. It should be noted that we did not show the overall sex-specific rate without controlling for poverty

TABLE 9.3

Patient-Based Comorbidities for 10 Diseases by Sex and Poverty: 2005–2011

	Males			Females		
	Low Pvt	**Mid-Pvt**	**High Pvt**	**Low Pvt**	**Mid-Pvt**	**High Pvt**
Heart disease	1351.4	1428.9	1560.4	989.4	1095.9	1262.1
Cancer	418.0	426.8	453.5	308.9	322.9	349.1
CLRD	542.0	661.2	845.1	**561.8**	**661.6**	**931.9**
Stroke	304.2	309.2	362.8	256.4	259.3	322.2
Unintentional injuries	1073.2	1102.2	1258.8	1047.2	1084.1	1268.1
Alzheimer's disease	66.0	58.1	64.0	**81.1**	**71.4**	**79.9**
Diabetes	576.7	645.0	801.9	445.0	570.5	831.3
Kidney diseases	550.7	520.8	691.7	368.1	365.0	517.8
Influenza and pneumonia	474.2	582.7	689.2	394.9	504.0	606.3
Suicide	31.7	37.5	64.5	**55.0**	**62.1**	**91.9**

Note: Rates are age standardized per 100,000 per year. Diseases are based on one to nine diagnosis codes and E-codes for each patient. Pvt, poverty level. Low pvt, census tracts with <5% people under the poverty level; mid-pvt, census tracts with >5% and <15% people under the poverty level; high pvt, census tracts with >5% people under the poverty level.

in Table 9.3, but the average of 3 rates is very close to the overall rate, normally within 2 or 3 per 100,000. Finally, even though the rates follow a general SES gradient, with a higher rate for those from high-poverty neighborhoods, the gaps between males and females vary in both directions. The standard rate for diabetes increased drastically for the high-poverty group for both males and females, but the magnitude of the increase was much greater for females than for males. As a result, the female rate was higher than the male rate for this group. It should be mentioned that type 1 diabetes (not shown) is more common among females than males (based on International Classification of Diseases, Ninth Revision [ICD-9] codes 250.01, 250.03, 250.11, and 250.13). This implies that poor neighborhoods had a lot more females with type 1 diabetes. Overall, gender differences in disease prevalence were dominant by higher rates for males than for females, and the SES was not the driver of these differences.

9.3 USING PREVALENCE DATA TO ASSESS DISEASES MORE COMMON AMONG FEMALES

Due to the emphasis on gender difference, we purposefully sought diseases that were more common among women (Table 9.4). We first selected three autoimmune diseases in which women are biologically more vulnerable than men: multiple sclerosis (MS), lupus, and rheumatoid arthritis (Barsky et al., 2001). Women were about three times more likely than men to be diagnosed with MS when admitted to a hospital. Females were also at least four times more likely than males to have lupus and rheumatoid arthritis, respectively. In addition, all incidents had a poverty gradient, regardless of sex; people from a low-income neighborhood were more likely to have MS, lupus, and rheumatoid arthritis. Even though these sex differences are well known, there are benefits in studying them. For example, sex differences in MS could be due to the effects of sex hormones (estrogen and testosterone) or sex chromosomes (XX or XY). Pregnancy reduces relapses by 80%, significantly more than most of the currently available drug therapies. Noticing the effect of pregnancy on MS pathogenesis, laboratories have produced an estriol medication that has shown some efficacy in clinical trials (Smith-Bouvier et al., 2008). In this regard, we attempted to add the SES dimension, which might help to reduce sex-based disease disparities.

TABLE 9.4

Annualized Prevalence for Selected Disease by Sex and Poverty

	Males			Females		
	Low Pvt	Mid-Pvt	High Pvt	Low Pvt	Mid-Pvt	High Pvt
MS	6.5	7.9	9.9	22.9	24.5	26.6
Lupus	5.5	6.7	8.3	30.5	31.9	34.2
Rheumatoid arthritis	30.6	34.0	39.0	68.1	76.2	85.6
Asthma	47.5	60.4	66.0	113.0	143.2	167.5
Dysthymic disorder	0.9	1.7	2.3	11.2	11.8	11.0
Eating disorder	30.7	25.0	26.4	123.5	120.9	124.7
Migraines	176.8	212.1	287.4	321.8	368.4	545.6
Chronic pain	33.2	40.9	53.2	46.4	62.3	88.1
Obesity	230.7	270.0	282.9	310.9	426.1	564.2
STD/STI	1.7	3.0	4.3	5.2	5.3	13.6

Note: Rates are age standardized per 100,000 per year. Diseases are based on one to nine diagnosis codes and E-codes for each patient. Pvt, poverty level. Low pvt, census tracts with <5% people under the poverty level; mid-pvt, census tracts with >5% and <15% people under the poverty level; high pvt, census tracts with >5% people under the poverty level.

Recall that we included CLRD in Table 9.2, where females had a moderately higher rate than males. Since asthma is included in CLRDs, we took asthma only in Table 9.4 and found that the rate for females on average was about 1.4 times higher than that for males. This result suggests that the male–female difference in CLRD was mainly contributed by asthma.

The next three diseases in Table 9.4 belong broadly to mental and neurological disorders. In general, females are more vulnerable than males to psychological risk factors. Compared to males, females are more likely to experience stressful life events and psychological stressors (Leach et al., 2008). Specifically, we selected persistent depressive (dysthymic) disorder, eating disorder, and migraine. The average standard rates (over poverty) of dysthymic disorder (depression), eating disorder, and migraine for females were 7, 4.5, and 1.8 times higher, respectively, than the rates for males. Some depression can be attributed to chronic conditions. Although there is no sex difference in diabetes prevalence, women with type 2 diabetes had twice the odds of developing depression as did men, and the odds were significantly higher (Anderson et al., 2001). Among the three diseases, depression also had marked SES gradients; the poorer the neighborhood, the higher the standard rates. Depression is a risk factor for a host of diseases, such as high blood pressure, insomnia, substance abuse, and suicide, and some of those diseases were also found to be more prevalent among women.

We know that lower SES is associated with poorer health due to persistent exposure to hardship and stress and poorer access to primary care. We also know that diseases such as asthma, chronic pain, obesity, and sexually transmitted diseases and infections (STDs and STIs) have SES gradients and are more common among women. However, we do not know if, or how, sex interacts with SES for these diseases. In the absence of interaction, the SES gradients should follow the same proportional effects or gaps between males and females in disease incidence. If the gap between males and females is small for one SES group and larger for another SES group, then SES interacts with sex. One way to check the proportionality is to use the rate ratio between males and females by SES group. If the rate ratios are the same, then sex effects are the same across SES groups.

Figure 9.1 shows rate ratios based on Table 9.4. On the one hand, the rate ratios for rheumatoid arthritis, asthma, and migraines were almost flat along neighborhood SES, suggesting these sex differences were almost purely biologically constructed. On the other hand, we observed some SES gradients in rate ratio. MS, lupus, and dysthymic disorder had SES gradients favoring the poor: the

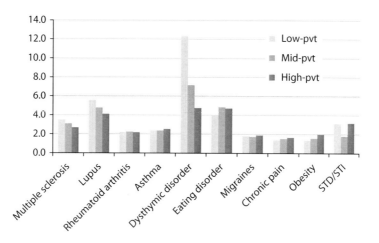

FIGURE 9.1 Female-to-male prevalence ratios by poverty for selected diseases.

higher the poverty level, the smaller the female and male gap, suggesting that they had either some behavioral (dysthymic disorder) or diagnosis (lupus) disparities. Eating disorder, chronic pain, and obesity had SES gradients that were less favorable to the poor: those in high-poverty neighborhoods had greater female–male gaps than those in low-poverty neighborhoods, suggesting that high-poverty neighborhoods exert stronger sex effects than low-poverty neighborhoods. Compared to men, women are more likely to work part-time, have lower wages, or work from home, all of which may create barriers to their access to quality health care, which in turn may increase their onset of comorbidity. If they happen to live in poor neighborhoods, diseases with known SES effects would take an extra toll on their health.

9.4 ASSESSING HOSPITAL PROCEDURE DISPARITIES

If sex-specific comorbidities are highly associated with SES, we would expect SES gradients in procedures to be much less apparent, because of variations in access to care and affordability. As early as 1975, Wennberg and Gittelsohn showed that geographic variation in surgical procedure per capita could be double or triple. The study was expanded later in the Dartmouth Atlas study of a Medicare population, which identified a list of procedures likely to vary by region. The list was coded according to primary and secondary procedures with some control for diagnosis codes. We found that searching for the first two (primary and secondary) yielded almost identical results to searching for all six procedures on the Dartmouth study list. Hence, we opted to used six procedures to search for the list. In addition, we added knee and hip replacements, as well as spinal fusion. Both knee and hip replacements have been negatively associated with SES in England (Judge et al., 2010), and those with low SES had less access to knee replacement services (Yong et al., 2004). Spinal fusion is related to back pain.

First, let's examine sex difference in procedures (Table 9.5, first six columns). In general, males tended to have more procedures per 100,000 population than females. Males were more likely to have a resection for colon cancer, lower-extremity revascularization, abdominal aortic aneurysm repair, coronary angiography, coronary artery bypass grafting, percutaneous coronary interventions, and aortic or mitral valve replacement; with the exception of resection for colon cancer, all of these sex differences in procedures were strongly aligned with sex-specific differences in heart disease.

We should also mention that we included a number of procedures that are not available for both sexes, and thus cannot be used for comparing sex differences. Nevertheless, we can assess procedure disparities by SES, so that we do not have to exclude those procedures listed on the Dartmouth Atlas. For example, radical prostatectomy is for prostate cancer, and it should not be included in gender

TABLE 9.5

Standardized Hospital Procedures for Selected Diseases

	Low Poverty		Midpoverty		High Poverty		H/L	H/L
	Male	Female	Male	Female	Male	Female	Male	Female
Back surgery	191.3	185.2	171.9	173.2	146.5	169.5	0.77	0.92
Spinal fusion	102.3	118.2	97.3	112.8	84.8	106.2	0.83	0.90
Abdominal aortic aneurysm repair	21.2	4.8	22.4	4.1	20.3	5.2	0.96	1.09
Hip replacement, total and partial	129.1	152.9	121.2	155.4	104.3	143.0	0.81	0.94
Arthroplasty knee	201.9	277.5	193.9	293.9	**156.0**	278.5	0.77	1.00
Mastectomy for cancer		24.6		27.9		24.6		1.00
Resection for colon cancer	22.8	20.2	22.9	22.5	23.1	22.7	1.01	1.12
TURP for BPH	52.6		45.1		46.6		0.88	
Lower-extremity revascularization	19.1	8.1	19.4	9.7	25.5	12.6	1.33	1.56
Cholecystectomy	87.3	121.4	89.8	129.9	92.8	151.0	1.06	1.24
Coronary angiography	180.9	108.7	209.0	134.1	227.4	167.7	1.26	1.54
Coronary artery bypass grafting	112.3	33.7	116.8	38.8	101.0	40.4	0.90	1.20
Percutaneous coronary interventions	252.5	97.9	250.2	111.0	240.4	130.4	0.95	1.33
Radical prostatectomy	50.5		52.0		38.5		0.76	
Aortic and mitral valve replacement	32.2	14.3	33.7	18.9	28.6	18.3	0.89	1.28

Source: The Dartmouth Institute. Variation in surgical procedures. The Dartmouth Atlas of Health Care. http://www.dartmouthatlas.org/pages/variation_surgery_2. Accessed October 2014.

Note: These procedures were selected according to the Dartmouth Atlas. TURP for BPH, transurethral resection of the prostate for benign prostatic hyperplasia; H/L, sex-specific ratio between high-poverty and low-poverty rates.

disparity for obvious reasons. We found that those in poor neighborhoods were much less likely to get this procedure. If we look at prostate cancer from the Nebraska Cancer Registry data, we find that the incidence rate in high-poverty neighborhoods was higher than that in midpoverty neighborhoods. By connecting the tabulated information from two sources, we were able to infer that a high prostate incidence rate in poor neighborhoods was associated with a low prostatectomy rate.

To help interpret the data, we included two columns in Table 9.5 that divide the high-poverty rate by the corresponding low-poverty rate for males and females separately. The results show that SES interacted significantly with sex in terms of performed procedures. For males, 10 out of 14 standard procedure rates were higher in low-poverty neighborhoods than in high-poverty neighborhoods, compared to 3 out of 13 for females. For example, 129 patients per 100,000 individuals had a hip replacement in a low-poverty neighborhood; the corresponding number for a high-poverty neighborhood was 104. This result is in stark contrast to comorbidity-based analysis (Table 9.3): poor neighborhoods had a much higher disease incidence rate than well-off neighborhoods. In other words, a high incidence of hospitalization may not lead to a high incidence of procedures, especially for males. It should also be noted that females tend to be more likely to develop arthritis and osteoporosis. Since these diseases are major risk factors for hip fractures and knee deterioration, more women used knee replacement and hip replacement services. However, those in high-poverty neighborhoods had fewer of those procedures while having higher comorbidities related to arthritis and osteoporosis, suggesting that either those conditions were less severe or there was an access to care issue in these neighborhoods.

9.5   ASSESSING MEASUREMENT CONSISTENCY

There are two measurement issues that deserve our attention: (1) To what extent does the number of diagnoses used in calculating comorbidities affect the results? (2) How do the results differ if the unit of analysis is incidence rather than individual patients? We raised the first issue because we used 9 diagnosis codes from the 7 years of data, rather than the 25 diagnosis codes from 2007 to 2011. We raised the second issue because most HDD are record based without patient-level identifiers at the state level. Consequently, only incidence can be calculated. To answer question (2), we only need to run a record-based comorbidity analysis and compare the results with those from Table 9.3. To answer question (1), we would have to use 2007–2011 data, which have more diagnosis codes.

As mentioned in the introduction, the morbidity–mortality paradox suggests that women have more comorbidities than men, at least in later age (Green and Pope, 1999; Case and Paxson, 2005). Table 9.6 replicates the analysis in Table 9.1 with 2007–2011 data and compares numbers of comorbidities using

TABLE 9.6
Number of Comorbidities by Age, Sex, and Poverty: 2007–2011

Age Group	Male			Female		
	Low Pvt	Mid-Pvt	High Pvt	Low Pvt	Mid-Pvt	High Pvt
Based on 9 Diagnosis Codes						
<15	2.8	2.7	2.8	2.5	2.5	2.6
15–24	4.6	4.7	4.8	4.5	4.6	4.7
25–34	5.2	5.2	5.4	4.6	4.7	4.8
35–44	5.6	5.9	6.2	5.6	5.6	5.9
45–54	6.5	6.6	6.9	6.4	6.4	6.8
55–64	7.2	7.2	7.4	7.1	7.1	7.4
65–74	8.0	7.8	7.9	7.8	7.7	7.8
75–84	8.4	8.3	8.3	8.3	8.1	8.2
Age ≥ 85	8.7	8.5	8.5	8.6	8.3	8.3
Total	6.3	6.3	6.5	6.2	6.1	6.3
Based on 25 Diagnosis Codes						
<15	2.9	2.8	3	2.6	2.6	2.7
15–24	5.1	5.2	5.4	4.7	4.8	4.9
25–34	6	6	6.4	4.9	4.9	5.1
35–44	6.7	7.1	7.7	6.3	6.4	6.9
45–54	8.2	8.3	9.3	7.7	8	8.9
55–64	9.6	9.7	10.6	9.2	9.4	10.3
65–74	11.5	11.2	11.7	10.8	10.6	11.3
75–84	13.2	12.7	12.9	12.2	11.8	12.2
Age ≥ 85	14.4	13.2	13.1	13	12.2	12.3
Total	8.6	8.5	8.9	7.9	7.9	8.3

Note: The number of comorbidities is person based. If a patient has three hospitalizations, then the one with the largest number of comorbidities during the 5-year period is selected. Pvt, poverty level. Low pvt, census tracts with <5% people under the poverty level; mid-pvt, census tracts with >5% and <15% people under the poverty level; high pvt, census tracts with >5% people under the poverty level.

25 versus 9 diagnosis codes. The total 1.08 million records between 2007 and 2011 represent about 0.57 million unique patients. In other words, each patient visited a hospital about twice in the 5-year period. Note also that patient population frequencies by sex were 39.33% for males and 60.67% for females, which are in proportion to hospitalization incidents by sex. In terms of directional effects, the results of the 9 diagnoses versus the 25 diagnoses were the same in terms of sex and SES differences. The addition of 16 diagnosis codes led to about two more comorbidities on average than for the 9 diagnosis codes (lower and upper panels of Table 9.6). All the differences between the 25 diagnosis codes and the 9 diagnosis codes in the first age group were within 0.2. The discrepancies increased in older age groups, to 4.6 on average. In general, discrepancies among females were narrower than among males, 1.8 versus 2.3 on average, respectively. We can conclude that the discrepancies between the two groups of diagnosis codes in the number of comorbidities are large enough if the concern is the number of comorbidities. However, if the concern is gender or SES difference, the 9 diagnosis codes yielded results with identical directional effects as the 25 diagnosis codes.

We now answer assessment question 2: How do the results differ if the unit of analysis is based on records rather than individual patients? It is straightforward to use all the records to reproduce Table 9.3 for standard incidence rates. As shown in Table 9.7, even though the standardized

TABLE 9.7
Standardized Hospitalization Rates: 10 Diseases by Sex and Poverty

Variable	Low Pvt	Mid-Pvt	High Pvt	Overall
Males				
Heart disease	3,218.3	3,345.3	3,687.6	3,391.5
Cancer	827.6	834.2	884.9	842.9
CLRDs	1,194.1	1,439.0	1,885.3	1,487.9
Stroke	493.3	488.7	581.5	508.9
Unintentional injuries	1,600.5	1,590.9	1,858.3	1,646.7
Alzheimer's disease	99.8	85.8	94.2	89.7
Diabetes	1,310.4	1,514.0	1,927.9	1,563.7
Kidney diseases	1,072.4	999.8	1,375.0	1,091.6
Influenza and pneumonia	707.2	858.7	996.5	857.6
Suicide	39.4	45.4	79.2	52.4
At-risk population	186,582	461,535	222,946	871,063
Females				
Heart disease	2,293.7	2,472.1	3,044.6	2,556.3
Cancer	607.2	630.6	714.4	642.7
CLRDs	1,171.5	1,354.1	**2,136.5**	**1,491.9**
Stroke	407.2	399.2	507.6	423.0
Unintentional injuries	1,580.6	1,586.8	**1,943.5**	**1,658.6**
Alzheimer's disease	**121.5**	**103.9**	**116.4**	**109.1**
Diabetes	1,043.3	1,345.5	**2,098.4**	1449.6
Kidney diseases	697.7	692.5	1,083.1	775.3
Influenza and pneumonia	555.1	705.7	879.6	711.1
Suicide	**67.3**	**77.9**	**114.4**	**84.7**
At-risk population	190,556	471,163	225,923	887,642

Note: Rates are age standardized per 100,000 per year. Diseases are based on nine diagnosis codes and the primary E-code for each hospitalization. Pvt, poverty level. Low pvt, census tracts with <5% people under the poverty level; mid-pvt, census tracts with >5% and <15% people under the poverty level; high pvt, census tracts with >5% people under the poverty level.

incidence rates are certainly higher than the corresponding prevalence, most directional effects pertaining to the male–female comparison and SES gradients are similar to those in Table 9.3. However, there are some exceptions. The CLRD rate for females in low- and midpoverty neighborhoods was much less than that for males. The unintentional injury rate for females in midpoverty neighborhoods was just slightly lower than that for males. These exceptions reflect minor male–female differences in hospital care-seeking behaviors. By and large, if one does not have individual patient identification variables, it is fine to use incidence-based results when comparing gender differences.

9.6 CHAPTER SUMMARY

In this chapter, we attempted to use age-standardized rates to assess gender differences in diseases and hospital procedures. In the context of the morbidity–mortality paradox in demography, we showed that hospitalization data in general do not support the paradox. Males had more comorbidities than females, and more males than females were hospitalized for the leading causes of death. Moreover, males had higher rates than females for 7 out of 10 leading causes of death comorbidities: heart disease, cancer, stroke, unintentional injuries, diabetes, kidney diseases, and influenza and pneumonia. Females were more likely to have three comorbidities: Alzheimer's disease, CLRDs, and suicide.

In addition, we looked at neighborhood SES within gender difference. We found that SES gradients are often in proportion to sex difference. Patients from high-poverty neighborhoods had more comorbidities for both males and females; the only exceptions were the last two age groups, where patients from high-poverty neighborhoods tend to have fewer comorbidities for both sexes. In all 10 leading causes of deaths, the hospital-based prevalence was higher in high-poverty neighborhoods than in low-poverty neighborhoods. With few exceptions (e.g., Alzheimer's disease and kidney diseases), the above statement is also applicable to mid- to low-poverty disease-specific comparisons.

In the evaluation of measurement consistency, we found that (1) using 9 diagnosis codes and 25 diagnosis codes would have almost identical directional effects in gender difference, and (2) using incidence rate and prevalence by and large also generated the same directional effects.

REFERENCES

Anderson, R.J., K.E. Freedland, R.E. Clouse, and P.J. Lustman. 2001. The prevalence of comorbid depression in adults with diabetes: A meta-analysis. *Diabetes Care* 24(6): 1069–1078.

Barsky, A.J., H.M. Peekna, and J.F. Borus. 2001. Somatic symptom reporting in women and men. *Journal of General Internal Medicine* 16(4): 266–275.

Case, A. and C. Paxson. 2005. Sex differences in morbidity and mortality. *Demography* 42(2): 189–214.

Green, C.A. and C.R. Pope. 1999. Gender, psychosocial factors and the use of medical services: A longitudinal analysis. *Social Science and Medicine* 48(10): 1363–1372.

Hunt, K. 2002. A generation apart? Gender-related experiences and health in women in early and late mid-life. *Social Science and Medicine* 54: 663–676.

Judge, A., N.J. Welton, J. Sandhu, and Y. Ben-Shlomo. 2010. Equity in access to total joint replacement of the hip and knee in England: Cross sectional study. *British Medical Journal* 341: c4092.

Kulminski, A.M., I.V. Culminskaya, S.V. Ukraintseva, K.G. Arbeev, K.C. Land, and A.I. Yashin. 2008. Sex-specific health deterioration and mortality: The morbidity–mortality paradox over age and time. *Experimental Gerontology* 43(12): 1052–1057.

Leach, L.S., H. Christensen, A.J. Mackinnon, T.D. Windsor, and P. Butterworth. 2008. Gender differences in depression and anxiety across the adult lifespan: The role of psychosocial mediators. *Social Psychiatry and Psychiatric Epidemiology* 43(12): 983–998.

Redondo-Sendino, Á., P. Guallar-Castillón, J.R. Banegas, and F. Rodríguez-Artalejo. 2006. Gender differences in the utilization of health-care services among the older adult population of Spain. *BMC Public Health* 6: 155.

Smith-Bouvier, D.L., A.A. Divekar, M. Sasidhar, S. Du, S.K. Tiwari-Woodruff, J.K. King, A.P. Arnold, R.R. Singh, and R.R. Voskuhl. 2008. A role for sex chromosome complement in the female bias in autoimmune disease. *Journal of Experimental Medicine* 205(5): 1099–1108.

The Dartmouth Institute. Variation in surgical procedures. The Dartmouth Atlas of Health Care. http://www.dartmouthatlas.org/pages/variation_surgery_2. Accessed October 2014.

Verbrugge, L.M. and D.L. Wingard. 1987. Sex differentials in health and mortality. *Women and Health* 12(2): 103–145.

Wennberg, J. and A. Gittelsohn. 1975. Health care delivery in Maine I: Patterns of use of common surgical procedures. *The Journal of the Maine Medical Association* 66(5): 123–130, 149.

Yong, P.F., P.C. Milner, J.N. Payne, P.A. Lewis, and C. Jennison. 2004. Inequalities in access to knee joint replacements for people in need. *Annals of the Rheumatic Diseases* 63(11): 1483–1489.

10 Rural–Urban Disparities in Hospitalization

10.1 INTRODUCTION

Similar to many rural areas in the United States, young people in rural Nebraska leave for urban areas, while many older Nebraskans are aging in place. The relatively small proportion of the young and middle-aged causes problems of intergenerational care of the elderly, who might have to rely more on themselves or institutions for informal and formal care. Compared to rural areas in the southern United States, rural areas in Nebraska have a relatively small number of African Americans and other minorities. Although income level is relatively low compared to urban areas, not many places in rural Nebraska have extreme poverty, which is defined as 20% or more of the population living under the federal poverty line.

In this chapter, Douglas, Sarpy, and Lancaster Counties are considered to be in urban areas, while the remaining counties are in rural areas. This definition corresponds to the definition of metropolitan counties prior to 2000, requiring that a metropolitan area must have at least 100,000 residents if its largest urbanized area has <50,000 people. In addition, all major urban hospitals are located in these three counties. Based on this definition, we have 923,726 (52.5%) and 834,979 (47.5%) populations, respectively, for urban and rural areas. For the 7-year period from 2005 to 2011, Nebraska hospitals had 694,116 (50.5%) and 680,465 (49.5%) urban and rural hospitalizations, respectively. The above number seems to suggest that rural residents, on the whole, had more hospitalizations than urban residents, but this may not necessarily be true due to differences in age distribution, among others, between rural and urban areas.

Table 10.1 lists population age structure by three poverty levels (<5%, 5%–20%, and >20%) and sex for both urban and rural counties. It shows that rural Nebraska had much higher percentages of older people, age 65 and older. In the case of females residing in neighborhoods with >20% of residents under the poverty line, the share of the elderly age 75+ was doubled, compared to their male counterparts in urban areas. Thus, differences in age distributions between rural and urban areas would lead us to expect greater hospitalization rates in rural areas if the age effect were not controlled. However, if we expect that people from high-poverty areas are more likely to be hospitalized, then it is the urban areas that had greater shares of high-poverty populations for both males and females. The two forces—age structure and poverty level—seem to work against each other in the context of rural–urban differences.

In addition, greater distance to a hospital is often cited as an important barrier to rural health care. A study in British Columbia, Canada (Lin et al., 2002), found that the hospitalization rate in each census enumeration area followed a bell curve. It was relatively low at the city center. The rate climbed rapidly as the distance from the central grew, until about 10–15 kilometers; then the rate gradually declined as the distance continued to get farther away from the center. Since the distance to the hospital for each patient–hospital pair has already been built into the public health data infrastructure, it is easy to present the average distance to the hospital for rural and urban patients for documentation.

Studies in the United States have found that many quality of care indicators for critical access hospitals in rural areas were worse than those for their urban counterparts (Lutfiyya et al., 2007). However, a study by Ross et al. (2008) found that hospital remoteness was not necessarily linked to 30-day readmissions and mortality for three critical care conditions: acute myocardial infarction (AMI), heart failure, and pneumonia. Nevertheless, real and perceived quality issues drive many rural patients to bypass rural hospitals to seek health care in urban hospitals. Since 1992, the federal

TABLE 10.1
Population Age Structure by Sex and Urban Status

	Rural Poverty Level			Urban Poverty Level		
	<5%	5%–20%	>20%	<5%	5%–20%	>20%
			Female			
Age <35	40.5	41.3	50.3	47.0	49.1	58.4
35–44	12.4	11.6	11.0	15.0	12.5	11.9
45–54	17.3	15.2	13.2	15.2	13.9	12.1
55–64	13.4	12.8	10.5	12.4	11.0	8.5
65–74	8.5	8.9	6.4	5.6	6.5	4.7
Age ≥75	7.9	10.2	8.6	4.8	6.9	4.3
Total N	40,244	335,030	44,606	150,312	218,595	98,855
	(9.6)	(79.8)	(10.6)	(32.1)	(46.7)	(21.1)
			Male			
Age <35	42.6	44.1	55.1	49.1	53.5	59.1
35–44	12.5	11.9	11.3	15.2	12.9	12.9
45–54	17.1	15.7	12.4	15.1	13.4	12.7
55–64	14.4	13.2	10.3	11.8	10.4	8.7
65–74	7.8	8.2	5.9	5.5	5.4	4.0
Age ≥75	5.6	7.0	5.0	3.3	4.4	2.6
Total N	40,079	331,166	43,854	146,503	210,851	98,610
	(9.7)	(79.8)	(10.6)	(32.1)	(46.2)	(21.6)

Note: Poverty distributions by urban status are in parentheses. Three counties are considered urban: Douglas, Sarpy, and Lancaster. All other counties are rural.

government has used Medicare reimbursements and designation of rural critical access hospitals to help finance rural hospitals and provide high-quality hospital care to rural patients. However, bypassing rural hospitals to seek care in urban hospitals has been persistent. We wanted to assess rural patients whose hospitalizations were in urban hospitals by the listed diseases and conditions. Previous studies on rural Medicare patients suggested that younger, male Medicare beneficiaries were more likely to be rural bypassers (Adams et al., 1991). Some studies over specific geographical areas found that rural residents had a propensity to bypass a rural hospital even after controlling for hospital characteristics (Escarce and Kapur, 2009). However, the proportion of bypassers was not high ranging from 18% to 64% (Hall et al., 2010). We could presume that patients with functional limitations tended to be nonbypassers, but functional limitations cannot be measured from the hospital claims data.

The rest of the chapter is organized as follows. In Section 10.2, we provide an analytical approach. In Section 10.3, we present main surveillance results that include rural–urban differences in about 40 hospitalization categories, rural hospital bypassing rates, and distance to hospitals. We then present a case study of rural–urban differences in injuries in Section 10.4. Finally, we summarize the chapter with some concluding remarks.

10.2 OUR APPROACH TO MODEL RURAL–URBAN DIFFERENCE

In this chapter, we offer both surveillance and analytical approaches to assess rural–urban health disparities. We used the same hospital discharge data (HDD) of 2005–2011 for incidence, the ACS 2006–2010 for at-risk-population, and odds ratio presentations. Note that we use the term *incidence* to indicate that records, rather than patients, are the unit of analysis. Since rural–urban differences in health status and outcomes are mostly studied in the contexts of access to care, we also put an emphasis on this area. Hence, most of our discussions are based on the principal diagnosis to admissions rather than comorbidities. We used the first two diagnosis codes, or primary and secondary diagnoses, as the principal diagnosis for admissions. We used the same 40 diseases or conditions as in Chapter 8. If a listed disease is found in the first and second diagnosis codes, then we attributed the admission to the disease. We chose this strategy because quite often, secondary diagnosis is an important reason for admission. For instance, a fall injury may be the primary reason for seeking hospital care, but the patient might be admitted and then treated more for heart disease. However, to bridge the findings from Chapter 8 and this chapter, we also used the same analytical approach from Chapter 8 by searching nine diagnosis codes for a comorbidity condition (see Appendix 10A).

In Chapter 8, we estimated a log-rate model using age and sex as control variables. In Chapter 9, we surveyed sex difference in disease and used sex explicitly in standard hospitalization estimations. In both chapters, we also included poverty level as a unique dimension for socioeconomic status (SES). In this chapter, we add rural–urban status as another dimension to estimate a five-way log-rate model (age, sex, urban status, disease status, and poverty level).

$$\log\big(\text{Disease}(x)\big) = \text{Age} + \text{Sex} + \text{Poverty} + \text{Urban status} + \text{Offset}\big(\log\text{of population}\big) \quad (10.1)$$

where disease(x) is the comorbidity incidence or frequency indexed by age group, poverty, and urban status. Age group has six categories, and poverty has five categories. We only present odds ratios for urban with the referent being rural. This model can help us answer the following question: When controlling for age, sex, and poverty level, which place of residence had higher hospitalization rates for selected diseases and conditions? We estimate this model for each disease or condition.

Furthermore, we need to model neighborhood SES effects for both rural and urban areas explicitly. There are two ways to achieve this objective. First, we could do a two-way interaction model between urban status and poverty level. However, this method requires knowledge about categorical data analysis that is not common in a public health practice setting. Alternatively, we could simply estimate two separate models, one for urban areas and the other for rural areas. The model outputs for the latter are more straightforward and easier to understand than model outputs from the interaction method. We opted to use the second method by estimating the following model twice, one for urban areas and one for rural areas:

$$\log\big(\text{Disease}(x)\big) = \text{Age} + \text{Sex} + \text{Poverty} + \text{Offset}\big(\log\text{of population}\big) \quad (10.2)$$

In this way, model estimates for poverty level (five levels) only need to refer to one main effect, and we used poverty group 1 (<5%). In addition, we still need to estimate the model in Equation 10.2 for 40 diseases and for each place of residence category. We realize that something has to give when presenting the results. Therefore, we present the contrast between the poorest neighborhood category (poverty population >20%) and the least poor (<5%), even though all poverty levels were included in our estimations.

10.2.1 Distance to Hospital

Since all patients within the Nebraska HDD were geocoded to the master address index, and all hospitals were geocoded and updated for biopreparedness, it was straightforward to calculate distance to hospital. As we built public health data infrastructure, we had already calculated the great circle distance or crow flying distance to the hospital for each patient and attached this to the enhanced Nebraska HDD data. Originally, we attempted to calculate network distance for more than a million records and found that it was not feasible to efficiently (computational and geographic information system [GIS] resource use) determine road classes and other travel criteria for calculating the shortest network distance calculation. However, we did sample 1% randomly and found the ratios between road network distance and great circle distance to be 1.19 and 1.07 for urban and rural areas, respectively.

10.2.2 RURAL HOSPITAL BYPASSER

Given that each patient and hospital location is known (during the data production stage), it was straightforward to assign each patient as a bypasser or not based on the rural–urban county dichotomy. First, we assigned each hospital to be an urban or rural hospital. If a hospital is in an urban county, it is an urban hospital; otherwise, it is a rural hospital. Since each resident is already assigned with urban and rural status, we can tell if a resident is a bypasser or not. A rural bypassing patient is defined as having to bypass a nearby rural hospital to seek care in an urban hospital.

10.3   RURAL–URBAN HOSPITALIZATION DISPARITY SURVEILLANCE RESULTS

Table 10.2 lists odds ratios for the 40 diseases and conditions based on the models in Equations 10.1 (first column) and 10.2 (the next two columns). The first column shows odds ratios of urban versus rural. Hence, the interpretation is that controlling for age and poverty, urban patients are more likely to have most of the surveyed diseases. Six principal diagnoses causing more hospitalization for rural residents were pneumonia, coronary atherosclerosis, nonspecific chest pain, fracture of neck of femur, appendicitis and other appendiceal conditions, and other fracture. In addition, three diagnoses—urinary tract infections, biliary tract disease, and gastrointestinal hemorrhage—had no rural–urban difference. The three diagnoses that urban patients were most likely to be admitted for were acute and unspecified renal failure, hypertension (HP) with complications and secondary HP, and respiratory failure, insufficiency, and arrest, and they were all at least 50% more often. The three diagnoses that urban patients were least likely to be admitted for were appendicitis and other appendiceal conditions, nonspecific chest pain, and pneumonia, and they were at least 10% less often. Finally, there was a small tendency that urban residents were more likely than rural residents to have preventable hospitalizations. Previous literature suggests that rural residents tended to have more preventable or ambulatory care sensitive condition (ACSC) hospitalizations due to lack of primary care (Stranges and Stocks, 2010). This was not the case in Nebraska during the study period.

The last two columns of Table 10.2 present SES odds ratios contrasting poverty group 5 (>20%) and poverty group 1 (<5%) separately for urban and rural areas. The results for urban areas showed that patients from the poorest neighborhoods were more likely to be admitted to hospitals for most diagnoses. Similar to the comorbidity model in Chapter 8, diagnoses responsible for hospitalizations were clustered around mental health (e.g., schizophrenia and other psychotic disorders, substance-related disorders, and alcohol-related disorders) and ambulatory sensitive conditions such as diabetes, HP, and asthma. While rural areas had much weaker and fewer significant SES effects, among the 17 significant SES effects, those clustered around mental health were strong. In addition, 12 odds ratios had inverse SES effects, or those residing in the poorest rural neighborhoods were less likely to be admitted than those from well-off rural neighborhoods. The top three effects were for secondary malignancies, syncope, and osteoarthritis, and they were all 20% less often. Implicitly, the opposite rural SES disparities suggest that rural patients living in the poorest areas were either less likely to have these conditions or less likely to seek hospital care due to distance and other barriers (e.g., access to quality care).

There were some examples where distance could be a critical factor for the rural poor. AMI requires timely critical care, while renal disease requires dialysis services that are more accessible in urban areas. Both AMI and renal failure were less likely for the rural poor than for the rural well-off neighborhoods, and both were more common in urban areas and for the urban poor. In Chapter 8, we showed that AMI in poor neighborhoods (poverty group 5) was 37% more likely

TABLE 10.2

Rural and Urban Differences in Principal Diagnoses for Hospitalization

Diagnosis	Urban/Rural	Poverty Level 5 vs. Level 1	
		Urban	Rural
All hospitalizations	1.08 (1.08, 1.09)	1.57 (1.56, 1.58)	1.04 (1.03, 1.05)
Pneumonia	0.86 (0.84, 0.87)	1.87 (1.81, 1.93)	1.09 (1.04, 1.14)
Osteoarthritis	1.04 (1.01, 1.06)	0.9 (0.86, 0.93)	0.75 (0.7, 0.8)
Congestive heart failure	1.09 (1.07, 1.11)	1.88 (1.81, 1.95)	1.06 (1.01, 1.12)
Septicemia (except in labor)	1.4 (1.36, 1.43)	1.76 (1.68, 1.85)	0.86 (0.79, 0.93)
Mood disorders	1.32 (1.29, 1.34)	2.39 (2.32, 2.47)	1.62 (1.53, 1.72)
Cardiac dysrhythmias	1.02 (1, 1.04)	1.21 (1.17, 1.26)	0.95 (0.9, 1)
COPD and bronchiectasis	1.09 (1.08, 1.11)	2.64 (2.55, 2.74)	1.18 (1.12, 1.25)
Back problems	1.06 (1.04, 1.09)	1.04 (0.99, 1.09)	0.84 (0.78, 0.91)
Skin and subcutaneous tissue infections	1.03 (1.01, 1.06)	2.19 (2.09, 2.3)	1.17 (1.09, 1.27)
Coronary atherosclerosis	0.95 (0.93, 0.97)	1.39 (1.34, 1.45)	0.94 (0.89, 1)
Urinary tract infections	1.01 (0.99, 1.04)	1.81 (1.74, 1.88)	1 (0.94, 1.06)
Nonspecific chest pain	0.87 (0.84, 0.89)	2.48 (2.35, 2.63)	1.19 (1.09, 1.29)
AMI	1.07 (1.04, 1.1)	1.64 (1.55, 1.73)	0.91 (0.83, 0.99)
Acute cerebrovascular disease	1.23 (1.19, 1.26)	1.8 (1.7, 1.91)	0.94 (0.86, 1.03)
Diabetes mellitus with complications	1.34 (1.3, 1.38)	3.22 (3.04, 3.4)	1.32 (1.19, 1.46)
Biliary tract disease	0.98 (0.95, 1.01)	1.38 (1.3, 1.47)	0.96 (0.87, 1.05)
Fluid and electrolyte disorders	1.04 (1.02, 1.06)	1.7 (1.65, 1.75)	1.11 (1.06, 1.17)
Asthma	1.08 (1.05, 1.12)	2.68 (2.54, 2.83)	1.61 (1.46, 1.78)
Schizophrenia and other psychotic disorders	1.31 (1.27, 1.35)	8.02 (7.5, 8.58)	2.9 (2.58, 3.27)
Acute and unspecified renal failure	1.9 (1.85, 1.94)	2.13 (2.05, 2.22)	0.8 (0.74, 0.86)
Respiratory failure, insufficiency, arrest	1.57 (1.53, 1.61)	2.46 (2.35, 2.58)	1.02 (0.94, 1.11)
Gastrointestinal hemorrhage	0.98 (0.95, 1.01)	1.84 (1.73, 1.96)	0.96 (0.87, 1.05)
Intestinal obstruction without hernia	0.96 (0.93, 0.98)	1.26 (1.19, 1.34)	1.12 (1.03, 1.22)
Pancreatic disorders (not diabetes)	1.07 (1.03, 1.11)	2.11 (1.97, 2.26)	1.11 (0.99, 1.25)
Diverticulosis and diverticulitis	1.06 (1.02, 1.1)	1.1 (1.01, 1.19)	0.86 (0.76, 0.97)
Fracture of neck of femur (hip)	0.92 (0.89, 0.96)	1.23 (1.14, 1.34)	0.82 (0.74, 0.91)
Epilepsy, convulsions	1.15 (1.11, 1.19)	2.6 (2.44, 2.76)	1.25 (1.12, 1.4)
Appendicitis and other appendiceal conditions	0.88 (0.84, 0.92)	1.08 (0.99, 1.17)	1.01 (0.89, 1.15)
Alcohol-related disorders	1.04 (1.01, 1.08)	3.21 (3.02, 3.41)	2.89 (2.59, 3.23)
Fracture of lower limb	1.08 (1.04, 1.13)	1.77 (1.64, 1.91)	0.87 (0.76, 0.99)
Other nervous system disorders	1.39 (1.35, 1.43)	1.53 (1.45, 1.62)	1.05 (0.95, 1.16)
Syncope	1.21 (1.16, 1.26)	1.38 (1.26, 1.5)	0.79 (0.68, 0.92)
Substance-related disorders	1.33 (1.28, 1.39)	3.36 (3.13, 3.62)	1.8 (1.56, 2.08)
Intracranial injury	1.16 (1.1, 1.21)	1.76 (1.61, 1.91)	0.99 (0.85, 1.14)
HP with complications and secondary HP	1.62 (1.56, 1.68)	3.08 (2.89, 3.28)	1 (0.88, 1.13)
Other fractures	0.96 (0.93, 1)	1.25 (1.16, 1.35)	0.87 (0.78, 0.98)
Intestinal infection	1.11 (1.06, 1.16)	1.4 (1.29, 1.51)	0.94 (0.83, 1.07)
Secondary malignancies	1.25 (1.21, 1.29)	1.23 (1.17, 1.31)	0.79 (0.72, 0.87)
Anemia	1.06 (1.03, 1.09)	1.76 (1.66, 1.86)	0.93 (0.85, 1.02)
ACSCs	1.01 (1, 1.02)	2.03 (1.99, 2.07)	1.14 (1.11, 1.17)

Note: Reason for admission is based on the primary and secondary diagnosis codes only. Except for Douglas, Sarpy, and Lancaster Counties, all counties were rural.

to be found among hospitalized patients than those coming from poverty group 1 neighborhoods. However, in terms of seeking care for AMI, in Table 10.2, we found that urban residents in poverty group 5 were 64% more likely, while rural residents were 9% less likely to have AMI hospitalizations. This result has a number of implications. It could be that the rural poor were less likely to have an AMI than the rural "rich." It could also be that the rural poor who had AMIs did not go to the hospital, either rural or urban, which would cause selection bias in our HDD.

In the surveillance context, we intended to document the proportions of rural bypassing incidents by some poverty levels. Table 10.3 shows proportions of bypassers for rural patients. The bypassing (rural hospital) rates ranged from 12.9% to 46.5%, with an overall rate of 26.3%. Diagnoses that had above 40% bypassing rates include AMI, coronary atherosclerosis, secondary malignancies, respiratory failure, insufficiency, arrest, back problems, and intracranial injury. Diagnoses that had <15% bypassing rates include pneumonia, appendicitis and appendiceal conditions, urinary tract infections, asthma, chronic obstructive pulmonary disease (COPD), and bronchiectasis. It is obvious that high bypassing rates are for more serious diseases and injury, while the low bypassing rates are for less serious diseases. Note that these bypassing rates are revealed rates. Not included were patients who might wish to bypass a rural hospital but cannot afford travel and other costs associated with bypassing hospital care.

When we compared the bypassing rates between the poorest and the well-off neighborhoods in rural areas, we found that patients from the well-off rural neighborhoods tend to have much higher bypassing rates. The overall bypassing rate for the well-off neighborhood category was 47.68% compared to 18.69% for the poorest neighborhoods. The top bypassing rate was 72%, which was for back problems (spondylosis, intervertebral disk disorders, and other back problems) in the least poor neighborhood category, which was followed by intracranial injury (67%) and secondary malignancies (66%). For the poorest neighborhood category, the lowest three are appendicitis and appendiceal conditions (4.82%), asthma (9.18%), and pneumonia (9.91%), respectively. In all select diagnoses, the bypassing rates were about 20% higher in the least poor than in the poorest neighborhood category. It almost made us wonder if the huge bypassing rate difference was caused by a locational difference between the rich and poor neighborhoods, as many least poor people live in exurbia and, in our context, could be counties adjacent to urban counties. However, we performed some sensitivity analysis by including four adjacent counties that form the greater Omaha–Lincoln metropolitan area. The result does not make much difference, suggesting that the income barrier, rather than the distance, caused SES differences in bypassing rates.

To further confirm our interpretation of poverty effects, we used mean crow flying distance to the hospital for each diagnosis (Table 10.4). As expected, the mean distance to the hospital for urban patients, overall, was 6 miles, compared to 29 miles for rural patients. For rural patients, a long distance to the hospital was associated with those diseases that had a greater rural hospital bypassing rate. This implies that if a disease condition required more coordinated hospital care, rural residents were willing to travel a greater distance to seek the care, which often meant urban hospital care. In fact, when we used the overall distance to the hospital for rural patients (first column of Table 10.4) to correlate with the overall bypassing rate, we had a 0.91 ($p < 0.01$) correlation coefficient. This suggests that distance had a very strong influence on the bypassing rates; the greater the distance, the greater the bypassing rate.

However, the distance and bypassing rate relationship in the context of disease severity cannot be used to explain differences in bypassing rates between poverty groups 1 and 5. The low-poverty group 1 (<5%), which had a greater bypassing rate (see Table 10.3) and greater mean distance (22 miles) than the high-poverty group (18-mile distance), also had greater bypassing rates regardless of disease or condition codes (see Table 10.3). If a short distance to the hospital is associated with a greater bypassing rate, then poverty group 5 would have had higher bypassing rates than poverty group 1.

We present surveillance results for rural–urban difference in readmission in Table 10.5, which uses the same modeling strategy presented in Table 10.3. The first column is urban–rural difference

TABLE 10.3

Rural Hospital Bypassers by Poverty Status for Rural Patients

Primary and Secondary Diagnoses	Overall		Poverty <5%		Poverty >20%	
	%	Total N	%	Total N	%	Total N
All hospitalizations	26.33	680,465	47.68	57,823	18.69	65,045
Pneumonia	12.91	57,518	28.14	3,806	9.91	4,206
Osteoarthritis	32.38	53,710	54.99	2,393	20.77	1,642
Congestive heart failure	25.58	71,465	40.59	3,131	18.22	3,244
Septicemia (except in labor)	29.37	17,225	41.09	1,280	23.07	1,088
Mood disorders	33.56	66,465	61.93	1,584	23.28	3,084
Cardiac dysrhythmias	27.30	95,542	42.54	2,948	19.51	2,748
COPD and bronchiectasis	14.68	72,940	35.25	2,397	10.18	2,673
Back problems	45.86	36,333	72.05	1,481	33.42	1,197
Skin and subcutaneous tissue infections	18.24	19,872	41.05	1,201	15.96	1,410
Coronary atherosclerosis	43.94	112,075	62.91	2,650	34.80	2,256
Urinary tract infections	14.25	46,420	28.02	2,027	10.57	2,194
Nonspecific chest pain	18.49	18,801	38.04	970	13.77	1,060
AMI	41.90	15,525	57.43	1,184	32.27	1,001
Acute cerebrovascular disease	25.01	11,735	45.90	1,011	19.66	941
Diabetes mellitus with complications	22.16	21,197	50.37	671	16.82	898
Biliary tract disease	27.16	15,012	47.78	902	19.12	884
Fluid and electrolyte disorders	15.11	99,598	35.05	2,990	10.60	3,492
Asthma	14.42	28,626	39.35	648	9.18	1,100
Schizophrenia and psychotic disorders	21.21	12,091	42.15	363	16.87	1,144
Acute and unspecified renal failure	38.43	24,656	55.23	1,405	31.27	1,084
Respiratory failure, insufficiency, arrest	45.58	23,715	57.82	1,119	36.26	1,095
Gastrointestinal hemorrhage	18.17	13,762	34.78	854	13.97	809
Intestinal obstruction without hernia	18.93	19,089	38.92	1,015	14.50	1,117
Pancreatic disorders (not diabetes)	22.54	7,523	43.03	509	10.22	558
Diverticulosis and diverticulitis	17.83	13,551	40.88	565	11.51	469
Fracture of neck of femur (hip)	23.53	9,598	33.94	766	21.30	676
Epilepsy, convulsions	30.10	16,693	52.26	532	22.58	713
Appendicitis and appendiceal conditions	13.70	4,657	41.29	402	4.82	477
Alcohol-related disorders	18.93	17,554	48.08	416	10.34	1,277
Fracture of lower limb	27.78	6,764	51.93	491	24.21	442
Other nervous system disorders	32.68	38,781	56.87	735	25.52	776
Syncope	18.11	8,246	40.32	372	13.90	295
Substance-related disorders	23.31	11,961	48.91	274	15.10	596
Intracranial injury	46.46	5,492	67.25	345	40.32	372
HP with complications and secondary HP	37.53	33,698	54.33	497	31.42	487
Other fractures	25.81	10,252	50.16	620	18.34	567
Intestinal infection	18.13	7,676	36.49	485	16.94	484
Secondary malignancies	44.06	16,823	65.54	1,007	33.47	738
Anemia	19.89	67,325	37.47	942	12.60	897
ACSCs	16.67	224,503	36.43	8,617	12.54	9,983

Note: Reason for admission is based on the primary and secondary diagnosis codes only. Except for Douglas, Sarpy, and Lancaster Counties, all counties were rural.

TABLE 10.4

Average Travel Distance (Miles) to Hospital by Disease and Residence Location

Primary and Secondary diagnoses	Overall		Poverty <5%		Poverty >20%	
	Rural	Urban	Rural	Urban	Rural	Urban
All hospitalizations	29	6	22	7	18	4
Pneumonia	20	5	14	7	10	4
Osteoarthritis	32	6	24	7	19	5
Congestive heart failure	26	5	19	7	14	4
Septicemia (except in labor)	34	5	18	7	19	4
Mood disorders	32	6	30	10	30	6
Cardiac dysrhythmias	31	5	19	7	17	4
COPD and bronchiectasis	24	5	15	6	9	4
Back problems	36	6	38	8	32	5
Skin and subcutaneous tissue infections	24	5	17	7	12	4
Coronary atherosclerosis	36	5	32	7	30	4
Urinary tract infections	23	5	12	7	10	4
Nonspecific chest pain	23	5	16	7	12	3
AMI	43	5	29	6	27	4
Acute cerebrovascular disease	28	5	21	7	17	4
Diabetes mellitus with complications	29	5	22	7	18	4
Biliary tract disease	32	6	19	7	21	5
Fluid and electrolyte disorders	25	5	14	7	12	4
Asthma	28	6	18	7	11	4
Schizophrenia and psychotic disorders	27	6	26	10	26	6
Acute and unspecified renal failure	38	5	22	7	26	4
Respiratory failure, insufficiency, arrest	39	5	26	7	27	4
Gastrointestinal hemorrhage	23	5	14	6	13	4
Intestinal obstruction without hernia	29	6	16	7	10	5
Pancreatic disorders (not diabetes)	30	5	19	7	13	4
Diverticulosis and diverticulitis	22	5	14	6	10	4
Fracture of neck of femur (hip)	29	5	12	6	16	4
Epilepsy, convulsions	33	6	26	9	26	5
Appendicitis and appendiceal conditions	20	6	17	7	7	4
Alcohol-related disorders	31	6	26	11	19	5
Fracture of lower limb	34	6	20	7	26	5
Other nervous system disorders	33	6	26	8	27	5
Syncope	24	5	15	6	12	3
Substance-related disorders	32	7	20	11	23	6
Intracranial injury	46	7	37	9	33	6
HP with complications and secondary HP	38	5	24	7	26	3
Other fractures	32	6	25	7	19	5
Intestinal infection	25	6	15	7	14	4
Secondary malignancies	41	6	34	8	33	5
Anemia	30	5	19	7	15	4
Preventable ACSCs	27	5	16	7	12	4

Note: Travel distance is based on Euclidean distance. Except for Douglas, Sarpy, and Lancaster Counties, all counties were rural.

TABLE 10.5
Model-Based 30-Day Readmission Estimates (Odds Ratios) by Rural–Urban and Poverty Level (5 vs. 1)

Principal Diagnosis	Urban vs. Rural	Urban Pvt5	Rural Pvt5
Pneumonia	1.119 (1.029, 1.217)	1.076 (0.905, 1.279)	1.027 (0.812, 1.298)
Congestive heart failure, nonhypertensive	1.132 (1.046, 1.225)	1.158 (0.988, 1.359)	0.919 (0.729, 1.16)
Septicemia (except in labor)	1.233 (1.039, 1.463)	0.928 (0.675, 1.277)	0.77 (0.448, 1.324)
Mood disorders	1.129 (1.023, 1.247)	1.591 (1.352, 1.871)	0.996 (0.729, 1.362)
Cardiac dysrhythmias	1.022 (0.918, 1.138)	1.521 (1.197, 1.932)	1.068 (0.79, 1.443)
COPD and bronchiectasis	1.146 (1.046, 1.257)	1.34 (1.115, 1.611)	1.106 (0.838, 1.459)
Skin and subcutaneous tissue infections	1.072 (0.893, 1.287)	1.598 (1.132, 2.257)	1.535 (0.915, 2.576)
Nonspecific chest pain	1.19 (0.87, 1.627)	1.831 (1.045, 3.209)	1.125 (0.342, 3.7)
AMI	0.852 (0.659, 1.102)	1.567 (0.93, 2.642)	3.225 (1.545, 6.735)
Acute cerebrovascular disease	0.998 (0.764, 1.303)	0.495 (0.291, 0.844)	0.84 (0.435, 1.623)
Diabetes mellitus with complications	1.077 (0.891, 1.303)	1.497 (1.074, 2.088)	1.893 (0.945, 3.793)
Biliary tract disease	0.81 (0.638, 1.029)	1.131 (0.713, 1.793)	1.026 (0.497, 2.118)
Asthma	0.913 (0.718, 1.16)	1.535 (1.003, 2.349)	1.066 (0.479, 2.371)
Schizophrenia and other psychotic disorders	1.02 (0.861, 1.208)	1.469 (1.067, 2.022)	1.157 (0.658, 2.035)
Acute and unspecified renal failure	1.24 (1.078, 1.427)	1.234 (0.984, 1.548)	0.686 (0.422, 1.115)
Respiratory failure, insufficiency, arrest	1.425 (1.208, 1.681)	1.549 (1.183, 2.03)	1.361 (0.816, 2.27)
Gastrointestinal hemorrhage	0.937 (0.739, 1.189)	1.614 (0.985, 2.644)	0.961 (0.486, 1.901)
Fracture of lower limb	1.487 (1.001, 2.209)	0.49 (0.242, 0.992)	0.721 (0.233, 2.228)
HP with complications and secondary HP	1.02 (0.788, 1.32)	1.872 (1.217, 2.879)	5.53 (1.201, 25.456)
Other fractures	0.597 (0.367, 0.969)	0.582 (0.184, 1.843)	1.34 (0.405, 4.437)
Intestinal infection	1.403 (1.102, 1.787)	1.052 (0.649, 1.705)	0.89 (0.413, 1.917)
Anemia	0.829 (0.64, 1.074)	1.091 (0.595, 2.002)	1.168 (0.545, 2.503)
Avoidable hospitalization	1.011 (0.961, 1.064)	1.373 (1.241, 1.519)	1.105 (0.953, 1.28)
Complication of device	1.195 (1.033, 1.381)	1.154 (0.893, 1.492)	1.165 (0.731, 1.858)
Complications of surgical procedures	1.018 (0.895, 1.157)	1.414 (1.114, 1.796)	1.088 (0.759, 1.559)

Note: Only significant results are included.

controlling for age, sex, and poverty. Columns 2 and 3 are poverty contrasts for separate urban and rural readmission models controlling for age and sex. Overall, urban patients were more likely to be readmitted for 12 diagnoses, while they had a reduced chance for "other fracture." There were 13 significant diagnoses for the poverty contrast in urban areas, all in the expected directional effects except the fracture of lower limb. In contrast, only three diagnoses were significant in rural areas (AMI, complications of surgical procedures, and HP with complications and secondary HP).

To briefly summarize, we found that urban residents were more likely to be hospitalized for the listed diagnoses. The urban SES effects were mostly consistent with the overall SES disparities in hospitalization found in Chapter 8. The SES effects for rural areas were mixed. Diagnoses related to mental health and preventable ambulatory sensitive conditions tended to have SES effects consistent with those in urban areas. Some reversed SES effects require further studies. Some diagnoses that had a reverse SES effect, such as AMI or renal failure, can be explained by lack of quality and timely care for the poor in rural hospitals. One may ascribe some of the differences to the composition effects, as there are not many census tracts in rural areas that are extremely poor (>20% of people under poverty) and very well off (<5%). However, it could also be that some broader income differences are at work too. An income of $20,000 for a family of four has quite different meanings for rural and urban families.

10.4   CASE STUDY: RURAL–URBAN INJURY SURVEILLANCE

Many risk factors in rural areas are different from factors in urban areas (Hartley, 2004), yet few studies have systematically surveyed along the poverty line. Injuries, intentional or unintentional, are the top five leading causes of death in the United States. It is known that injuries are common in rural areas, especially nonfatal injuries. Previous studies have examined motor vehicle injury in Nebraska (Zhang and Lin, 2013). However, besides motor vehicle crash (MVC) injuries, little is known about other injuries between rural and urban areas in Nebraska. We conducted this case study for three reasons. First, disease surveillance showed that rural areas had a slightly higher crude hospitalization rate. When age, sex, and poverty variables were put into the surveillance model, rural areas were less likely to have hospitalizations for most disease and diagnosis conditions. Most people treat age effects as confounders that cause a distortion of the association between an exposure, in our case rural–urban, and an outcome, in our case a disease-specific hospitalization rate. If we purely treated age as a confounder, then we would simply acknowledge the effect from the fact that rural areas had substantially more elderly people (shown in Table 10.1), who were more likely to be hospitalized for a set of diseases due to the age effect. In public health practice, age effects should be explicitly documented, even though our main interest is rural–urban difference. In this way, people can target specific age groups for intervention.

The second reason for this case study is the use of E-codes, or external injury codes. So far, we have not used E-codes, which are quite different from International Classification of Diseases, Ninth Revision (ICD-9) codes. The external cause of injury codes (E-codes) describe the force that causes the injury (e.g., fall, motor vehicle traffic [MVT] accident, or poisoning) and the intent of the injury, intentional or unintentional (e.g., if the injury is inflicted purposefully). In other words, an E-code describes how an injury occurred and if it was accidental or intentional. Although a patient can be admitted for heart attack, the cause might be a fall injury. E-codes, which are also based on ICD codes, can be found separately in the HDD. E-codes can have six digits, including the beginning E and 1 decimal place. In practice, the decimal is often omitted, leaving only five digits, with the first four digits being most frequently used. In addition, a secondary or third E-code also identifies the place of injury (i.e., the four-digit place of occurrence E-code is E849, which ignores the last digit). E-codes are used to classify injury incidents by mechanism (e.g., MVC, fall, struck by or against, adverse medical or drug effects) and intent (e.g., unintentional, self-inflicted, assault, or undetermined). Documenting injuries via E-codes can help area health agencies identify major injuries and design injury prevention strategies. It should be mentioned even though other fractures can be a primary diagnosis for hospitalization. An E-code associated with the fracture injury can never be used as the primary diagnosis. In this regard, E-codes are supplemental to ICD-9 diagnosis codes.

The third reason for this case study is to broaden injury surveillance not only for injury categories, but also for SES. Injuries, according to E-codes, can be divided into more than 20 major categories. According to the Centers for Disease Control and Prevention (CDC) injury matrix, major unintentional or accidental injuries include cut or pierce, drowning or submersion, fall, fire or flame, hot object or substance, firearm, machinery, MVT, pedal cyclist or pedestrian and other transport, natural environment, bites and stings, struck by or against, poisoning, and other specified accidental injuries. Major intentional injuries include assaults or purposely inflicted injury, adverse effects due to drugs, adverse effects due to medical care (e.g., medical procedures as the cause of an abnormal reaction or later complication), suicide or self-inflicted, and assault. In the current surveillance, we classify them into 10 categories and assess them according to rural–urban and neighborhood SES jointly. Since prior assessments rarely bring neighborhood SES and rural–urban dimensions together at the population level, surveillance along these two lines that considers the whole state over a sufficient period of time may shed new insight into preventing injury hospitalization.

Tiesman et al. (2007) used the National Health Interview Survey to examine injuries by place of residence. They used county-based nine-level urban influence codes (UICs) to group respondents into large urban (population >1 million), small urban (population <1 million), large rural (adjacent to a metro or UICs 3–6), and small rural (UICs 7–9). Using injury episodes or incidence, the study found that small rural residents had a 26% higher injury rate than large urban residents. The study, however, did not differentiate injury categories. Coben et al. (2009) examined rural–urban differences in nine injury categories: unintentional (fall, motor, poisoning), self-inflicted (poisoning, cut or pierce, firearm), and assault (struck by or against, firearm, cut or pierce). They used the Nationwide Inpatient Sample (NIS) of the Healthcare Cost and Utilization Project (HCUP) and the same four rural–urban categories as Tiesman et al. It was found that patients in the small rural category had the highest risks of unintentional injuries contributed mainly by MVCs; those in the small rural category also had the highest risk of self-inflicted injuries, while the large urban category had the highest risk in assault injuries. Since the largest metro area in Nebraska has a population of <500,000, the urban area in our definition is small urban in these studies. The current study would not be able to further divide rural into two categories due to sample size and potential confidentiality concerns, so we kept the two categories that we used in this chapter. For this reason, we also want to put Tieman et al.'s results in reference to small urban, where the elevated rate for small rural would be 16% higher. If we combine large rural and small rural into rural, similar to our study, the elevated rate would be 12%. The elevated rates for unintentional injury in general and fall in particular would be 23% and 12%, respectively, in Coben et al.'s study.

Since the national studies found fairly consistent results, we did not want to use local data to make a comparison. Rather, we wanted to see how age and poverty may affect the incidence rate. As mentioned earlier, crude disease rates tend to be higher in rural areas, but the age-adjusted rates were higher in urban areas. To show age effects on the rural–urban difference in injuries, we ran three models for 10 injury categories, with the first 7 being unintentional. In model I, we included controls for sex and then evaluated rural–urban differences. In model II, we added age to model I, and in model III, we added poverty to model II. During the study period, there were 222,253 hospitalizations that were associated with the primary E-codes, and they accounted for 16.17% of total in-state residents' hospitalizations. We used 10 injury categories (Table 10.6), where "all unintentional" overlaps with categories from MVT to poisoning. The all unintentional category, together with the self-inflicted, assault, and adverse effects—drugs and medical injury categories, accounts for 99.4% of 222,253 hospitalizations. In terms of specific injury categories, fall is the single most important unintentional injury category, whereas adverse effects due to drug or medical care is the most dominant injury category. However, this latter category is rarely considered injury from the public health point of view because it does not have a clear indication of whether it is accidental.

The results from model I showed that 6 out of the 10 categories were much less likely to happen in urban areas than in rural areas. The opposites included accidental poisoning, self-inflicted injuries, and assaults. However, when we introduced model II, 7 out of the 10 categories were much more likely to happen in urban areas. What remain unchanged included MVT injuries, other traffic-related injuries (cyclist, pedestrian, or other transportation), and natural environment, bites and stings, struck by or against. For example, all unintentional injuries were 19% less likely in urban than in rural areas according to model I, and they were 9% more likely in model II; fall injuries were 27% less likely in urban areas in model I, and they were 11% more likely in model II. The reversal in rural–urban effects in model II is expected, as rural areas had much higher proportions of older age categories. It implicitly suggests that if we want to reduce injury hospitalization, we need to bring down injuries associated with aging. Since the inclusion of poverty in model III yielded more or less the same effects as in model II, we will not describe the results here, but leave them as references for later use.

We suspected that age distributions between rural and urban areas play an important role in altering rural–urban effects in Table 10.6. In Table 10.7, we made age effects explicitly based on model III outputs in Table 10.6 for each injury group. Recall that the effects of MVT injury were consistent with and without controlling for age; this is because the age effects were similar at both ends. If we were to use the middle-age 45–54 category as the reference, the age effects would be symmetrically similar toward both younger and older age groups. Since urban areas have more

TABLE 10.6
Urban–Rural Odds Ratios for E-Codes by Different Control Effects

	% of Total	Model I	Model II	Model III
All unintentional	38.71	0.81 (0.8, 0.82)	1.09 (1.07, 1.1)	1.1 (1.08, 1.12)
MVT	3.23	0.82 (0.78, 0.86)	0.83 (0.8, 0.87)	0.84 (0.8, 0.89)
Other traffic related	1.13	0.49 (0.46, 0.54)	0.51 (0.47, 0.55)	0.53 (0.48, 0.58)
Fall	21.24	0.73 (0.72, 0.75)	1.11 (1.09, 1.13)	1.12 (1.09, 1.14)
Unintentional–cut or pierce	0.38	0.99 (0.86, 1.14)	1.01 (0.88, 1.16)	1.02 (0.88, 1.19)
Natural environment, bites, struck by	2.26	0.73 (0.69, 0.78)	0.84 (0.79, 0.89)	0.83 (0.78, 0.88)
Unintentional—poisoning	1.55	1.2 (1.12, 1.28)	1.35 (1.26, 1.44)	1.36 (1.27, 1.46)
Self-inflicted	3.64	1.55 (1.48, 1.62)	1.38 (1.32, 1.44)	1.38 (1.32, 1.45)
Assault	1.16	2.82 (2.58, 3.09)	2.63 (2.4, 2.88)	2.37 (2.15, 2.6)
Adverse effects—drugs and medical	55.89	0.99 (0.98, 1)	1.3 (1.29, 1.32)	1.32 (1.3, 1.33)

Note: % of total, percent of total 222,253 injury-related hospitalization incidents. Model I includes sex and urban indicator only. Model II includes age group, sex, and urban indicator. Model III includes age group, sex, poverty level, and urban indicator. Each injury group was modeled separately.

TABLE 10.7
Age Effects in Odds Ratio for 10 Injury Groups

Injury Categories	15–24	25–34	35–44	45–54	55–64	65–74	75+
Unintentional	2.20	1.93	2.54	3.45	4.84	9.87	34.51
MVT injury	5.85	3.78	3.41	3.16	3.27	3.80	5.93
Other traffic related	1.69	1.35	1.58	1.95	1.78	1.80	1.74
Fall injury	1.40	1.63	2.95	5.28	9.80	25.05	112.71
Unintentional—cut or pierce	2.49	2.33	2.72	2.48	2.33	3.20	3.54
Natural environment, bites, struck by	1.45	1.03[a]	1.29	1.48	1.77	2.64	5.66
Unintentional—poisoning	1.83	1.57	2.26	3.28	2.96	4.11	6.82
Self-inflicted	12.17	8.28	8.87	6.31	2.43	1.28	1.06[a]
Assault	6.52	4.44	3.78	2.78	1.12[a]	0.90[a]	0.63
Adverse effects—drugs and medical	1.36	2.06	3.56	6.08	10.24	21.32	33.45

Note: The age referent is age 0–14. Model estimates were based on age group, sex, poverty level, and urban indicator. Each injury group was modeled separately.

[a] Not significant at $p < 0.05$.

younger age population, and rural areas have more older population, the effects along age tend to cancel each other out symmetrically according to differences in rural–urban population age distributions. Consequently, the odds ratios for models I and II are very similar for MTV in Table 10.6. Injuries due to fall, on the other hand, were extremely disproportionally attributed to aging. In the age groups 55–65, 65–74, and 75+, the odds ratios disproportionally increased from 9.80 to 112.71, which essentially represents age effects, which explains the rural–urban effect reversals between models I and II in Table 10.6. We should mention that fall is the most commonly seen unintentional injury. The 10 times or more injuries in older age categories (age 55+) suggest that we really need to concentrate on where older people live, especially in rural areas, for fall prevention. In contrast to fall, self-inflicted injury is a commonly cited intentional injury, and it had an age effect tipping toward young age groups, where urban areas tend to have more for both those injuries and younger age populations. Controlling for age in model II substantially reduced their effects, especially for self-inflicted injuries. This suggests that we need to concentrate on younger age groups, especially in urban areas, for intentional injury prevention. Between the two extremes—fall and self-inflicted injuries—specific age effects in odds ratios can help injury prevention programs design age-specific intervention programs that are often pooled with other control variables, rather than targeted intervention variables.

Having investigated rural–urban and age effects, we now turn our attention to poverty effects at the neighborhood level. We estimated Equation 10.3 for rural and urban effects separately, thus producing two sets of estimates for poverty levels:

$$\log(\text{Disease}(x)) = \text{Age} + \text{Sex} + \text{Poverty} + \text{Offset}(\log \text{ of population}) \qquad (10.3)$$

The results are shown in Table 10.8. First, most effects from poverty between poverty groups 2–5 and poverty group 1 were significant for urban areas, but not for rural areas. In most cases, the odds

TABLE 10.8
Poverty Effects in Odds Ratios in Rural and Urban Areas

Injury categories (E-codes)	Rural				Urban			
	Pvt2	Pvt3	Pvt4	Pvt5	Pvt2	Pvt3	Pvt4	Pvt5
Unintentional injury	1.03	1.02	1.04*	0.96	1.10*	1.21*	1.33*	1.59*
Unintentional—cut or pierce	1.05	0.87	1.02	0.96	1.56*	1.26	1.66*	1.92*
MVT injury	1.12	1.04	1.01	0.81*	1.09	1.16*	1.22*	1.72*
Other traffic-related injury	1.08	1.09	1.26*	0.65*	1.01	0.86	1.02	0.93
Fall injury	1.05	1.04	1.07*	1.04	1.07*	1.14*	1.22*	1.41*
Unintentional–poisoning	1.11	1.33*	1.36*	1.45*	1.39*	1.44*	2.09*	3.22*
Natural environment, bites, struck by	1.02	0.93	0.93	0.77*	0.94	1.05	1.16	1.53*
Self-inflicted injury	1.25*	1.28*	1.60*	1.61*	1.28*	1.39*	1.68*	2.16*
Assault injury	1.12	1.41*	1.99*	1.93*	1.44*	1.94*	4.68*	8.34*
Adverse effects—drugs and medical	0.96*	0.89*	0.92*	0.90*	1.09*	1.24*	1.41*	1.46*

Note: Pvt, poverty group. All poverty groups are in reference to poverty group 1 with <5% people under the poverty line.
*Significant at $p < 0.05$.

ratios were >1, suggesting that poorer neighborhoods had more injury hospitalizations. Few injuries in rural areas tend to be less likely from the poorest group, and they include MTV, other traffic related, and natural environment, bites and stings, and struck by or against. Compared to well-off neighborhoods, those in the poorest neighborhoods were less active, or they lacked transportation to be outside for those injuries in rural areas.

Finally, the odds ratios for adverse effects in drug or medical care were <1 in rural areas suggesting that poverty group 1 (<5%) was more likely compared to all other poverty group. In other words, well-off neighborhoods in rural areas were disadvantaged in adverse drug or medical care effect injuries. The reverse is true for urban areas, where all odds ratios were in favor of the well-off neighborhoods. Residents of poverty groups 2–5 were all were more likely than group 1 to experience adverse drug or medical care injuries, controlling for age and sex in urban areas. As shown earlier, the adverse effect category accounts for almost 56% of all E-code-based injuries. Although they are not part of traditional accidental and intentional categories, it might be worth looking into diagnoses together with these late adverse effects due to drug, surgery, and other medical care.

To briefly summarize, we demonstrated in this case study the importance of age effects in the evaluation of rural–urban differences in injury hospitalization. Most injuries are preventable. By focusing on age effects, we can intervene more effectively in certain age groups that have the greatest risks. Examples include the elderly for fall injuries, the younger than middle age for self-inflicted injuries, the college age for assault injuries, and the college age and 75+ age groups for MVT injuries. In addition, we found that SES effects in injuries were generally higher in poor neighborhoods, and the effects were much stronger in urban areas than in rural areas.

10.5   CHAPTER SUMMARY

In this chapter, we have examined rural–urban differences for common hospitalizations. We mentioned that the age distribution favored the urban residents, while the SES distribution slightly favored the rural residents. It seemed that the SES effect outweighed the age effect. In addition, in urban areas, there were more primary care physicians available in close proximity to many patients, but the SES gradients were much stronger than those in rural areas, which had at least a geographic access issue. In most diagnoses, urban residents were more likely to be hospitalized than rural residents. While low SES was associated with high hospitalization incidence in urban areas for many diagnoses, the effects were mixed in rural areas. In rural areas, two clusters of diseases that were related to mental health and preventable hospitalization due to ACSCs were negatively associated with SES, while some diseases that require timely and quality care tended to be positively associated with SES (the higher the SES, the greater the incidence rates).

Location factors play an important role in differentiating access to care in rural areas. Depending on poverty groups and diagnoses, rural hospital bypassing rates range from 5% to 72%. Traumatic brain injury, back problems, and cancer saw the highest bypassing rates, while COPD and bronchiectasis pneumonia, and asthma had the lowest bypassing rates. For documentation purposes, we also included distance traveled. Traumatic brain injury, AMI, and cancer patients in rural areas traveled the greatest distances; patients with diagnoses of diverticulosis and diverticulitis, appendicitis and appendiceal conditions, and pneumonia for admissions traveled the least distances. Empirical studies showed that rural residents in Nebraska have more frequent ambulance transfers than urban residents from a secondary to a tertiary hospital, more volunteer than paid emergency medical services (EMS) responders, more EMS by a helicopter, and greater EMS scene-to-facility distance (Grossman et al., 1997; Mueller, 1999). Moreover, a previous study showed that the average travel distance to medical care for rural residents was twice that of urban residents (Probst et al., 2007). Our current surveillance showed that the average distance to hospitals for rural residents was at least four times that of urban residents.

In the case study of injury admissions, we showed the importance of explicitly including age effects. Given that it is common to control for age effects, people sometimes tend to forget or ignore exploiting age effects for public health intervention. Older age groups should be targeted for MVC and fall injury prevention, while young age group should be focused on for MVT injury, self-inflicted injury, and assault injury prevention. Although not often seen in injury surveillance, we included adverse drug effects for hospitalization to show its importance. More than 50% of E-code hospitalizations were related to adverse effects—drugs and medical care. As prescription drug abuse becomes more common, future E-code surveillance should include adverse effects—drugs and medical care in different categories and separately evaluate them. In this part of the study, we did not extend to it injury-related disability and mortality. Sihler and Hemmila (2009) reported that rural residents were more likely to be discharged disabled, which could be due to injury severity that could be constructed from ICD-9 codes. It is possible to look into rehabilitation, other follow-up care (Zwerling et al., 2005), and mortality outcomes due to injury, which are the limitations of the current studies.

APPENDIX 10A: COMMODITY-BASED RURAL–URBAN COMPARISON

To make a model comparable to the Chapter 8 incidence reports, odds ratios for 40 diagnoses in comorbidities are listed in Table 10.A1, which is a companion to Table 10.3. The first column contrasts urban versus rural. Controlling for age and poverty, urban patients were more likely to have most of the surveyed diseases, except pneumonia, fracture of neck of femur, and appendicitis and other appendiceal conditions, which were more likely to be found among rural patients. In addition, three disease conditions—other fractures, biliary tract disease, and nonspecific chest pain—were not significant. Diseases at least 50% more likely to be found among urban patients than rural patients were substance-related disorders, HP with complications and secondary HP, respiratory failure, insufficiency, arrest, and acute and unspecified renal failure.

The results for poverty group 5 in urban areas (Table 10.A1, column 2) resembled corresponding odds ratios in Chapter 8 (Table 8.1), suggesting that results without place of residence stratification were primarily contributed by urban neighborhoods—the poorer the neighborhoods, the greater the odds of having a diagnosed disease or condition. In contrast, poverty effects in rural areas were mixed. Only 12 odds ratios had consistent results between urban and rural areas, all of which showed weaker effects compared to their urban counterparts; 13 out of 41 odds ratios were not significant, and 15 odds ratios had opposite directional effects compared to urban areas.

The 12 odds ratios consistent between urban and rural areas include mood disorders, COPD and bronchiectasis, skin and subcutaneous tissue infections, nonspecific chest pain, diabetes mellitus with complications, asthma, schizophrenia and other psychotic disorders, pancreatic disorders (not diabetes), epilepsy and convulsions, alcohol-related disorders, substance-related disorders, and preventable hospitalization. This group has some diseases in common. Diabetes and asthma are parts of preventable hospitalization. Mood disorders, schizophrenia and other psychotic disorders, alcohol-related disorders, substance-related disorders, and epilepsy are all related to mental health. Implicitly, SES disparities in these diseases and conditions permeate both urban and rural areas, with a weaker effect in rural areas.

The 15 odds ratios having opposite place of residence effects include osteoarthritis, septicemia (except in labor), cardiac dysrhythmias, coronary atherosclerosis, AMI, fluid and electrolyte disorders, acute and unspecified renal failure, diverticulosis and diverticulitis, fracture of neck of femur, syncope, HP with complications and secondary HP, other fractures, intestinal infection, secondary malignancies, and anemia. A straightforward interpretation is that compared to the well-off neighborhood (poverty group 1), rural patients living in the poorest areas were either less likely to have these conditions or less likely to seek hospital care associated with these conditions.

TABLE 10.A1

Rural and Urban Differences in Comorbidity Incidence

Diagnosis	Urban/Rural	Poverty Level 5 vs. Level 1	
		Urban	Rural
All hospitalizations	0 (1.57, 1.56)	0 (1.04, 1.03)	1.04 (1.03, 1.05)
Pneumonia	1 (1.94, 1.89)	0 (1.04, 1)	1.04 (1, 1.08)
Osteoarthritis	2 (1.13, 1.09)	0 (0.85, 0.82)	0.85 (0.82, 0.89)
Congestive heart failure	3 (1.93, 1.88)	0 (0.99, 0.96)	0.99 (0.96, 1.03)
Septicemia (except in labor)	4 (1.87, 1.8)	0 (0.94, 0.88)	0.94 (0.88, 1)
Mood disorders	5 (2.13, 2.09)	0 (1.29, 1.25)	1.29 (1.25, 1.34)
Cardiac dysrhythmias	6 (1.36, 1.33)	0 (0.87, 0.84)	0.87 (0.84, 0.89)
COPD and bronchiectasis	7 (2.41, 2.36)	0 (1.09, 1.06)	1.09 (1.06, 1.13)
Back problems	8 (1.41, 1.37)	0 (1, 0.95)	1 (0.95, 1.05)
Skin and subcutaneous tissue infections	9 (2.16, 2.08)	0 (1.12, 1.05)	1.12 (1.05, 1.19)
Coronary atherosclerosis	10 (1.56, 1.53)	0 (0.94, 0.91)	0.94 (0.91, 0.97)
Urinary tract infections	11 (1.88, 1.83)	0 (0.93, 0.89)	0.93 (0.89, 0.97)
Nonspecific chest pain	12 (2.51, 2.41)	0 (1.14, 1.06)	1.14 (1.06, 1.23)
AMI	13 (1.68, 1.59)	0 (0.9, 0.83)	0.9 (0.83, 0.97)
Acute cerebrovascular disease	14 (1.82, 1.73)	0 (0.93, 0.85)	0.93 (0.85, 1.01)
Diabetes mellitus with complications	15 (2.97, 2.86)	0 (1.15, 1.08)	1.15 (1.08, 1.23)
Biliary tract disease	16 (1.48, 1.41)	0 (0.95, 0.88)	0.95 (0.88, 1.02)
Fluid and electrolyte disorders	17 (1.87, 1.84)	0 (0.97, 0.94)	0.97 (0.94, 1)
Asthma	18 (2.46, 2.39)	0 (1.32, 1.25)	1.32 (1.25, 1.4)
Schizophrenia and other psychotic disorders	19 (7.29, 6.91)	0 (2.32, 2.13)	2.32 (2.13, 2.53)
Acute and unspecified renal failure	20 (2.17, 2.11)	0 (0.83, 0.79)	0.83 (0.79, 0.88)
Respiratory failure, insufficiency, arrest	21 (2.33, 2.25)	0 (0.99, 0.93)	0.99 (0.93, 1.05)
Gastrointestinal hemorrhage	22 (1.91, 1.81)	0 (0.95, 0.87)	0.95 (0.87, 1.03)
Intestinal obstruction without hernia	23 (1.34, 1.28)	0 (1.01, 0.94)	1.01 (0.94, 1.08)
Pancreatic disorders (not diabetes)	24 (2.34, 2.21)	0 (1.1, 0.99)	1.1 (0.99, 1.22)
Diverticulosis and diverticulitis	25 (1.11, 1.05)	0 (0.87, 0.8)	0.87 (0.8, 0.94)
Fracture of neck of femur (hip)	26 (1.26, 1.17)	0 (0.88, 0.8)	0.88 (0.8, 0.98)
Epilepsy, convulsions	27 (2.72, 2.61)	0 (1.32, 1.23)	1.32 (1.23, 1.42)
Appendicitis and other appendiceal conditions	28 (1.09, 1)	0.05 (1.03, 0.9)	1.03 (0.9, 1.17)
Alcohol-related disorders	29 (3.64, 3.51)	0 (2.19, 2.05)	2.19 (2.05, 2.35)
Fracture of lower limb	30 (1.82, 1.7)	0 (0.9, 0.8)	0.9 (0.8, 1.01)
Other nervous system disorders	31 (1.73, 1.68)	0 (1.04, 0.99)	1.04 (0.99, 1.09)
Syncope	32 (1.41, 1.32)	0 (0.88, 0.79)	0.88 (0.79, 0.98)
Substance-related disorders	33 (4.28, 4.11)	0 (1.84, 1.7)	1.84 (1.7, 2)
Intracranial injury	34 (1.91, 1.78)	0 (0.97, 0.86)	0.97 (0.86, 1.1)
HP with complications and secondary HP	35 (2.49, 2.42)	0 (0.89, 0.85)	0.89 (0.85, 0.94)
Other fractures	36 (1.32, 1.24)	0 (0.88, 0.8)	0.88 (0.8, 0.96)
Intestinal infection	37 (1.54, 1.45)	0 (0.92, 0.83)	0.92 (0.83, 1.02)
Secondary malignancies	38 (1.34, 1.28)	0 (0.8, 0.75)	0.8 (0.75, 0.86)
Anemia	39 (1.89, 1.85)	0 (0.94, 0.9)	0.94 (0.9, 0.97)
Avoidable hospitalization	40 (2.06, 2.03)	0 (1.09, 1.07)	1.09 (1.07, 1.11)

Note: Comorbidities are based on search of nine diagnosis codes. Except for Douglas, Sarpy, and Lancaster Counties, all counties were rural.

REFERENCES

Adams, E.K., R. Houchens, G.E. Wright, and J. Robbins. 1991. Predicting hospital choice for rural Medicare beneficiaries: The role of severity of illness. *Health Services Research* 26(5): 584–612.

Coben, J.H., H.M. Tiesman, R.M. Bossarte, and P.M. Furbee. 2009. Rural–urban differences in injury hospitalizations in the U.S., 2004. *American Journal of Preventive Medicine* 36(1): 49–55.

Escarce, J.J. and K. Kapur. 2009. Do patients bypass rural hospitals? Determinants of inpatient hospital choice in rural California. *Journal of Health Care for the Poor and Underserved* 20(3): 625–644.

Grossman, D.C., A. Kim, S.C. MacDonald, P. Klein, M.K. Copass, and R.V. Maier. 1997. Urban-rural differences in prehospital care of major trauma. *Journal of Trauma: Injury, Infection and Critical Care* 42(4): 723–729.

Hall, M.J., J. Marsteller, and M.F. Owings. 2010. Factors influencing rural residents' utilization of urban hospitals. National Health Statistics Report No. 31. Hyattsville, MD: National Center for Health Statistics.

Hartley, D. 2004. Rural health disparities, population health, and rural culture. *American Journal of Public Health* 94(10): 1675–1678.

Lin, G., D. Allen, and M. Penning. 2002. Examining distance effects on hospitalizations using GIS: A study of three health regions in British Columbia, Canada. *Environment and Planning A* 34(11): 2037–2053.

Lutfiyya, M.N., D.K. Bhat, S.R. Gandhi, C. Nguyen, V.L. Weidenbacher-Hoper, and M.S. Lipsky. 2007. A comparison of quality of care indicators in urban acute care hospitals and rural critical access hospitals in the United States. *International Journal for Quality in Health Care* 19(3): 141–149.

Mueller, K.J. 1999. Health status and access to care among rural minorities. *Journal of Health Care for the Poor and Underserved* 10(2): 230–249.

Probst, J., S. Laditka, J. Wang, and A. Johnson. 2007. Effects of residence and race on burden of travel for care: Cross sectional analysis of the 2001 US national household travel survey. *BMC Health Services Research* 7(1): 40.

Ross, J.S., S.-L.T. Normand, Y. Wang, B.K. Nallamothu, J.H. Lichtman, and H.M. Krumholz. 2008. Hospital remoteness and thirty-day mortality from three serious conditions. *Health Affairs* 27(6): 1707–1717.

Sihler, K.C. and M.R. Hemmila. 2009. Injuries in nonurban areas are associated with increased disability at hospital discharge. *Journal of Trauma* 67(5): 903–909.

Stranges, E. and C. Stocks. 2010. Potentially preventable hospitalizations for acute and chronic conditions, 2008. HCUP Statistical Brief No. 99. Rockville, MD: Agency for Healthcare Research and Quality, November.

Tiesman, H., C. Zwerling, C. Peek-Asa, N. Sprince, and J.E. Cavanaugh. 2007. Non-fatal injuries among urban and rural residents: The National Health Interview Survey, 1997–2001. *Injury Prevention* 13(2): 115–119.

Zhang, Y. and G. Lin. 2013. Disparity surveillance of nonfatal motor vehicle crash injuries. *Traffic Injury Prevention* 14(7): 697–702.

Zwerling, C., C. Peek-Asa, P.S. Whitten, S. Choi, N.L. Sprince, and M.P. Jones. 2005. Fatal motor vehicle crashes in rural and urban areas: Decomposing rates into contributing factors. *Injury Prevention* 11(1): 24–28.

11 Racial Disparities in Hospitalization

11.1 INTRODUCTION

So far, we have not touched racial disparities, because it would be the most difficult task when race and ethnicity variables are not available from the Nebraska hospital discharge data (HDD). The Nebraska HDD had race and ethnicity fields, but about 93% of the values were missing for race. Even among the nonmissing race values, many are categorized as U, or unknown. That is why we know little about racial disparities in hospitalization in Nebraska. However, racial and ethnicity disparities are too important to not report. Both the federal and state governments officially acknowledge the existence of health and health care disparities between the racial and ethnic minority populations and White Americans. The health of minority populations has consistently lagged behind that of Whites. Racial and ethnic minority groups tend to be socially and economically disadvantaged, and they also have low health literacy. These factors together affect minority patients' access to care in many ways, including preventive care, ambulatory care, and in- and post-hospital care. Nationally, the rate of preventable hospitalizations for minorities is twice that for Whites. African Americans are substantially less likely to have coronary artery bypass surgery than Whites. Not only are African Americans more likely to have heart disease, diabetes, renal disease, and chronic obstructive pulmonary disease (COPD), but they also experience more severe forms of these diseases.

One of the four overarching goals of Healthy People 2020 is "to achieve health equity, eliminate disparities and improve the health of all groups." Taking this initiative, the U.S. Department of Health and Human Services (DHHS) started to assess health disparities in the U.S. population by tracking rates of death, chronic and acute diseases, injuries, and behavior risk factors among racial groups. Since hospitalization is a major data source to identify commodities, we decided to add race information, so that it can be used to assess racial disparities in hospitalization. We first identified major databases within the state Division of Health, and they included the driver's license database, cancer registry data, trauma registry data, mortality data, birth record database, and motor vehicle crash data. Second, we standardized the race and ethnicity variables. For data improvement on race, we identified a common denominator: White, Black, Asian, American Indian, and Other. Note that ethnicity was not included because the driver's license database did not have a separate field for ethnicity.

The rest of the chapter is divided into three parts. First, we briefly summarize the data linkage project that provided race information from other, secondary data sources. Second, we provide a general racial disparity assessment using HDD, similar to what we did in previous chapters. Third, we provide a multivariate data analysis by adding race to the acute myocardial infarction (AMI) analysis that we did in Chapter 17.

11.2 USING MULTIPLE DATA SOURCES TO GENERATE THE RACE VARIABLE FOR HDD

As mentioned in Section 11.1, the race variable in the Nebraska HDD was unusable because 93% of the values were missing. Since the Nebraska DHHS often links various datasets to the Nebraska HDD, we were able to systematically harvest race and ethnicity variables from those datasets that had been linked to HDD. Datasets that have been used for linkage include the cancer registry, trauma registry, birth (including mother's and father's race and ethnicity) and death records, and driver's license—all of them have a program-specific purpose. We already mentioned the data linkage schema in Chapter 5. Each record linkage for program data was based on the name (first and last), sex, date of birth, and residential address based on the probabilistic linkage methodology described in Chapter 4. After each individual linkage, we created a consolidated race variable.

There were a number of steps to consolidate different race variables into a single race variable. First, it was necessary to decide a race data quality scale. Since 7% of the race records were valid, we keep them without any manipulation. To update records for the remaining missing values, we used the following rank order. Race information from the cancer registry took priority due to its superior data quality, which was then followed by death records, birth records, and both mother's and birth father's race. We then have race variables from the trauma registry, and finally driver's license data. It should be noted that driver's license data had the largest number of records, but no information about those 14 years old or younger, and it had a sizable "other." For this reason, we listed it as having low-quality race information.

The second step of consolidating race is to find the common race categories. While the cancer registry and vital records use five race categories, the trauma registry and driver's license data only have four race categories. For this reason, we settled on four race categories: White, Black, Asian, and American Indian.

However, there were still 155,782 (10.59%) records that could not be linked to any program dataset. In order to conduct population-based studies, it is necessary to impute about 10.59% of the records that could not be found in any public health database. In this final process, we used census tract information by making two assumptions: (1) age and sex are not important in imputation, and (2) the racial composition of census tracts could represent the racial composition of those patients without race information who resided there. Based on these two assumptions, the imputation algorithm is straightforward. We can simply assign an individual race based on the racial mix of the census tract of each hospital patient without considering age and sex. Assuming that there are 100 missing values in a census tract that has 80% Whites, 10% Blacks, 5% Asians, and 5% American Indians, then the probability-based imputation would generate approximately 80 Whites, 10 Blacks, 5 Asians, and 5 American Indians. We used the 2010 population census DP-1 table ("Profile of General Population and Housing Characteristics: 2010") to obtain the proportion of each single race for each census tract, and assigned race to each remaining missing race value according to the census tract–based imputation scheme.

Table 11.1 shows the results of the single-race variable compilation from the linked program data (first column), the imputed census population (second column), and the 2010 census population profile. It seems that the linked race profile is more consistent with the 2010 census profile.

TABLE 11.1

Race Distributions in Percent between the Linked and Imputed

	Linked	Imputed	Combined	Nebraska Population
White	84.819	81.794	84.599	86.1
African American	8.646	7.945	8.595	4.5
American Indian	0.740	0.077	0.692	1
Asian	1.465	1.592	1.474	1.9
Other	4.331	8.592	4.639	6.5
Total N	1,273,316	99,536	1,372,852	1,826,341

Note: Linked records are from public health databases; imputed records are from the census. The table is for in-state patients only. The total number of in-state and out-of-state patient records is 1,485,131. The total number of records without race is 155,782. The total number of in-state patient records without race is 99,536. The last column includes race composition for the 2010 Nebraska population.

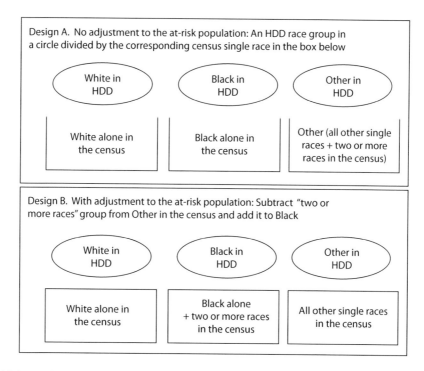

FIGURE 11.1   Design of race-specific incidence rate calculation. Circles contain incidence and boxes contain at-risk population.

A major difference between the imputed and the linked data is "other." The imputed data had 8.6% for this category, while the linked data had 4.3%. However, since only 7.25% of hospitalizations had imputed race, they are unlikely to make much difference in terms of disease incidence rates, which was confirmed by our preliminary analysis. In the following, we used the combined race as the numerator, which included both geographically imputed and individually linked files (third column in Table 11.1).

As for the denominator or at-risk population, we wanted to use poverty-specific race populations from the 2010 census. To this end, we extracted percentage of people under the poverty line from the ACS 2007–2011 data and linked it the 2010 census data at the census tract level. The next step was to deal with multiple races in the census data. For the Whites, the Census includes White alone or with other combinations. Hence, the HDD race and the census race are not 100% compatible even for the same racial group. For this reason, it was necessary to make some adjustment about racial groups in the census data so that they are compatible with the HDD racial categories. We show adjustment strategies in Figure 11.1, while leaving the details in Table 11.A1. Design A in the figure assumes that Whites and Blacks in the HDD file align perfectly with single race from the census files. This design would inflate race-specific hospitalization rates because it does not take two or more races into consideration in the Black or White at-risk population. In design B, we moved "two or more races" to the Black population group. This alternative design assumes that all people who checked two or more races had at least one race being Black. This oversimplification follows the so-called "one drop of blood rule" and will definitely deflate the Black hospitalization rate, but it would generate conservative estimates for Blacks.

11.3   PATIENT-BASED ASSESSMENT FOR MAJOR COMORBIDITIES

With the above methodological introduction, we are ready to conduct disparity assessments. Since the uncertainties introduced by the race mismatch between the at-risk population and hospitalized patients, we want to establish some bounds for the assessments. Table 11.2 attempts to provide some sensitivity analysis by moving and not moving the 2.2% of the at-risk population from the Other race category to Black. For ease of comparison, we used the same approach as in Chapter 9 for sex differences in major diseases, which present data in age-standardized rates. To refresh our memory, the table provides hospital-based surveillance or calculation of comorbidities for 10 common causes of death. The surveillance was conducted by searching up to 10 diagnoses, and if one of the diagnoses met the disease definition, then the visit records a comorbidity condition. Since much survey

TABLE 11.2

Standard Comorbidity Prevalence by Sex and Race for Common Causes of Death

	White		Black		Other	
Comorbidities	Male	Female	Male	Female	Male	Female
			Panel A			
Heart disease	1296	960	3697	3049	1319	1137
Cancer	409	312	613	482	200	160
Chronic lower respiratory diseases	284	230	873	823	308	300
Stroke	610	619	1682	2025	565	665
Unintentional injuries	458	426	1635	1296	538	463
Alzheimer's disease	87	82	228	250	67	94
Diabetes	581	511	1958	2065	832	912
Kidney diseases	489	328	1741	1348	573	480
Influenza and pneumonia	530	440	1217	1151	510	509
Suicide	31	21	117	52	41	30
			Panel B			
Heart disease	1296	960	3076	2561	1611	1403
Cancer	409	312	512	402	244	195
Chronic lower respiratory diseases	284	230	730	694	377	371
Stroke	610	619	1364	1616	699	818
Unintentional injuries	458	426	1287	1054	661	578
Alzheimer's disease	87	82	186	213	85	119
Diabetes	581	511	1625	1713	1004	1107
Kidney diseases	489	328	1439	1132	700	595
Influenza and pneumonia	530	440	979	939	640	643
Suicide	31	21	91	41	50	37
			Rate Ratios			
Heart disease	1.00	1.00	0.83	0.84	1.22	1.23
Cancer	1.00	1.00	0.84	0.84	1.22	1.22
CHRONIC lower respiratory diseases	1.00	1.00	0.84	0.84	1.22	1.24
Stroke	1.00	1.00	0.81	0.80	1.24	1.23
Unintentional injuries	1.00	1.00	0.79	0.81	1.23	1.25
Alzheimer's disease	1.00	1.00	0.82	0.85	1.26	1.26
Diabetes	1.00	1.00	0.83	0.83	1.21	1.21
Kidney diseases	1.00	1.00	0.83	0.84	1.22	1.24
Influenza and pneumonia	1.00	1.00	0.80	0.82	1.25	1.26
Suicide	1.00	1.00	0.78	0.78	1.21	1.24

Note: Comorbidities are standardized per 100,000 by search up to 10 diagnoses. Black and White in panel A are based on one race only. Black in panel B includes Black alone and two or more race categories.

data are reported by prevalence, we changed our approach a little by using prevalence too. However, in order to be consistent with previous tables in Chapter 9 for gender differences, we also included an incidence table (Table 11.A3).

Results shown in panels A and B of Table 11.2 are based on designs A and B in Figure 11.1, respectively. Let's first establish some general comorbidity profiles by race from panel A. With the exception of Other males, prevalence was much higher among racial minorities than Whites regardless of sex. In many diseases, Black rates doubled those for Whites. For instance, the heart disease rate for Black males was 3,697 per 100,000, in contrast to 1,296 for White males. Comparing Other males with White males, there was no clear pattern. White males had higher rates for cancer, stroke, and Alzheimer's disease, while Other males had higher rates for heart disease, COPD, and diabetes. In other words, males of other races did not necessarily have higher disease prevalence than White males according to design A.

When comparing panels A and B in Table 11.2, one can use a rate ratio by dividing the standard rates in panel B by the standard rates in panel A. Since there was no adjustment for Whites, all the rate ratios equal 1. For Blacks, the rate ratios suggest that all the standard rates reduced by 15% or more after the adjustment in panel B. In contrast, the standard rates for Other increased by 21% on average after the adjustment, which is enough to change the directional effect for heart diseases by sex. One could argue that the adjustment should be made to White rather than Black. If we did that, the standard rates for White after adjustments would decrease by a fraction of 1% for both sexes (not shown), which would not change any directional effects between Black and White. Hence, the slight underestimates of Black rates and overestimates of Other rates are acceptable. For this reason, we use the adjusted at-risk population from now on.

We know that racial minorities are more likely to be poor than Whites. Due to small population sizes for Blacks and Others, we opted to use three poverty groups without gender stratification. These three groups are <5%, 5%–15%, and >15%, representing the low-poverty, midpoverty, and high-poverty census tracts, respectively. In addition to prevalence, we present poverty levels by race in rate ratios, with the reference rate being low poverty (Table 11.3). The incidence rates and rate ratios are in Appendix 11B (Table 11.A3). If a rate ratio is >1, the rate of interest is greater than the rate of the referent, and we expect that all rate ratios are >1. The results were broadly consistent with our expectations; a number of rate ratios were <1, or against our expectation. For Whites, poor neighborhoods tended to be associated with higher comorbidity rates. All rate ratios were above 1 except Alzheimer's and kidney diseases for midpoverty. In addition, the rate ratios for high-poverty census tracts seem to be greater than those from midpoverty census tracts. The general pattern for Blacks and Other is likewise similar to that of Whites. However, Blacks were less likely to have unintentional injuries and suicide. It was possible that those who had injuries or suicide attempts in the midpoverty group did not end up in hospitals.

11.4 PREVALENCE, READMISSION, AND MORTALITY FOR MAJOR HOSPITALIZATIONS

11.4.1 Prevalence

Normally, prevalence is appropriate for comorbidity surveillance, and incidence is suitable for hospitalization. However, there is some merit to using prevalence based on the primary diagnosis. First, the primary diagnosis basically is the reason for admission. Second, a model-based rate can be derived for how many people per 1000 residents were admitted for a particular diagnosis. Third, it removes potential outliers, or those patients who had been repeatedly hospitalized for the same diagnosis due to swing beds or other reasons. Finally, we could use the same number of observations to conduct surveillance for prevalence, readmission, and mortality. For this reason, we used primary diagnosis based on individual patients. If two admissions had the identical primary diagnosis for the same patient, the first one was used.

TABLE 11.3

Standard Comorbidity Prevalence and Prevalence Ratios by Race and Poverty

Disease	White			Black			Other		
	Low Pvt	Mid-Pvt	High Pvt	Low Pvt	Mid-Pvt	High Pvt	Low Pvt	Mid-Pvt	High Pvt
				Prevalence					
D1	1034	1109	1236	2395	6698	2426	1422	1756	1738
D2	330	350	387	447	627	518	226	233	229
D3	245	247	290	449	1373	564	304	438	402
D4	495	593	784	1060	3016	1546	686	846	898
D5	394	451	512	1285	2449	985	512	609	711
D6	88	82	93	154	375	165	113	114	117
D7	449	535	677	1381	3673	1638	863	1059	1343
D8	384	380	484	939	2187	1319	575	619	806
D9	372	484	593	574	1899	934	533	696	701
D10	20	28	28	99	101	49	47	51	38
				Rate Ratio					
D1	1.00	1.07	1.19	1.00	2.80	1.01	1.00	1.23	1.22
D2	1.00	1.06	1.17	1.00	1.40	1.16	1.00	1.03	1.01
D3	1.00	1.01	1.19	1.00	3.06	1.26	1.00	1.44	1.32
D4	1.00	1.20	1.58	1.00	2.84	1.46	1.00	1.23	1.31
D5	1.00	1.14	1.30	1.00	1.91	0.77	1.00	1.19	1.39
D6	1.00	0.93	1.06	1.00	2.44	1.07	1.00	1.01	1.04
D7	1.00	1.19	1.51	1.00	2.66	1.19	1.00	1.23	1.56
D8	1.00	0.99	1.26	1.00	2.33	1.40	1.00	1.08	1.40
D9	1.00	1.30	1.60	1.00	3.31	1.63	1.00	1.31	1.32
D10	1.00	1.39	1.40	1.00	1.02	0.50	1.00	1.09	0.81

Note: D1, heart disease; D2, cancer; D3, chronic lower respiratory diseases; D4, stroke; D5, unintentional injuries; D6, Alzheimer's disease; D7, diabetes; D8, kidney diseases; D9, influenza and pneumonia; D10, suicide. Low-pvt census tracts, <5% people under the poverty level; mid-pvt census tracts, >5% and <15% people under the poverty level; high-pvt census tracts, >5% people under the poverty level; low pvt, census tracts with <5% people under the poverty; mid-pvt, census tracts >5% and <15% people under the poverty line; high pvt, census tracts with >5% people under the poverty line.

We used the first admission for prevalence, readmission, and mortality assessments according to following steps:

1. Select diseases based on the primary diagnosis.
2. Save hospitalization records for each diagnosis to a separate file.
3. Sort file by master patient ID and admission date in each diagnosis file.
4. If the patient was only admitted one time, select the record as the indexed event; otherwise, select the first admission record as the indexed event.
5. Delete duplicate records by retaining the indexed event, and proceed with age- and sex-adjusted model-based estimation and presentation.

It should be pointed out that even though the assessment is patient based, a patient can still appear multiple times based on primary diagnosis. If the patient had diabetes as the primary diagnosis in one hospitalization and asthma as another, then the same patient can appear twice in two different assessment models: one for diabetes and one for asthma.

As shown in Table 11.4, except for the diagnosis of appendicitis and other appendiceal conditions, all odds ratio were significantly greater than 1 for Blacks. Rank-ordered odds ratios from more than 7 to at least 3 were found for hypertension (HP) with complications and secondary HP, schizophrenia and other psychotic disorders, diabetes mellitus with complications, acute and unspecified renal failure, alcohol-related disorders, intracranial injury, and nonspecific chest pain. Nationally, the preventable hospitalization rate for Blacks was about twice that for Whites. The number in Nebraska is about 1.96. Moving to the Other race column, we found that 33 out of 40 odds ratios were significant. However, the effects between Other and White were much smaller than those between Black and White. For comparison, we attached patient-based comorbidity diagnoses in Appendix 11A (Table 11.A3).

Sometimes, an initial surveillance leads to some other surveillance questions. A case in point is asthma. From Table 11.4, Blacks were 2.22 (95% confidence interval [CI] 2.044–2.406) times as likely as Whites to seek hospital care for asthma. But if we just restricted inpatient care through emergency department admission, we would see an odds ratio of 3.1 for asthma, or three times as likely for Blacks as for Whites. Here, all we needed to do was to restrict admission type to emergency department for a subset of data, run the same model, and inspect the odds ratios descriptively. Sometimes, one needs to compare two subsets statistically. For example, the state public health agency tends to compare disease trends over time by race. One recent trend is a decline in standard mortality due to heart disease. The 5-year mortality from 2004 to 2008 compared to from 2009 to 2013 declined by 13.4% for Whites and 10.3% for Blacks. However, for 7 years of data, it is hard to show trends. Knowing the caveat, we simply demonstrate how to conduct a model-based intertemporal assessment.

1. Break the data into 2005–2007 and 2009–2011 and leave the middle year of 2008 out.
2. Use the same model as in Table 11.4, and select those that relate to heart disease.
3. Compile two time periods of data—time 1 (2005–2007) and time 2 (2007–2011)—together with age, sex, and race in a four-way constancy table.
4. Add an interaction term to race, so that the time 1 effect is the main effect of Black versus White, and time 2 effects are from the interaction term and are relative to the main effects.
5. The model output would have the main effects of age group, sex, race, and time, and the interaction effect of time × race.

Here, we run a single model for each selected disease. The results are shown in Table 11.5. In the interpretation of the model-based estimates, the main effect of time itself needs to be considered in addition to the interaction term. As shown in Table 11.5, time 1 effects were all significant, while four out of five time 2 effects were significant. Time 2 effects suggest that Black–White disparities had widened for the four diseases: cardiac dysrhythmias, coronary atherosclerosis, nonspecific chest pain, and AMI. This small exercise had consistent mortality time effects in Chapter 7.

11.4.2 Readmission

Even though racial disparities in readmissions have been documented, there are two apparent gaps in the literature: (1) too few studies documenting readmission disparities for a host of diagnoses and (2) poverty is not often controlled when investigating readmission. Given an increased emphasis on readmissions as a quality measure, it is imperative to conduct race- and poverty-based surveillance so that public health can play some role in the hospital readmission reduction program that reduces payments to hospitals with high readmission rates for selected diseases. If racial and socioeconomic status (SES) disparities exist, then the program might penalize those with the least resources.

TABLE 11.4

Racial Disparities (Odds Ratios) in Hospitalization by Primary Diagnosis

Primary Diagnosis	Black	CI: 95%	Other	CI: 95%
All hospitalizations	1.508	1.497–1.519	1.341	1.331–1.351
Pneumonia	1.556	1.490–1.625	1.256	1.194–1.322
Osteoarthritis	2.246	2.156–2.339	0.777	0.721–0.837
Congestive heart failure	2.919	2.762–3.084	1.614	1.481–1.758
Septicemia (except in labor)	2.445	2.300–2.599	1.373	1.260–1.497
Mood disorders	1.976	1.904–2.050	**0.997**	0.950–1.045
Cardiac dysrhythmias	2.613	2.484–2.750	**1.086**	0.997–1.183
COPD and bronchiectasis	2.412	2.258–2.577	1.223	1.103–1.356
Back problems	1.689	1.601–1.783	0.837	0.776–0.902
Skin and subcutaneous tissue infections	2.087	1.967–2.214	1.321	1.228–1.421
Coronary atherosclerosis	2.411	2.297–2.530	1.262	1.174–1.356
Urinary tract infections	2.124	1.998–2.257	1.605	1.491–1.727
Nonspecific chest pain	3.228	3.074–3.390	1.887	1.768–2.013
AMI	2.552	2.408–2.705	1.375	1.263–1.497
Acute cerebrovascular disease	2.957	2.786–3.138	1.677	1.537–1.830
Diabetes mellitus with complications	3.746	3.517–3.989	2.051	1.888–2.228
Biliary tract disease	1.788	1.674–1.909	1.451	1.350–1.559
Fluid and electrolyte disorders	1.949	1.840–2.066	1.295	1.203–1.393
Asthma	2.218	2.044–2.406	1.471	1.334–1.622
Schizophrenia and other psychotic disorders	4.305	4.024–4.605	1.509	1.367–1.667
Acute and unspecified renal failure	3.723	3.474–3.991	1.746	1.566–1.947
Respiratory failure, insufficiency, arrest	2.379	2.186–2.589	1.210	1.068–1.370
Gastrointestinal hemorrhage	2.711	2.526–2.909	1.709	1.552–1.881
Intestinal obstruction without hernia	2.194	2.042–2.358	1.125	1.014–1.248
Pancreatic disorders (not diabetes)	2.901	2.676–3.145	1.718	1.554–1.901
Diverticulosis and diverticulitis	2.126	1.947–2.321	1.060	0.930–1.207
Fracture of neck of femur (hip)	2.004	1.839–2.185	**0.954**	0.819–1.111
Epilepsy, convulsions	2.149	1.974–2.339	1.294	1.165–1.437
Appendicitis and other appendiceal conditions	**0.986**	0.903–1.076	**1.033**	0.951–1.122
Alcohol-related disorders	3.547	3.242–3.880	1.655	1.471–1.862
Fracture of lower limb	2.406	2.244–2.580	1.393	1.273–1.523
Other nervous system disorders	2.102	1.919–2.303	**0.994**	0.874–1.132
Syncope	2.762	2.515–3.035	1.303	1.124–1.511
Substance-related disorders	2.717	2.407–3.066	**1.061**	0.890–1.266
Intracranial injury	3.404	3.155–3.673	1.755	1.585–1.945
HP with complications and secondary HP	7.768	7.155–8.433	3.600	3.185–4.069
Other fractures	2.675	2.467–2.899	1.305	1.156–1.472
Intestinal infection	1.385	1.261–1.521	**0.981**	0.876–1.098
Secondary malignancies	1.361	1.204–1.538	0.575	0.470–0.704
Anemia	2.791	2.528–3.081	1.689	1.474–1.936
Avoidable hospitalization	1.969	1.918–2.021	1.349	1.306–1.394

Note: All odds ratios are patient–diagnosis pairs rather than record–diagnosis pairs. If a patient had asthma hospitalizations, only one is counted. If a confidence interval crosses 1, it is not statistically significant. All race categories are derived from data linkage with other data sources. All odds ratios are in reference to Whites. Odds ratios in bold were not significant.

FIGURE 6.1 Poverty MOE to poverty rate ratio by census tracts in Nebraska: 2007–2011. The rate is the ratio of MOE divided by the poverty rate within each census tract. If the rate is greater than 100, the MOE is greater than the estimated poverty rate. The box below the state map is Omaha and its adjacent area. (From 2007–2011 ACS 5-year data for poverty.)

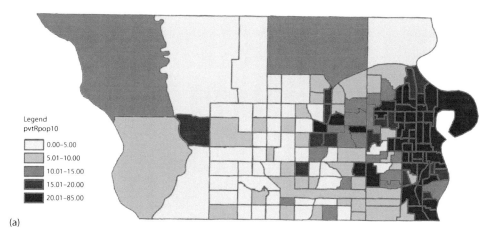

(a)

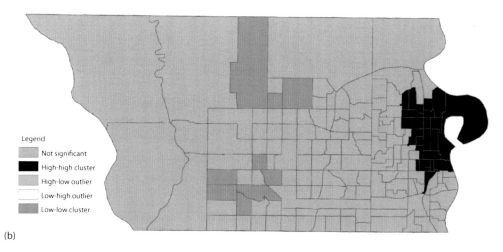

(b)

FIGURE 6.2 (a) Poverty rates in Douglas County (Omaha): 2006–2010. (b) Poverty rate clusters in Douglas County (LISA): 2006–2010.

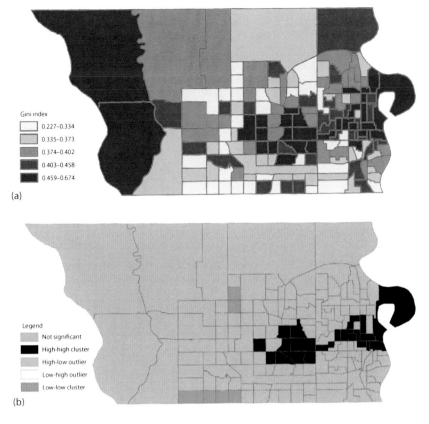

FIGURE 6.3 (a) Gini indices in Douglas County: 2006–2010. (b) Gini index clusters in Douglas County (LISA): 2006–2010.

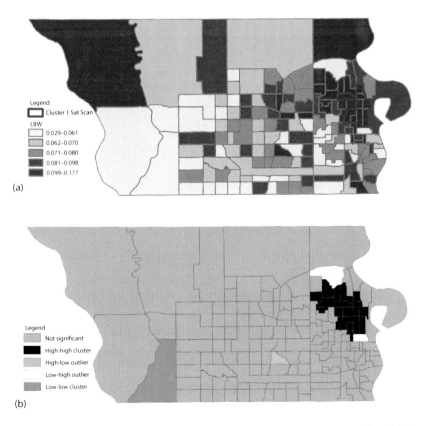

FIGURE 6.4 (a) Low-birth-weight rate in Douglas County (Omaha): 2005–2011. (b) Low-birth-weight rate clusters (LISA) in Douglas County: 2005–2011.

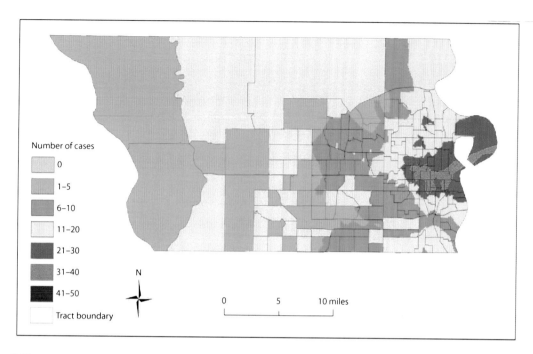

FIGURE 6.5 Smoothed HIV cases in Douglas County: 1983–2010.

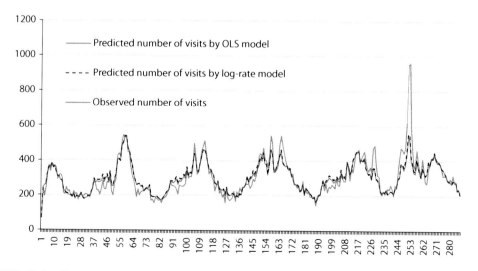

FIGURE 12.4 Observed and predicted ILI ED visits for age 5 or younger.

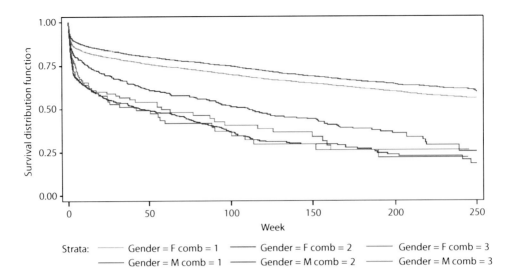

FIGURE 16.1 Kaplan–Meier curves by sex and comorbidities.

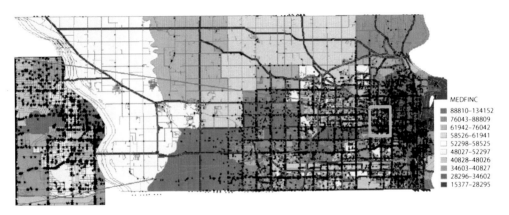

FIGURE 18.1 Nonhighway traffic crashes and median family incomes by census tracts in Douglas County, Nebraska.

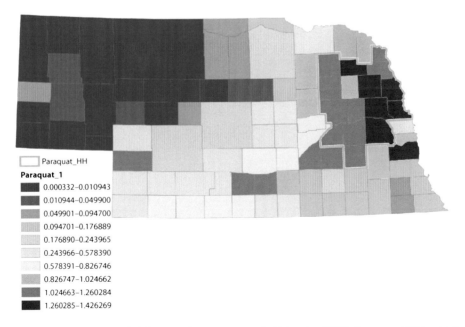

FIGURE 20.3 Paraquat usage in kilograms by census tract in Nebraska: 2005. Paraquat_HH is a high-value cluster based on the LISA adjusted for multiple testings. (Paraquat usage concentrations are from Wan, N., *Applied Geography* 56: 99–106, 2015.)

TABLE 11.5
Black–White Disparities in Heart Disease–Related Diagnoses

Heart Disease–Related Diagnoses	Main Effect: T1		T2 Relative to T1	
	OR	95% CI	OR	95% CI
Congestive heart failure	2.385	2.182–2.606	0.965	0.895–1.040
Cardiac dysrhythmias	1.818	1.665–1.985	1.217	1.138–1.303
Coronary atherosclerosis	1.743	1.616–1.879	1.119	1.040–1.203
Nonspecific chest pain	2.294	2.126–2.475	1.294	1.204–1.390
AMI	1.741	1.567–1.934	1.196	1.105–1.294

Note: T1, 2005–2007; T2, 2009–2011. The main effect is the referent, and T2 effects are relative to T1 or the interaction effect.

TABLE 11.6
Black vs. White Readmission with and without Controlling for Poverty

Diagnosis	Not Accounting for Poverty		Accounting for Poverty	
	OR	95% CI	OR	95% CI
Mood disorders	1.195	1.084–1.319	1.157	1.047–1.280
Cardiac dysrhythmias	0.868	0.746–1.010	0.852	0.731–0.993
Nonspecific chest pain	1.351	1.162–1.570	1.308	1.120–1.527
AMI	0.825	0.662–1.029	0.782	0.624–0.978
Schizophrenia and other psychotic disorders	1.304	1.121–1.517	1.242	1.064–1.450
Gastrointestinal hemorrhage	1.200	1.005–1.433	1.147	0.955–1.378
Pancreatic disorders (not diabetes)	1.235	1.000–1.525	1.184	0.953–1.470
Alcohol-related disorders	1.542	1.217–1.954	1.568	1.226–2.006
Syncope	1.419	1.086–1.854	1.357	1.030–1.788
Intracranial injury	0.545	0.389–0.763	0.552	0.393–0.775
Avoidable hospitalization	1.078	1.001–1.162	1.038	0.961–1.121
Complication of device	1.151	1.015–1.307	1.120	0.984–1.276

Note: OR, odds ratios from a logistic regression controlling for age and sex. Only significant odds ratios are shown among 42 diseases and conditions.

We ran two sets of models of whether a patient was readmitted within 30 days of discharge. Model I generates estimates based on the 42 diseases and conditions used in Chapter 8 (Table 11.6, columns 1 and 2). Model II generates model-based estimates from the same 42 diseases and conditions by adding an additional control of neighborhood poverty level (Table 11.6, columns 1 and 2). Only significant odds ratios from either models were retained. In total, 12 diseases had significant odds ratios from either model, and they were from 24 model runs, or 12 from each model.

The results without controlling for poverty showed that 10 diseases had a significant Black–White difference in readmissions. Black patients were more likely to be readmitted to the hospital within 30 days for 9 out of 10 significant results. The widest disparities were in alcohol-related disorders, syncope, and nonspecific chest pain, where the latter two may relate to heart disease. Another cluster of diseases were around mental health, as schizophrenia and other psychotic disorders, mood disorders, and alcohol-related disorders could all be caused by mental health, for which

Blacks were more likely to experience readmissions. The only significant disease with reduced odds was intracranial injury (mostly from traumatic brain injury). It is also worth pointing out that preventable hospitalizations tend to be less likely for Blacks than Whites in readmissions.

We mentioned in Chapter 8 that neighborhood poverty plays a significant role in readmission. Here, poverty also mediates through race. When the poverty variable is introduced into the model, which had five levels (<5%, 5%–10%, 10%–15%, 15%–20%, >20%), only eight disease conditions were significant. The top three conditions in Model I that had the widest disparities remain the top three in Model II. The most significantly elevated risks in Model I had some reduction in risk, four of them from significant to nonsignificant (confidence intervals cross 1). In particular, gastrointestinal hemorrhage, pancreatic disorders (not diabetes), preventable hospitalization conditions, and complication of device were all from elevated risk to no difference between Blacks and Whites. In addition, two diseases—cardiac dysrhythmias and AMI—were from nonsignificant to significant, and both had a reduced risk of readmission. It is worth noting that preventable hospitalization became nonsignificant, suggesting that a significant Black–White difference in readmission for ambulatory care sensitive conditions (ACSCs) was mainly due to neighborhood SES. In sum, accounting for neighborhood poverty level attenuates Black–White disparities in readmission.

A major limitation of the surveillance-based readmission analysis is that it lacks further controls. Obviously, we should control disease severity and comorbidities, but different primary diagnoses require different sets of comorbidities, and it is impractical to set up different comorbidity sets for different diagnoses with comparable prognoses. Likewise, we did not control for type of hospitals. It is difficult to control for hospital factors, as some hospitals might not have admitted any particular disease-specific patient due to hospital-level factors. As we automatically went through each disease, we would have to devise some hospital-level controls that may not work for all types of diseases. In this regard, we treated all hospitals in the same way, which is clearly not the case. Another control that we could have is surgery procedure (Girotti et al., 2014). However, including it would lead us deeper into health care delivery that public health cannot do too much. We should also remember that poverty level is an area adjustment. This implies that poor SES is associated with less optimal care, or poor SES may lead to few options for home care that predispose the individual patient to readmission. Despite these caveats, we still believe that surveillance should not have many controls, because additional control points are also the entry points for public health interventions.

11.4.3 Mortality

We can assess mortality in the same way we do readmission. In addition to all the limitations for the readmission assessment, there were far fewer death cases in the sample. We restrict our analysis to White and Black patients only due to the small number of cases. To be cautious, we only used in-hospital mortality or those whose discharge status was "expired." We surveyed the 40 diagnoses and reported only those with significant differences between Black and White patients.

We first ran a logistic regression for in-hospital mortality with a dichotomous variable of alive or died (Table 11.7, first two columns). We found that except syncope, all other significant disease conditions were at least 30% less likely to die in the hospital for Black patients than White patients. Even though disease severity could not be controlled for, the results in general suggested that Black patients might have less severe symptoms for the primary diagnosis than White patients, because a priori, Blacks were more likely to have higher age- and sex-adjusted mortality from death records for a host of diseases.

To reconcile the logistic regression approach of in-hospital mortality and the higher overall mortality for many diseases from vital statistics reports, we also ran a log-rate model that requires at-risk population (Table 11.7, last two columns). Here, White in-hospital deaths are modeled by the White at-risk population, and Black in-hospital deaths are modeled by the Black at-risk population; the odds ratio between the two were reported. Since nine age groups, two sex groups, and two race groups require 36 cells in the contingency table, the number of zero cells becomes a concern.

TABLE 11.7

Logistic and Log-Rate Regression Results for Black vs. White In-Hospital Mortality

	Logistic Regression		Log-Rate Model	
	OR	95% CI	OR	95% CI
Pneumonia	0.54	0.39–0.75	1.14	0.83–1.56
Congestive heart failure	0.38	0.22–0.63	1.01	0.60–1.68
Acute cerebrovascular disease	0.68	0.54–0.87	1.29	0.88–1.88
Acute and unspecified renal failure	0.34	0.22–0.54	NS	NS
Respiratory failure, insufficiency, arrest	0.71	0.55–0.91	1.57	1–2.45
Fracture of neck of femur (hip)	0.31	0.11–0.83	1.5	0.56–4.05
Syncope	4.43	1.18–16.60	NS	NS
Intracranial injury	0.48	0.34–0.67	1.82	1.31–2.53
Avoidable hospitalization	0.48	0.36–0.64	1.22	0.92–1.62

Note: Both logistic and log-rate models are age and sex adjusted. The at-risk population for the log-rate model is from the 2010 census. NS, no stable estimates.

Consequently, both acute and unspecified renal failure and syncope could not get stable parameter estimates. Other than these two, Blacks had higher mortality rates than Whites for all listed diagnoses.

11.5 CASE STUDY: RACIAL DISPARITY IN REHABILITATION AMONG ELDERLY AMI PATIENTS

Cardiac rehabilitation (CR) is a Class I recommendation for the management of AMI in many national and international societies, such as the American College of Cardiology and the American Heart Association (AHA). CR intends to help patients regain autonomy and improve regular physical activities. CR is a secondary prevention program to detect disease progression and modify behavioral risk factors. CR interventions include exercise training, physical activity counseling, tobacco cessation, nutritional counseling, weight management, aggressive coronary risk factor management, and psychosocial counseling. It is recommended that all hospitalized patients have CR if they have a qualifying cardiovascular event or diagnosis, such as AMI, coronary artery bypass graft (CABG) surgery, or percutaneous coronary intervention (PCI). From the rural–urban difference case study (Chapter 17), we found that CR is the strongest predictor of 1-year mortality for AMI patients, but only 38% of AMI patients had CR in outpatient settings. In this case study, we wanted to examine factors associated with CR among AMI patients. Based on the sample of AMI patients in Chapter 10, we added a race variable and treatment procedure variables, such as PCI and CABG surgery. Note that most AMI patients are elderly, so we restricted the sample to those age 65 and over. In this way, all the patients in the sample had either commercial insurance, Medicare, or Medicaid.

Baseline characteristics of patients by rehabilitation status are presented in Table 11.8. Patients having CR were typically younger, male, and White, and they were less likely to come from poorer neighborhoods. Patients having CR were more likely to have PCI and CABG procedures during the hospital stay. Patients were less likely to have CR if they had COPD, renal disease, cerebrovascular disease, and cancer. Finally, urban patients and rural hospital bypassers were more likely to have CR.

Since our main interest is racial disparities, and PCI and CABG patients were more likely to have CR, we also cross-tabulated race by these two procedures. The percentages of patients having

TABLE 11.8
AMI Outpatient Rehabilitation Rate among Age 65+

	No Rehab	With Rehab	N	p-Value
Overall	61.18%	38.82%	7646	Chi-square
Sociedemographics				
Age 65–69	45.68%	54.32%	1517	<0.0001
Age 70–74	48.63%	51.37%	1458	
Age 75–79	55.36%	44.64%	1391	
Age 80–84	65.23%	34.77%	1412	
Age 85–89	79.24%	20.76%	1079	
Age 90+	92.52%	7.48%	789	
Sex: Female	66.82%	33.18%	3626	<0.0001
Sex: Male	56.09%	43.91%	4020	
Race: White	60.55%	39.45%	7037	<0.0001
Race: Black	67.40%	32.60%	457	
Race: other	71.71%	28.29%	152	
Poverty (0%–4.99%)	53.16%	46.84%	1185	0.0004
Poverty (5%–9.99%)	59.06%	40.94%	2616	
Poverty (10%–14.99%)	64.41%	35.59%	2245	
Poverty (15%–19.99%)	63.21%	36.79%	685	
Poverty (20% or more)	68.20%	31.80%	915	
Treatment Procedures				
PCI (no)	71.52%	28.48%	4631	<0.0001
PCI (yes)	45.32%	54.69%	3015	
CABG (no)	64.63%	35.37%	6834	<0.0001
CABG (yes)	32.14%	67.86%	812	
Comorbidity				
COPD (no)	59.44%	40.56%	6331	<0.0001
COPD (yes)	69.58%	30.42%	1315	
Diabetes (no)	61.01%	38.99%	5886	0.5699
Diabetes (yes)	61.76%	38.24%	1760	
Renal disease (no)	59.66%	40.34%	6753	<0.0001
Renal disease (yes)	72.68%	27.32%	893	
HP (no)	63.82%	36.18%	3090	<0.0001
HP (yes)	59.39%	40.61%	4556	
Cerebrovascular disease (no)	60.39%	39.61%	7183	<0.0001
Cerebrovascular disease (yes)	73.43%	26.57%	463	
Cancer (no)	60.64%	39.36%	7389	<0.0001
Cancer (yes)	76.65%	23.35%	257	
Locational Factors				
Rural	62.65%	37.35%	4554	0.0014
Urban	59.02%	40.98%	3092	
Not a rural bypasser	63.33%	36.67%	5828	<0.0001
Rural bypasser	54.29%	45.71%	1818	

Note: A rural bypasser is a rural resident who went to an urban hospital for care.

CABG were 10.63, 10.5, and 10.53, respectively, for Whites, Blacks, and Other races; the percentages of patients having PCI were 39.35, 39.82, and 42.11, respectively. These percentages indicate that Blacks were equally likely to have PCI or CABG procedures as White patients. Based on the facts that (1) Black patients were less likely to have CR and (2) PCI and CABG patients were more likely to have CR, Black patients having either PCI or CABG were much more less likely to have CR.

Given the importance of hospital procedures, we intend to run two models, one without procedures and the other with procedures. Preliminary results suggest that CABG had some correlation with other variables, such as rural bypassers. We therefore did not include it in the final analysis. Table 11.9 displays the final multivariate analysis using the backward-selecting method. The multivariate results essentially retained univariate results. In particular, Black patients were 38% less

TABLE 11.9

Logistic Regression on Factors Associated with Having AMI Rehabilitation

	Model I without PCI Indicator		Model I with PCI Indicator	
	OR	95% CI	OR	95% CI
Age (ref age 65–69)				
Age 70–74	0.915	0.789–1.061	0.915	0.828–1.120
Age 75–79	0.712	0.612–0.829	0.712	0.677–0.921
Age 80–84	0.486	0.416–0.568	0.486	0.459–0.629
Age 85–89	0.232	0.193–0.279	0.232	0.220–0.320
Age 90+	0.074	0.056–0.099	0.074	0.070–0.126
Sex (male vs. female)	1.284	1.160–1.422	1.250	1.127–1.386
Race (ref White)				
Black	0.624	0.503–0.774	0.652	0.524–0.811
Other races	0.546	0.376–0.793	0.549	0.376–0.803
Poverty (ref <5%)				
Poverty (5%–9.99%)	0.869	0.745–1.013	0.851	0.730–0.992
Poverty (10%–14.99%)	0.706	0.600–0.83	0.688	0.587–0.807
Poverty (15%–19.99%)	0.734	0.595–0.905	0.720	0.582–0.890
Poverty (20% or more)	0.562	0.462–0.683	0.570	0.468–0.695
PCI (ref no)			2.156	1.944–2.391
Comorbidity (ref no)				
COPD	0.563	0.491–0.645	0.592	0.516–0.680
Diabetes	0.862	0.765–0.970	0.883	0.783–0.996
Renal disease	0.611	0.517–0.724	0.678	0.572–0.804
HP	1.268	1.143–1.406	1.207	1.087–1.341
Cerebrovascular disease	0.549	0.438–0.687	0.609	0.485–0.765
Cancer	0.434	0.319–0.591	0.477	0.349–0.651
Location Factors				
Rural bypassers (yes vs. no)	1.494	1.308–1.707	1.319	1.141–1.525
Urban vs. rural	1.32	1.168–1.493		
Distance to hospital			0.998	0.997–1.000

Note: Positive AMI rehabilitation is determined by searching for two CPT codes (93797 and 93798) in the outpatient visits. If one of the codes is found, rehab is yes; otherwise, it is no. PCIs are defined according to the Dartmouth Atlas: International Classification of Diseases (ICD) procedure codes 00.66, 36.06, 36.07, and 36.09.

likely to have CR than White patients, and patients of Other races fared even worse at 45% less likely. In Model II, we include PCI and found that having PCI increased the odds of rehabilitation by more than 100%. In the meantime, what used to be a significant rural–urban place of residence factor dropped from the final model, suggesting that having PCI explained some rural–urban differences in rehabilitation. The introduction of PCI in Model II also reduced the odds ratio for the bypasser effect from 49% (odds ratio [OR] 1.494, 95% CI 1.308–1.707) to 32% (OR 1.319, 95% CI 1.141–1.525) more likely.

In this case study, we assessed racial disparities in CR. We concluded that reduced odds in rehabilitation for Black and Other race patients were independent of many risk factors, such as age, sex, SES, in-hospital procedures, comorbidities, and place of residence. A previous study by Johnson et al. (2004) in the United Kingdom showed that patient's educational level is an important predictor of rehabilitation; those having an education beyond high school were much more likely to enroll in a CR program. In a U.S. multicenter perspective study, those avoiding health care because of cost were also less likely to participate in CR (Parashar et al., 2012), whereas Whites and those with a high school diploma were more likely to participate in CR at 6 months after AMI. We used neighborhood poverty as a proxy measure, and it showed that area poverty level is an independent factor associated with CR, which is consistent with Paarashar et al.'s findings. Another important factor is underreferral of CR among minorities at discharge (Aragam et al., 2011); however, an early and smaller sample study of the Get With the Guidelines program on post-AMI referral and enrollment into a CR program showed that underenrollment was a significant barrier (Mazzini et al., 2008). Although we could not directly test referral issues, information from treatment suggests that minority patients were either underenrolled or underreferred. Cultural differences could be a reason for lack of referral and less awareness of CR benefits. A study by Barber et al. (2001) found that the ability to speak English is a strong predictor of lack of CR referrals. Since many Hispanic patients and some Black refugee patients do not speak English in Nebraska, language barriers could also be a factor (Gregory et al., 2006; Sanderson et al., 2007).

Although it was not our main interest, locational factors also play some role in having CR care. Urban residents were 32% more likely to have CR, while rural bypassers were 49% more likely. Since rural bypassers who sought AMI care were already in big urban areas, it was a lot easier for them to stay longer to be enrolled into a CR program. There is certainly some interaction between place of residence and procedures; as we found once PCI was introduced into a multivariate model, the place of residence factor become nonsignificant. Likewise, we observed a quite substantial reduction in odds ratio for the bypasser effect when PCI was introduced.

11.6   CHAPTER SUMMARY AND CONCLUDING REMARKS

Racial disparity surveillance should be at the forefront of hospital-based surveillance efforts. In Nebraska, however, race was not included in most HDD. For this reason, we had to derive the race variable from multiple public health databases and government records. This was part of the data infrastructure building effort that involved a half-dozen public health datasets and more than 1.5 million records. Without this effort, we would not be able to conduct any racial disparity surveillance based on HDD. As the Affordable Care Act is fully implemented, race information will be collected, so that we will not have to conduct such a large data infrastructure project again. Nevertheless, our project helped to establish baseline data by race for the whole state of Nebraska, beginning in 2005.

Another consideration of race-based assessment is program needs. Although many disease-oriented programs, such as cardiovascular disease (CVD), cancer, and injury, are interested in both incidence and prevalence, the state health disparity program was more interested in prevalence. Questions such as racial disparity in having diabetes or HP are important and conceptually more intuitive for minority health disparity reports. Although some of these questions can be obtained from the Behavioral Risk Factor Surveillance System (BRFSS), the BRFSS only had limited disease-related questions, and the BRFSS is very costly if a state wants to add a minority panel to produce reliable race-specific estimates for local health departments, which are about the size of three or five counties combined, in most cases. Such a geographic unit would not permit small-area analysis and potentially small-area-targeted interventions. For this reason, we adopted the approach of patient-based estimates rather than record-based estimates when engaging disparity surveillance.

Besides a small degree of uncertainty about race linkage that we have not yet been able to independently evaluate, we devoted a fair amount of space to evaluate at-risk population uncertainties due to multiple races from the 2010 census. Here, we wanted to produce conservative estimates for incidence rates for Blacks by pooling all the two or more races into Blacks. In so doing, we inflated the rates for both Whites and Other races. Since the population size for Other races is not large, the potential inflation could be large.

In assessing rates between Whites and Blacks, one needs to establish some criteria. First, the racial profiles between our hospitalization data and census population data should be comparable, and this was true after adjusting the multiracial group in the census data. Second, hospital-derived incidence and prevalence should be much smaller than survey-derived rates. Since we do not have clinically validated estimates within the state, we used the National Health Interview Survey data reported by Beckles and Chou (2013). Using diabetes in panel B of Table 11.4 as an example, the incidence rates for Whites and Blacks were, on average, 1.25% and 4.23%. The corresponding prevalence rates from the National Health Interview Survey were 6% and 10.9% based on the 2006 estimates. Since the incidence rate in the hospital data is higher than the prevalence rate, the number we had was at least reasonable. Third, disease-specific hospitalization disparities may help in the evaluation process. Diabetes is an ACSC, and Blacks tend to be more often hospitalized for it than Whites. For this reason, the hospital-based relative rate between Whites and Blacks should be much higher than the survey-based rates. The relative hospitalization rate was 230% higher among Blacks than among Whites. Nationally, diabetes prevalence ranged from 82% to 66% higher among Blacks than among Whites between 2006 and 2010. Finally, there are many hospital-based research findings that we could tap into. However, this would require a metadata analysis, as most research-based study designs tend to include many more control variables than what we had in this chapter. We did not calculate age-specific disease rates, and we suspect that some very high rates for Blacks might be due to the age effect. For instance, the national study by Crosby et al. (2013) showed that age-specific suicide rates for non-Hispanic Blacks were the highest between age 15 and 29 years. If the Nebraska rates followed the same distribution, they would contribute significantly to the higher standard suicide rates among Blacks.

Race tends to interact with many disparity dimensions. For this reason, there are more interactions that need to be considered in the racial disparity surveillance. Our conservative calculations of standard hospitalization rates showed that most diagnoses among Blacks doubled those for Whites. Decomposing race effects by neighborhood SES would enlarge the Black–White disparity for the midpoverty group and the White–Other races disparities for the high-poverty groups. Our model-based estimates showed that Blacks were not only more likely to be admitted, but also more likely to be readmitted to hospitals. Blacks also had higher mortality rates than Whites in major diseases, but they were less likely to die in the hospital than White patients.

At this point, we want to point out that mental health–related diseases and conditions persistently showed SES gradients. In Chapter 8, we found SES gradients with mental health–related conditions. In Chapter 9, we found clear SES gradients in suicide. In Chapter 10, we found SES gradients in self-inflict injury, and in this chapter, we again found SES gradients in suicide and mental health–related hospitalization. The founder of socioepidemiology, Durkheim showed that suicide was associated with social cohesion and suggested social intervention (Van Poppel and Day, 1996). After 100 years, we witness the same stubborn problem with the SES gradients, suggesting lack of social cohesion in high-poverty neighborhoods.

In our case study of CR for AMI patients, we found that Blacks and Other races were much less likely to participate in an outpatient CR program than Whites. This effect was independent of many risk factors, such as age, sex, SES, in-hospital procedures, comorbidities, and place of residence.

APPENDIX 11A: RACE ADJUSTMENT STRATEGIES USING CENSUS DATA

We noted that the Asian and American Indian populations were too small to have their own categories for stable incidence rates; we therefore grouped them with Other in the following racial disparity assessment. We used census 100% summary files or census DEC_10_SF1_P12 files A–G series to derive age group and sex for each single-race and multiple-race category. In particular, we used DEC_10_SF1_P12A for White alone and DEC_10_SF1_P12B for Black alone to derive single-race-alone at-risk populations for Blacks and Whites. We used census tables DEC_10_SF1_P12C to DEC_10_SF1_P12F (some other race alone) to combine all other single-race categories. We them used DEC_10_SF1_P12G (two or more races) to create the "two or more races" category that can be combined with Other (design A) or Blacks (design B) in Figure 11.1. Design A makes no attempt to adjust the at-risk population, in which Whites and Blacks in the HDD file are assumed to align nicely with single race from the census files. This design would inflate the race-specific hospitalization rate because it does not take two or more races into consideration in the Black or White at-risk population. In design B, we moved "two or more races," which accounted for 2.2% of the state population, to the Black population group. This alternative design assumes that all people who checked two or more races had at least one race being Black. This oversimplification will definitely deflate the Black hospitalization rate, but it will generate conservative estimates for Blacks.

In Table 11.A1, we assess denominators (at-risk population) by sex-specific age distributions for three racial groups: Whites, Blacks, and Other races, where Other races include other single races and two or more races. The three groups represent White alone, Black alone, and Other

TABLE 11.A1
Population Age Distributions by Race According to the 2010 Census

| | White | | Black | | Other | | | |
| | Single Race | | Single Race | | Single Race | | 2 or More Races | |
Percent	Male	Female	Male	Female	Male	Female	Male	Female
N	777,280	795,558	42,138	40,747	67,246	63,862	19,632	19,878
0–14	19.86	18.46	27.04	26.67	30.30	31.07	51.26	49.29
15–25	13.95	13.05	18.19	17.34	18.80	17.82	18.07	18.35
25–34	13.29	12.60	15.76	14.95	18.87	17.77	11.80	12.59
35–44	12.20	11.68	12.95	12.52	15.06	14.63	7.86	7.82
45–54	15.01	14.71	12.87	12.61	9.20	9.37	5.80	6.03
55–64	12.60	12.62	7.72	8.28	4.97	5.68	3.12	3.39
65–74	7.09	7.72	3.49	4.31	1.95	2.36	1.39	1.51
75–84	4.40	5.91	1.65	2.46	0.66	1.03	0.47	0.70
85+	1.61	3.26	0.33	0.87	0.18	0.28	0.23	0.32
Total	100.00	100.00	100.00	100.00	100.00	100.00	100.00	100.00
% of total N	42.56	43.56	2.31	2.23	3.68	3.50	1.07	1.09

Note: Blacks and Whites are based on one race only. Other, all other single-race groups and the two or more race group. Total N (Nebraska population) was 1,826,341 in 2010.

races that include Other race alone and two or more races. Together, they represented the total population, 1,826,341 in 2010. The last row has their distributions, with White males and females being 42.56% and 43.56%, respectively. Black males and females account for 2.31% and 2.23%, and Other races account for 9.3%. Here the 9.3% of Other races include 2.2% two or more races, 1.07% for males and 1.09% for females. As the table shows, racial minorities are heavily distributed in younger age groups, while Whites are more concentrated in older age groups. In all racial groups by sex, more than 30%–60% of populations were younger than 25, with age distributions progressively younger as we move from White to other races. The two or more racial groups under the Other category had the youngest population, with about 50% in the age 0–14 group. On the other hand, age 75+ accounted for <10% of populations in all racial groups, with Blacks and other racial groups being <4%. These age distributions suggest Whites were expected to have more hospitalizations due to the aging population.

Table 11.A2 lists patient admission distribution by age and race. Let's start with age distributions. Consistent with the notion of vulnerable populations of the very young and very old, hospital admissions were dominant by the two tails. In all racial groups by sex, patients younger than age 25 ranged from 24% to 52%, with age-specific admissions progressively toward higher percentages as we move from White to Other races. In the age group 0–14, hospital admissions for males in the Other race category accounted for 45.7% of this race group, and males tended to have 80% more hospital visits than females in this age group, suggesting predominant Hispanic patients in this category. On the other end of age spectrum, those age 75+ accounted for more than 27% male admissions, compared to 6% of the male population for Whites; the corresponding numbers for White females are 26% and 9% for hospital admissions and population, respectively. For Blacks and the Other race group, the gaps between age distributions of patient admissions and population are much smaller, but the ratios of the patient and population age distributions are similar to those of Whites.

We stress that the Blacks in HDD had 8.6%, or nearly twice the percentage of Blacks in the Nebraska population, while the Other race in HDD had 6.6% compared to 9.1% of other population.

TABLE 11.A2

Sex-Specific Patient Age Distributions by Race Groups

Age Group	White		Black		Other	
	Male	Female	Male	Female	Male	Female
N (1,372,852)	472,545	688,880	48,436	69,561	33,319	60,111
0–14	20.71	12.90	18.11	11.40	45.70	23.24
15–25	3.50	8.97	7.23	14.78	6.64	20.79
25–34	3.37	15.09	8.55	13.94	6.68	23.34
35–44	5.12	7.56	10.53	10.69	8.82	10.42
45–54	9.78	8.11	17.29	12.99	11.53	7.59
55–64	13.66	9.42	14.91	11.68	9.54	5.52
65–74	16.62	11.62	11.81	10.46	5.93	4.03
75–84	18.35	15.06	8.97	9.38	3.76	3.40
85+	8.91	11.26	2.61	4.69	1.40	1.67
	100.00	100.00	100.00	100.00	100.00	100.00
% of N (1,372,852)	34.4	50.2	3.5	5.1	2.4	4.4

Note: More than 92% of the race records were derived from other data sources. The total number of in-state hospital records (N) was 1,372,852 from 2005 to 2011.

TABLE 11.A3

Standard Comorbidity Incidence by Sex and Race for Common Causes of Death

Comorbidities	White		Black		Other	
	Male	Female	Male	Female	Male	Female
Panel A						
Heart disease	3045	2181	8599	7468	2822	2673
Cancer	806	621	1143	1026	379	295
Chronic lower respiratory diseases	451	358	1423	1337	506	450
Stroke	1334	1302	3671	4557	1167	1443
Unintentional injuries	555	523	1979	1593	644	568
Alzheimer's disease	133	123	384	394	97	139
Diabetes	1363	1214	4705	5492	1874	2174
Kidney diseases	942	630	3524	3030	1152	1052
Influenza and pneumonia	781	621	1719	1663	713	706
Suicide	32	21	121	53	42	32
Panel B						
Heart disease	3045	2181	7162	6300	3456	3305
Cancer	806	621	950	853	463	361
Chronic lower respiratory diseases	451	358	1196	1129	622	555
Stroke	1334	1302	3004	3683	1433	1767
Unintentional injuries	555	523	1563	1301	791	709
Alzheimer's disease	133	123	317	335	124	175
Diabetes	1363	1214	3901	4559	2267	2643
Kidney diseases	942	630	2911	2538	1401	1297
Influenza and pneumonia	781	621	1390	1363	891	889
Suicide	32	21	94	41	51	39
Rate Ratios of A to B						
Heart disease	1.00	1.00	0.83	0.84	1.22	1.24
Cancer	1.00	1.00	0.83	0.83	1.22	1.22
Chronic lower respiratory diseases	1.00	1.00	0.84	0.84	1.23	1.23
Stroke	1.00	1.00	0.82	0.81	1.23	1.22
Unintentional injuries	1.00	1.00	0.79	0.82	1.23	1.25
Alzheimer's disease	1.00	1.00	0.83	0.85	1.28	1.26
Diabetes	1.00	1.00	0.83	0.83	1.21	1.22
Kidney diseases	1.00	1.00	0.83	0.84	1.22	1.23
Influenza and pneumonia	1.00	1.00	0.81	0.82	1.25	1.26
Suicide	1.00	1.00	0.77	0.78	1.21	1.24

Note: Comorbidities are standardized per 100,000 by searching up to 10 diagnoses. Blacks and Whites in panel A are based on one race only. Blacks in panel B include Black alone and two or more race categories.

Hence, moving the "two or more races" group to Black in the at-risk population would move Black and Other at-risk populations closer to the corresponding patient visit distributions. Although it would be more reasonable to allocate some of the 2.2% of two or more race groups to White, some to Black, and some to the Other racial category, in the absence of a quick and easy method, we preferred design B in Figure 11.1.

APPENDIX 11B: COMPANION TABLES FOR TABLES 11.2 THROUGH 11.4

Table 11.A3 is a companion for Table 11.2. It compares incidence rates with (panel B) and without (panel A) two or more race adjustments. Without the adjustment, incidence rates among Blacks doubled those for Whites in most diseases. White males had higher rates for heart disease and cancer, while Other males had higher rates for stroke and COPD. The general conclusion is the same as in Table 11.4.

Table 11.B1 is a companion for Table 11.3. The results from incidence rates in this table were broadly consistent with those from prevalence (Table 11.3). For the Whites, poor neighborhoods tended to be associated with higher comorbidity rates. All rate ratios were above 1 except Alzheimer's disease for midpoverty. The general pattern for the Other race group is similar to that of Whites.

Table 11.B2 is a companion for Table 11.4. It simply reports results based on records rather than patient.

Again, the results were broadly consistent with those reported from Table 11.4.

TABLE 11.B1

Standard Comorbidity Incidence Rates by Race and Poverty

Disease	White			Black			Other		
	Low Pvt	Mid-Pvt	High Pvt	Low Pvt	Mid-Pvt	High Pvt	Low Pvt	Mid-Pvt	High Pvt
					Incidence				
D1	2,331	2,529	2,942	5,888	13,806	4,953	2,720	3,462	3,526
D2	650	690	774	913	875	908	494	353	428
D3	386	385	462	1,063	2,145	923	469	636	572
D4	1,019	1,241	1,781	2,262	5,309	2,969	1,251	1,476	1,773
D5	483	549	624	1,713	2,982	1,005	610	723	839
D6	132	123	141	343	601	267	118	168	157
D7	995	1,253	1,658	3,201	7,211	3,571	1,774	2,150	2,945
D8	713	723	959	2,126	3,742	2,531	1,088	1,142	1,615
D9	526	690	868	1,023	2,454	1,136	733	907	913
D10	21	29	28	101	105	51	47	54	39
					Rate Ratios				
D1	1.00	1.08	1.26	1.00	2.34	0.84	1.00	1.27	1.30
D2	1.00	1.06	1.19	1.00	0.96	0.99	1.00	0.71	0.87
D3	1.00	1.00	1.20	1.00	2.02	0.87	1.00	1.36	1.22
D4	1.00	1.22	1.75	1.00	2.35	1.31	1.00	1.18	1.42
D5	1.00	1.14	1.29	1.00	1.74	0.59	1.00	1.18	1.38
D6	1.00	0.93	1.07	1.00	1.75	0.78	1.00	1.42	1.33
D7	1.00	1.26	1.67	1.00	2.25	1.12	1.00	1.21	1.66
D8	1.00	1.01	1.34	1.00	1.76	1.19	1.00	1.05	1.48
D9	1.00	1.31	1.65	1.00	2.40	1.11	1.00	1.24	1.25
D10	1.00	1.38	1.38	1.00	1.04	0.50	1.00	1.15	0.84

Note: Companion table for Table 11.3. D1, heart disease; D2, cancer; D3, chronic lower respiratory diseases; D4, stroke; D5, unintentional injuries; D6, Alzheimer's disease; D7, diabetes; D8, kidney diseases; D9, influenza and pneumonia; D10, suicide. Low-pvt census tracts, <5% people under the poverty level; mid-pvt census tracts, >5% and <15% people under the poverty level; high-pvt census tracts, >5% people under the poverty level.

TABLE 11.B2
Racial Disparities (Odds Ratios) in Comorbidities

Comorbidities	Black	CI: 95%	Other	CI: 95%
All hospitalizations	1.508	1.497–1.519	1.341	1.331–1.351
Pneumonia	1.926	1.877–1.975	1.294	1.252–1.338
Osteoarthritis	2.542	2.48–2.605	1.007	**0.964–1.051**
Congestive heart failure	2.961	2.897–3.026	1.575	1.521–1.63
Septicemia (except in labor)	2.311	2.229–2.396	1.394	1.329–1.462
Mood disorders	2.562	2.521–2.603	1.090	1.066–1.115
Cardiac dysrhythmias	2.533	2.485–2.581	1.210	1.174–1.247
COPD and bronchiectasis	2.395	2.344–2.447	1.068	1.03–1.107
Back problems	2.305	2.244–2.369	1.077	1.035–1.12
Skin and subcutaneous tissue infections	2.285	2.206–2.366	1.345	1.285–1.407
Coronary atherosclerosis	2.718	2.672–2.766	1.363	1.326–1.4
Urinary tract infections	2.695	2.631–2.761	1.642	1.589–1.697
Nonspecific chest pain	3.710	3.59–3.834	1.943	1.856–2.035
AMI	2.718	2.6–2.84	1.488	1.394–1.588
Acute cerebrovascular disease	2.973	2.832–3.121	1.652	1.538–1.775
Diabetes mellitus with complications	3.890	3.784–4	2.241	2.16–2.325
Biliary tract disease	1.980	1.894–2.071	1.510	1.435–1.59
Fluid and electrolyte disorders	2.434	2.395–2.472	1.431	1.401–1.462
Asthma	3.115	3.045–3.187	1.479	1.434–1.526
Schizophrenia and other psychotic disorders	5.363	5.202–5.529	1.672	1.593–1.754
Acute and unspecified renal failure	3.628	3.529–3.73	1.643	1.573–1.717
Respiratory failure, insufficiency, arrest	2.495	2.417–2.574	1.192	1.137–1.25
Gastrointestinal hemorrhage	2.845	2.721–2.976	1.793	1.689–1.904
Intestinal obstruction without hernia	2.245	2.158–2.335	1.132	1.07–1.198
Pancreatic disorders (not diabetes)	3.454	3.294–3.621	1.889	1.777–2.007
Diverticulosis and diverticulitis	2.275	2.165–2.39	1.154	1.07–1.244
Fracture of neck of femur (hip)	2.266	2.109–2.435	0.948	**0.83–1.084**
Epilepsy, convulsions	2.744	2.657–2.835	1.384	1.325–1.445
Appendicitis and other appendiceal conditions	1.074	**0.992–1.163**	1.063	**0.984–1.148**
Alcohol-related disorders	4.232	4.125–4.342	1.899	1.835–1.965
Fracture of lower limb	2.711	2.566–2.866	1.492	1.388–1.604
Other nervous system disorders	2.615	2.553–2.678	1.236	1.194–1.279
Syncope	2.787	2.632–2.951	1.373	1.259–1.496
Substance-related disorders	4.551	4.43–4.676	1.753	1.688–1.82
Intracranial injury	3.154	2.98–3.337	1.535	1.421–1.659
HP with complications and secondary HP	4.886	4.776–4.998	2.466	2.382–2.554
Other fractures	2.916	2.772–3.067	1.349	1.251–1.455
Intestinal infection	1.686	1.588–1.79	1.092	1.012–1.178
Secondary malignancies	1.311	1.245–1.38	0.534	0.49–0.582
Anemia	2.788	2.736–2.841	1.766	1.723–1.81
Avoidable hospitalization	2.793	2.763–2.822	1.536	1.514–1.559

Note: Companion table for Table 11.4. Comorbidities are based on up to nine diagnoses. All odds ratios are patient diagnosis rather than record diagnosis specific. If a patient had asthma hospitalizations, only one is counted. If a confidence interval crosses 1, it is not statistically significant. All race categories are derived from data linkage with other data sources. All odds ratios are in reference to Whites. Bold means not significant.

REFERENCES

Aragam, K.G., M. Moscucci, D.E. Smith, A.L. Riba, M. Zainea, J.L. Chambers, D. Share, and H.S. Gurm. 2011. Trends and disparities in referral to cardiac rehabilitation after percutaneous coronary intervention. *American Heart Journal* 161: 544–551.

Barber, K., M. Stommel, J. Kroll, M. Holmes-Rovner, and B. McIntosh. 2001. Cardiac rehabilitation for community-based patients with myocardial infarction: Factors predicting discharge recommendation and participation. *Journal of Clinical Epidemiology* 54. 1025–1030.

Beckles, G.L. and C.F. Chou. 2013. Diabetes—United States, 2006 and 2010. *Morbidity and Mortality Weekly Report* (RR01): 1–15.

Crosby, A.E., L. Ortega, M.R. Stevens, and Centers for Disease Control and Prevention (CDC). 2013. Suicides—United States, 2005–2009. *Morbidity and Mortality Weekly Report Surveillance Summary* 62(Suppl. 3): 179–183.

Girotti, M.E., T. Shih, S. Revels, and J.B. Dimick. 2014. Racial disparities in readmissions and site of care for major surgery. *Journal of the American College of Surgeons* 218: 423–430.

Gregory, P.C., T.A. LaVeist, and C. Simpson. 2006. Racial disparities in access to cardiac rehabilitation. *American Journal of Physical Medicine and Rehabilitation* 85: 705–710.

Johnson, N., J. Fisher, A. Nagle, K. Inder, and J. Wiggers. 2004. Factors associated with referral to outpatient cardiac rehabilitation services. *Journal of Cardiopulmonary Rehabilitation* 42: 165–170.

Mazzini, M.J., G.R. Stevens, D. Whalen, A. Ozonoff, and G.J. Balady. 2008. Effect of an American Heart Association Get With the Guidelines program-based clinical pathway on referral and enrollment into cardiac rehabilitation after acute myocardial infarction. *American Journal of Cardiology* 101: 1084–1087.

Parashar, S., J.A. Spertus, F. Tang, K.L. Bishop, V. Vaccarino, C.F. Jackson, T.F. Boyden, and L. Sperling. 2012. Predictors of early and late enrollment in cardiac rehabilitation, among those referred, after acute myocardial infarction. *Circulation* 126(13): 1587–1595.

Sanderson, B.K., S. Mirza, R. Fry, J.J. Allison, and V. Bittner. 2007. Secondary prevention outcomes among black and white cardiac rehabilitation patients. *American Heart Journal* 153: 980–986.

Van Poppel, F. and L.H. Day. 1996. A test of Durkheim's theory of suicide—without committing the ecological fallacy. *American Sociological Review* 61: 500–507.

12 Using Emergency Department Data to Conduct Surveillance

12.1 INTRODUCTION

There are hundreds of millions of emergency department (ED) visits each year in the United States, and ED visits accounted for 50% of inpatient admission in 2009 (Morganti-Gonzalez et al., 2013). EDs are the only universally accessible source of outpatient care that is available 24/7. Many disease outbreaks of public health significance can be detected via ED data. However, ED data are traditionally used respectively for health service research, and their use in public health assessments has been infrequent. One major public health use of ED data early on was to prepare and track events, such as major sports (Weiss et al., 1988), heat waves (Rydman et al., 1999), and natural disasters (Lee et al., 1993; Greene et al., 2013). In those ad hoc assessments, ED data were either compiled by local public health agencies for a specific purpose or extracted by researchers, sometimes manually. Such event-based data analyses can hardly be incorporated into public health reporting systems due to unstandardized and laborious data compilation.

In recent years, most EDs have adopted an electronic medical records system, which provides a data source that can be used by many syndromic surveillance systems in real time or near real time. This trend, together with trends in new emerging infectious diseases and bioterrorism threats, puts ED data to the forefront of syndromic surveillance systems nationwide. Now, most states and local health departments have some kind of syndromic surveillance system that includes real-time reporting from EDs to public health agencies. Surveillance in real time often uses profile data from previous time units (year, month, week, day) as the referent. If the incidence within a time unit was excessive, it may set an alarm. There are many syndromic surveillance systems that are currently used by state and local health departments. Examples include the Early Notification of Community-Based Epidemics (ESSENCE) and the Centers for Disease Control and Prevention (CDC)-supported BioSense system. ESSENCE included 29 or more influenza-like illnesses (ILIs) based on International Classification of Diseases, Ninth Revision (ICD-9) diagnosis codes that can be linked or extracted directly from an ED electronic record system. After years of implementation, both technologies and practices in syndromic surveillance have matured enough, and they have become a routine part of surveillance activities in many local health agencies.

The primary purpose of syndromic surveillance is to detect and respond to potential public health emergencies. In this regard, syndromic surveillance detects disease signals that generally precede diagnoses or an outbreak. However, data collected from the system can be used to address public health issues during nonemergency situations. Buehler et al. (2004) listed eight public health functions that surveillance serves; besides (1) outbreak detection, the rest are not time sensitive: (2) supporting case finding, (3) estimating the impact of a disease, (4) tracing the natural history of a health condition, (5) determining the spread of illness, (6) generating hypotheses, (7) evaluating prevention and control measures, and (8) facilitating planning. These latter seven functions fit well with chronic disease surveillance functions using ED data. Massive data are collected daily through syndromic and other public health reporting systems, and they sit in various public health database systems that can be tapped for other public health assessments. This is perhaps the impetus of recent initiatives of incorporating syndromic surveillance infrastructure into chronic disease surveillance. One limiting factor of using ED data from surveillance systems is that the number of participating hospitals varies substantially due to extra staff workload that cannot be fully compensated. In addition, most hospitals only allow a limited set of diseases or conditions to be extracted. Even though the CDC has long developed the Data Elements for Emergency Department Systems (DEEDS), and has recently been included in stage 3 meaningful use, it takes time to trickle down into real-world surveillance systems.

Alternatively, one can look for opportunities to use ED outpatient claims data as a source of secondary data analysis. ED claims data are parts of outpatient data generated for patient treatment and billing. Diagnosis information and patient demographics can feed syndromic surveillance too, in the same way as from an ED, but a lot more simply and easily. With today's technology, it can be close to real time (daily or weekly). Within a public health agency, if we are willing to use ED data retrospectively, it is possible to update ED data on a monthly or seasonal basis. An advantage of using ED claims data is that data format and variable names tend to follow the national standards, and most data specialists within a public health agency can readily analyze these data if they have had experience with inpatient claims data. For those without experience, it would still not be difficult, as there are only a limited number of variables. One notable difference between inpatient claims data and outpatient claims data is that the former use procedure codes to document treatment, while the latter use correct procedure terminology (CPT) codes to document treatment. However, since most public health agencies do not deal with treatment, this difference poses very little challenge to data analysts at this point.

Compared to inpatient data, ED data tend to be more frequent, with a lot more records on any particular day. In addition, ED data tend to be more reflective of seasonal or other disease cycles than inpatient data, as the latter tend to be more schedule driven. In the context of meaningful use of clinical data, we expect that it will become more and more common to use ED data extracted either directly from hospitals or indirectly from claims data. In either case, public health surveillance specialists need to be prepared or trained for secondary data analysis, so that surveillance system data or secondary data can be used to address other public health issues. In some way, ED data are quite similar to ambulatory clinical data. Getting familiar with ED data analysis would prepare public health data analysts to conduct meaningful use of ambulation clinic data for surveillance. With future secondary data analysis needs in mind, we selected both infectious and chronic diseases from outpatient data. We chose ILI because it is one of the most commonly monitored syndromes in surveillance systems. We chose external injuries because they fit both disaster relief efforts and chronic disease preventions. We used ILI to examine potential population and geographic vulnerabilities, and we used external injuries to examine the potential effects of climate change in terms of extreme weather.

12.2 INFLUENZA AND POPULATION VULNERABILITY

Influenza can cause widespread illness, considerable loss of life, economic disruption, and long-term health and economic outcomes (Almond, 2006; Greene et al., 2006; Viboud et al., 2013). While the seasonal influenza is anticipated each year, its impacts on different population segments are less understood, especially at a local level. Timely evaluation of influenza-related morbidity is a priority for seasonal surveillance and pandemic preparedness. Syndromic surveillance for influenza is based on chief complaints and ICD-9 diagnosis to capture ILI before the laboratory-confirmed diagnosis (Travers et al., 2013). In syndromic surveillance, electronic reporting of ED chief complaints is available, but we do not use this due to lack of reporting hospitals. Since chief complaints and associated keywords are not available in outpatient claims data, we used 29 diagnosis codes embedded in ESSENCE to code ILI (Lombardo, 2003). In addition, we added the confirmed cases by the ICD-9 code 488 (influenza due to the identified avian influenza virus). This approach is appropriate for secondary data analysis because diagnosis codes are available from both claims data and syndromic surveillance systems.

We used outpatient claims data from the Nebraska Hospital Association, with admitting sources being the ED. This item was unfortunately dropped in the middle of 2010. For this reason, we only had data from January 1, 2005 to June 26, 2010. Individual ED visit data were aggregated by age group, rural–urban areas, and week ending on Saturday. We followed Olson et al.'s (2007) methodology with a minor twist. We used rates rather than disease counts, because sometimes not all EDs reported data, especially in the last few weeks of June in 2010. In addition, the first week in 2005 was not a full week. Consequently, count-based calculation would have wider variation than rate-based calculation. The ILI rate is defined as the number of ED ILI visits divided by the total ED visits within the state.

There was a total of 1,850,830 ED visits from the outpatient data between January 2005 and June 2010. Among them, 10.51% were ILI. In order to see if there was an urban–rural difference, the ILI rates were plotted by place of residence (Figure 12.1), where urban areas include Douglas, Sarpy, and Lancaster Counties. As one can see, both rural and urban areas followed almost identical ILI cycles, with a slightly higher rate for urban areas than in rural areas. In almost any given week, at least 5% of ED visits were due to ILI. Each year, there was an ILI peak around late January. Although the peaks seem higher each year after 2005, there is no clear secular trend. It is worth noting that the 2009 peak came rather early. Both rural and urban areas peaked between the second and fourth weeks of October, with an average rate of 28%. The 2009 H1N1 influenza pandemic occurred at a time when pandemic preparedness was geared toward H5N1 viruses. The mismatch of vaccine perhaps contributed to the early arrival of the pandemic, because seasonal influenza vaccine would not provide any significant protection against the 2009 H1N1 virus.

In Figure 12.2, we compare ILI cycles by four age groups. Due to potential small frequencies for calculating a stable rate, we only used four age groups: ≤ 5, 6–18, 19–64, and ≥ 65. It is very clear that not only different age groups had different rates, but also some age groups were quite different in terms of ILI cycles. The greatest contrast was between the age ≤ 5 and ≥ 65 groups. For the young age group, there were six peaks, even though the fifth peak, or the one in the 2008–2009 winter, was not as clear as others. For the old age group, the last two peaks (2008–2009, 2009–2010) were not apparent. In fact, there was no elevated rate that corresponded to the 2009 H1N1 peak in October. It was reported by the CDC that about one-third of adults older than 60 years of age had cross-reactive antibody against the 2009 H1N1 flu virus. Children, in contrast, had no existing cross-reactive antibody to the virus. These results suggest that separate models are necessary to evaluate excessive morbidity by age groups.

Children younger than 5 years and the elderly aged 65 years and older are most vulnerable, and they accounted for 7.2% and 13.5%, respectively, of the total Nebraska population. To demonstrate how one might assess excessive morbidity, we separately modeled the young and old age groups using Serfling regression (Serfling et al., 1967). Let $\omega = \pi/52.18$ represent the annual cycle, and

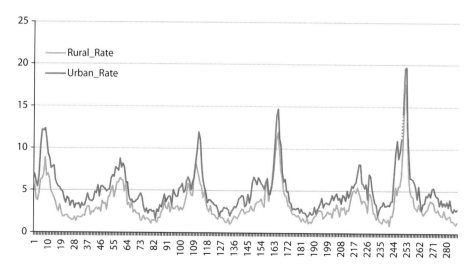

FIGURE 12.1   ILI ED visit rates in urban and rural Nebraska: 2005–2010.

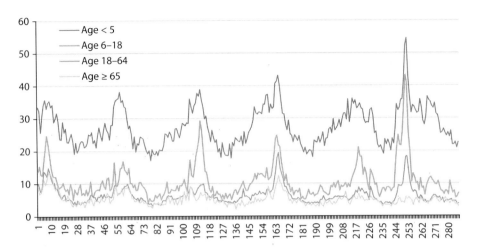

FIGURE 12.2 ILI ED visit rates by four age groups: January 2005–June 2010.

t index weeks from the beginning period to the end period; then a simple ordinary least-squares (OLS) Serfling regression model can be set up as

$$M_t = \alpha_0 + \beta_1 \cos(2\omega t) + \beta_1 \sin(2\omega t) + \gamma_2 \cos(4\omega t) + \gamma_2 \sin(4\omega t) + \delta_3 \cos(6\omega t) + \delta_3 \sin(6\omega t) + \varepsilon_\tau \quad (12.1)$$

where M_t is the expected rate of ED visits in week t. The t index is sequential week from week 1 in January 2005 to week 287 in June 2010. β_1 and γ_2 are annual sinusoidal and semiannual terms, respectively. In addition, we added a triannual term to capture the October peak in the 2009 H1N1 pandemic. Conceptually, it might be better without the triannual terms, as flu tends to vary annually and biannually. If empirical observation does not suggest a regular triannual term, it might not be appropriate to include it. From the model fitting perspective, if a triannual term was significant, it may reveal new insight. So the question of including it or not really depends on the purpose of the model. In addition, if a model fits better, it may reduce the expected excessive morbidity. Finally, if there was a secular trend, the t term can also be included. We did not include it because it was not significant.

The results of the Serfling curve are shown in Figure 12.3. All the coefficients were significant for the young age group, and they are listed below in parentheses:

$$M_t(\text{young}) = \alpha_0(27.406) + \beta_1 \cos(5.964) + \beta_1 \sin(1.892) + \gamma_2 \cos(-1.173) + \gamma_2 \sin(0.840)$$

$$+ \delta_3 \cos(-0.917) + \delta_3 \sin(0.993)$$

The R-squares for the unadjusted and adjusted models were 0.595 and 0.604, respectively. They suggest that the model fits well, driven primarily by seasonal flu. However, it appeared that the 2009 peak was not synchronized with the 2009 pandemic, as the model-based peak in 2009–2010 was at least 2 months behind.

To calculate excessive morbidity, one would simply use those coefficients to derive the expected rate for each week and calculate the rate difference between the expected and the observed. From the second week to the fourth week of October 2009, the excessive morbidity was 1232 for the young age group.

To fit the old age group, we slightly tweaked the model in Equation 12.1 by adjusting the triannual term to a quarterly term ($8\pi/52.18$). The parameter estimates are below.

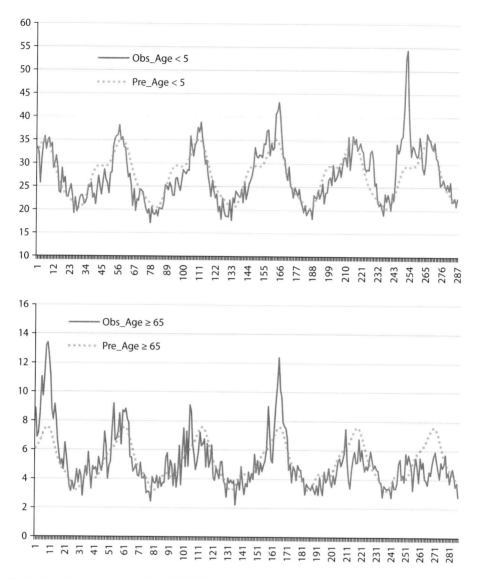

FIGURE 12.3 Observed and predicted ILI ED visit rates by two age groups.

$$M_t(\text{old}) = \alpha_0(5.094) + \beta_1\cos(1.233) + \beta_1\sin(1.216) + \gamma_2\cos(-0.339)$$

$$+ \gamma_2\sin(0.430) + \delta_3\cos(-0.31913)$$

With the exception of $\delta_3\cos$, all other terms were significant, with adjusted and unadjusted R-squares being 0.4775 and 0.4663, respectively.

As shown in Figure 12.3, lower panel, the Serfling curve for the elderly captured the seasonal flu trend over the study period. There were at least two predicted peaks where the observed ED visits for ILI were lower than expected. In 2009, for instance, the CDC declared an emergency regarding the 2009 H1N1 pandemic influenza, and there was a wide and intensive campaign to vaccinate vulnerable populations such as elderly and children. Although the vaccine was not that effective for children in terms of H1N1, it prevented expected seasonal flu peak. We can use the Serfling curve to calculate reduced morbidity, presumably due to the heavy vaccination campaign. During the 4-week period between January 31, 2010, and the end of February, the reduced ED visits for ILI were 124 for the elderly. Certainly, if we moved to October 2009, there would be a lot of excessive morbidity due to ILI.

Serfling regression models tend to capture seasonal flu trends, and they are quite useful for an initial morbidity assessment. However, they are not able to explain a pandemic outbreak, especially when the occurrence does not fit the regular cycle. In order to provide a better model fit, additional insights are often needed. We experimented with weather variables, school seasons (Neuzil et al., 2002), the reported H1NI cases, time, strain, and so forth. By including these variables, we could fit the young age group model to an R-square above 0.80%. However, with the exception of vaccination coverage (which was available for the elderly in BRFSS), we felt that putting explanatory variables into a Serfling regression would weaken the periodical nature of model prediction while strengthening modeled-based predictability.

Although it is easier to explain an OLS Serfling regression, the rate model does not consider visit volume. Since ILI visits are part of the denominator, weeks with smaller visit volumes may have some undue influence on the calculations of expected rates and excessive morbidity. When the rates are dominated by ILI visits, as seen in the 2009 peak season, the calculated ILI excessive morbidity may overestimate excessiveness. To assess this effect, we compared the OLS model with a log-rate model (Thompson et al., 2003). We used the same set of independent variables as in the model in Equation 12.1 for the young age group, with the offset term being the log of the total ED visits each week. The predicted number of ED visits during the 4-week period from January 31, 2010, to the end of February was 1142, or 90 less than 1232 from the OLS estimate. Although the mean difference between the OLS regression and log-rate regression was –3.61, which was significant according to the t-test ($p < 0.01$), such a difference is only a fraction of the overall means of 301.3461 and 304.9797 for the OLS and log-rate models, respectively. Hence, when the number of records is large, using OLS or log-rate models would yield very similar numbers of ILI visits, shown in Figure 12.4.

This brings us to the final point about estimated ILI rates. Visually comparing the predicted versus observed rates may provide a false impression in terms of predicted outbreak timing. If we used the rate estimates in the upper panel of Figure 12.3, we would pick the predicted 2009–2010 peak in early 2010. However, if we used the number of visits in Figure 12.4, we would pick October 2009 as the 2009–2010 peak. When visit volume is large during a few weeks time, a seemingly low visit rate would generate a high number of ILI visits. This problem is shared by all rate-based estimations,

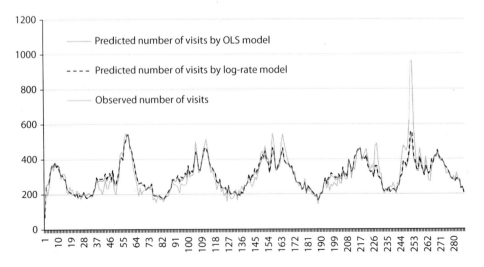

FIGURE 12.4 **(See color insert)** Observed and predicted ILI ED visits for age 5 or younger.

because it is just a visual effect. For this reason, one should always convert rates to expected numbers to pick expected ILI peaks over a period of time, as shown in Figure 2.4.

Another limitation at this point is that we were unable to assess population vulnerability by racial groups and neighborhood socioeconomic status (SES), as outpatient data do not have race and have not been geocoded. However, we know the ED is an important part of the safety net for vulnerable populations, data collected through the ED are likely to shed light on health disparities, and knowledge gained through ED disparity assessment will strengthen the safety net.

A recent SES study in New Haven was based on the surveillance system of Connecticut Emerging Infections Program (CT-EIP). It showed that adults residing in neighborhoods of lower SES were more likely to be hospitalized with influenza than the comparison groups (Tam et al., 2014). Such a study could be replicated using a chronic disease syndromic surveillance system. One concern often cited is that syndromic surveillance may not be population based, as only a handful of hospitals participated in the system. However, this should not be a big concern. Assuming that a hospital catchment area changes little within a year or two, one could simply evaluate population vulnerability by participating hospitals while not identifying a particular hospital to the general public.

12.3 LINKING WEATHER DATA TO HOSPITAL DATA

In the influenza analysis above, we used only time. Any climate-related association, such as seasonality, is implicit by week or month of a year. In this section, we look at the climate and weather explicitly, as many events relating to public health are weather related, such as hurricane, tornado, heat wave, and cold spell. In fact, medical professionals have long thought that weather and climate act both directly and indirectly on human health, triggering various physiological responses and physical behaviors (Drake, 1850; Sargent, 1982; Peterson, 1934). The Intergovernmental Panel on Climate Change (IPCC) reported that "many natural systems are being affected by regional climate changes, particularly temperature increases" (IPCC, 2007). The IPCC also pointed out potential global health effects from projected climate change–related exposures, such as (1) increases in malnutrition, and the effect of malnutrition on children's development; (2) increased deaths, disease, and injury due to heat waves, floods, storms, fires, and droughts; (3) increased frequency of cardiorespiratory diseases due to higher concentrations of ground-level ozone; and (4) altered spatial distributions of some infectious disease vectors.

In the United States, some potential climate change–related health effects have been identified, including respiratory problems, skin cancers, and cold and heat stroke due to exposure to extreme weather (Bernard et al., 2001; Greenough et al., 2001; McGeehin et al., 2001; Bunyavanich et al., 2003; Barnett et al., 2007; Liang et al., 2008). In addition, climate change will increase weather variability and the frequency of severe weather conditions (Meehl et al., 2009; Alley et al., 2003). Recent research findings have shown that extreme weather conditions, both hot and cold, are associated with excessive mortality and morbidity (Karl and Knight, 1997; Feigin et al., 2000; Ebi and Schmier, 2005), especially for those with cardiovascular disease (CVD) (Lin et al., 2009). Extreme weather, especially high temperature, is also linked to low birth weight in the United States (Lin and Zhang, 2012).

Figure 12.5, adapted from McMichael et al. (2006), highlights pathways from environmental climate effects to health effects in the context of inland areas, such as Nebraska. The leftmost boxes show that climate change can result from both anthropogenic greenhouse gas emissions and natural course of climate change. Mitigation can be entered between climate change factors and the three main environmental effects of climate change in the central section, while human adaptation is mainly between environmental effects and the health effects of the right-hand boxes. Many climate change–related health effects already exist, from those caused by extreme weather, which are relatively simple, to those caused by more complex processes of environmental degradation. With the exception of extreme events, most climate change–related health effects in Nebraska and other inland areas of the United States will likely be close to the bottom boxes and will be hard to measure (Mills, 2009). Consequently, those who live in inland areas are likely to be less concerned with climate change effects than are those who live in coastal regions.

In addition, the oval area in Figure 12.1 represents geographic and population vulnerabilities, which mediate between climate change and human health effects. Vulnerable geographic areas in Nebraska include (1) remote rural areas with scarce emergency response resources, (2) floodplains, and (3) other ecologically sensitive areas, such as the Sand Hills in Nebraska and the Ogallala Aquifer, one of the largest reserves of freshwater in the world. Vulnerable populations include children, older adults living alone, single mothers with children, and those living in poverty, and they tend to concentrate in a number of areas. Vulnerable geographic areas and vulnerable population segments are both more likely to be affected by the climate and weather effects listed in the right column of Figure 12.5.

In Nebraska, we do not have much evidence cited by IPCC in terms of climate change and its health effect. An analysis of temperature in the last 100 years showed a warming trend. As a result, the growing season in Nebraska in recent decades has started about 9 days earlier than in the 1950s (Feng and Hu, 2004), and winter wheat is also flowering 5–6 days earlier (Hu et al., 2006). Those trends may correspond to early pollen seasons in Nebraska, which in turn may increase the early onset of asthma and prolong asthma-sensitive seasons. Likewise, public health program data have rarely been collected for potential health impacts of climate change. However, many data, such as emergency medical services (EMS), hospital discharge data, and outpatient data, do have date of service and date of admission variables, and they can be linked to weather station data.

To demonstrate how we might utilize weather station data, we selected three urban counties in Nebraska that had high-quality airport weather stations. All stations had hourly data on temperature, humidity, wind speed, and precipitation in rain or snow. We generated four extreme weather indicator variables for 2005–2009: daily highest temperature greater than 90°F, daily lowest temperature less than 20°F, having snow, and having rain. We compiled the station data according to the three counties where they are located and summarized the data to one record per day per county.

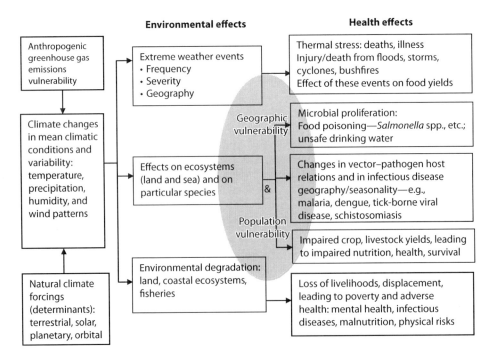

FIGURE 12.5   Main pathways of effects of climate change on population health.

We extracted ED data from outpatient data. In the three counties, we had 4,773,897 outpatient records from 2005 to 2009, among which 953,168 records were from the ED visits. Since each patient had the county of residence, we could link the weather station data within each county to each patient residing in the county by the date of the visit. All records were linked because there were no missing dates from weather station data or from ED data.

We selected external injuries as surveillance items and ran a set of age- and sex-adjusted logistic regressions to see if the 10 types of external injuries in Table 10.6 were significantly associated with any of four weather conditions. We used eight age groups (0–14 and then increment by 10 until age 85+) as controls. In Chapters 8 through 10, we surveyed hospitalization according to poverty levels. Instead of using poverty levels, we used four weather conditions as indicator variables.

The results show that motor vehicle crash (MVC) injuries were more likely to be associated with cold days (Table 12.1). When the lowest temperature hit 20°F, the odds of MVC ED visits increased by 24.5%. In addition, both rainy and snow days were associated with more MVC ED visits. All of the four weather conditions were significantly associated with other traffic-related injuries, such as bicycle and pedestrian. Both low-temperature and snow days reduced the odds by more than 65%, while high temperature and rainy days increased the odds. Fall injury ED visits were more likely to happen on low-temperature or snow days, and less likely happen to happen on rainy days. Finally, the remaining three injury categories (accidental cut or pierce; injuries due to the natural environment, bites or struck by or against; and assault injury) had similar directional effects. Low-temperature days or snow days had protective effects on them, while hot days and rainy days tended to have more of these injuries.

Weather-related ED visit surveillance can be extended in many ways. First, it can be carried out for any diagnoses suspected to be related to weather. We chose major injury categories because weather conditions are likely to be related, but such surveillance can be expanded to allergies and CVD, which are also deemed to be related to weather and climate. Second, since outpatient data are linked to inpatient data for those who were admitted to a hospital, we could track those ED patients to see which types of patients were more likely to be admitted for inpatient care. For instance, among 953,168 ED visits in the urban counties, 167,849 were eventually admitted to a hospital, with disproportionally more elderly and Medicare patients. Furthermore, since inpatient ED admissions are a subset of ED outpatient data, we could also analyze inpatient data for ED admission, because a hospital admission costs a lot more than an ED visit.

TABLE 12.1

Logistic Regressions for Injury ER Visits by Four Weather Variables

Type	Low Temperature, <20°F	High Temperature, >90°F	Rain	Snow
Motor vehicle traffic injury	1.245*	0.936	1.060*	1.191*
Other traffic-related injury	0.315*	1.379*	1.070*	0.327*
Fall injury	1.071*	1.018	0.944*	1.150*
Accidental cut or pierce	0.721*	1.311*	1.084*	0.947
Natural environment, bites, struck by	0.762*	1.260*	0.996	0.818*
Assault injury	0.868*	1.176*	1.071*	0.91

Note: Sample is based on counties with a population greater than 100,000 (Douglas, Sarpy, and Lancaster). Hospitalizations are based on Table 10.6. Only significant conditions were included. Diagnoses are based on the first E-code. Weather variables are based on airport weather stations within these counties. Rain or snow, any rain or snow during a day.

* Significant at $p < 0.05$.

Take fall injuries in urban areas as an example, Black patients were much less likely to be admitted to the ED for fall injuries than Whites; Black and White patients accounted for 3.3% and 6.9% of Black and White ED admissions, respectively. Following a similar line of reasoning, we found that patients in neighborhoods with >20% of the population under the poverty level (group 5) had 4.75% ED admissions attributable to them, as opposed to 7.15% of patients living in neighborhoods with <5% under the poverty level. Finally, our measures of extreme weather were crude, especially for rain and snow. As we refine weather condition indicators, we could also refine population segment, as people living with limited heating and cooling capacity are more vulnerable. As future climate change gathers more attention, weather and climate-related surveillance is likely to become more and more common.

A methodological limitation of using indicator variables to capture the date of events is that we do not know if there was an upward or downward trend after the date of events. If we used a crossover design to capture the trend of a few days after extreme weather, we would be able to conduct more rigorous analysis. However, each injury category is different, and thus a specific study design for each specific injury or disease in the surveillance set might be required, which would need institutional review board (IRB) approval and would be beyond the scope of surveillance and common public health data analysis.

12.4   CHAPTER SUMMARY AND CONCLUDING REMARKS

In this chapter, we have demonstrated two ways to use ED data. One is to evaluate population vulnerability by place of residence and age group. We found that place of residence has little effect, while the elderly had quite different seasonal variations from the younger age groups, especially for the H1N1 flu in 2009. Separate Serfling regressions showed greater excessive morbidity for the youngest age group than for the elderly. This type of analysis could be built into routine secondary data analysis based on either syndromic surveillance data or outpatient ED data, and the results can be used for program evaluation, such as an immunization campaign. The only critical issue is to digitally import ED data into a surveillance system.

Serfling regression models tend to capture seasonal trends, and they can be used for any diseases that have some cyclical incidence. However, a Serfling curve does not explain the observed pattern. In order to develop explanatory models, future studies should examine population vulnerability by race, SES, and locational factors. It is also possible to divide ILI into major influenza-related conditions (pneumonia, emphysema, asthma, other chronic obstructive pulmonary disease, and congestive heart failure) and upper respiratory infections (URIs) (common cold, simple influenza, sinusitis, tonsillitis, laryngitis, tracheitis, croup, epiglottitis, otitis media, and acute bronchitis) for planning ED resource allocations (Schull et al., 2005).

In the second part of the chapter, we experimented with a set of weather-related explanatory variables for chronic disease surveillance, which does not have to be in real time. We found that extreme weather conditions relate to a half-dozen injury categories. Obviously, both the weather conditions and injury categories used in the analysis need to be refined. Nevertheless, some reasonable explanations can be derived. For example, fall injuries were more likely to happen on cold or snowy days. These results were expected, and quantifying them would help to design fall prevention. In our integrated data system, we could trace fall patients through ED and inpatient admission by SES and racial characteristics. Linking data through a health care system certainly can provide more insights than otherwise separate data analyses within each system. Such an analytical model would be quite different from the traditional syndromic surveillance models commonly seen in infectious disease surveillance.

Given that climate change will increase the intensity and frequency of extreme weather, there is a need to develop weather–health indicators that can be used to track climate-induced health risks. Our experiment is just a beginning; future work needs long-term health outcome data and carefully designed weather–health relationships. The results would then be more meaningful to educate the general public about the effects of climate change on human health.

APPENDIX 12A: ORIGINAL LIST OF ILI SYNDROME IN ESSENCE

ICD-9 Code	Description
079.89	Viral infection NEC
079.99	Viral infection NOS
460	Nasopharyngitis, acute
462	Pharyngitis, acute
46400	Laryngitis, acute, without obstruction
464.10	Tracheitis, acute, without obstruction
464.20	Laryngotracheitis, acute, without obstruction
4650	Laryngopharyngitis, acute
465.8	Infectious upper respiratory, multiple sites, acute NEC
465.9	Infectious upper respiratory, multiple sites, acute NOS
4660	Bronchitis, acute
466.11	Bronchiolitis due to respiratory syncytial virus
466.19	Bronchiolitis, acute, due to other infectious organisms
478.9	Disease, upper respiratory NEC or NOS
4800	Pneumonia due to adenovirus
480.1	Pneumonia due to respiratory syncytial virus
480.2	Pneumonia due to parainfluenza
480.8	Pneumonia due to virus NEC
480.9	Viral pneumonia, unspecified
484.8	Pneumonia in other infectious disease NEC
485	Bronchopneumonia, organism NOS
486	Pneumonia, organism NOS
487	Influenza with pneumonia
487.1	Influenza with respiratory manifestation NEC
487.8	Influenza with manifestation NEC
490	Bronchitis NOS
780.6	Fever
784.1	Pain, throat
786.2	Cough

REFERENCES

Alley, R.B., J. Marotzke, W.D. Nordhaus, J.T. Overpeck, D.M. Peteet, R.A. Pielke, and L.D. Talley. 2003. Abrupt climate change. *Science* 299(5615): 2005.

Almond, D. 2006. Is the 1918 influenza pandemic over? Long-term effects of in utero influenza exposure in the post-1940 U.S. population. *Journal of Political Economy* 114: 672–712.

Barnett, A.G. 2007. Temperature and cardiovascular deaths in the US elderly: Changes over time. *Epidemiology* 18(3): 369–372.

Bernard, S.M., J.M. Samet, A. Grambsch, K.L. Ebi, and I. Romieu. 2001. The potential impacts of climate variability and change on air pollution-related health effects in the United States. *Environmental Health Perspectives* 109(Suppl. 2): 199–209.

Buehler, J.W., R.S. Hopkins, J.M. Overhage, D.M. Sosin, and V. Tong. 2004. Framework for evaluating public health surveillance systems for early detection of outbreaks. *Morbidity and Mortality Weekly Report Recommendations and Reports* 53(RR05): 1–11.

Bunyavanich, S., C.P. Landrigan, A.J. McMichael, and P.R. Epstein. 2003. The impact of climate change on child health. *Ambulatory Pediatrics* 3(1): 44–52.

Drake, D. 1850. *A Systematic Treatise, Historical, Etiological, and Practical, on the Principal Diseases of the Interior Valley of North America.* Cincinnati, OH: W. B. Smith, p. 447.

Ebi, K.L. and J.K. Schmier. 2005. A stitch in time: Improving public health early warning systems for extreme weather events. *Epidemiologic Reviews* 27: 115–121.

Feng, S. and Q. Hu. 2004. Changes in agro-meteorological indicators in the contiguous United States: 1951–2000. *Theoretical & Applied Climatology* 78: 247–264.

Greene, S.K., E.L. Ionides, and M.L. Wilson. 2006. Patterns of influenza-associated mortality among US elderly by geographic region and virus subtype, 1968–1998. *American Journal of Epidemiology* 163: 316–326.

Greene, S.K., E.L. Wilson, K.J. Konty, and A.D. Fine. 2013. Assessment of reportable disease incidence after Hurricane Sandy, New York City, 2012. *Disaster Medicine and Public Health Preparedness* 7: 513–521.

Greenough, G., M. McGeehin, S.M. Bernard, J. Trtanj, J. Riad, and D. Engelberg. 2001. The potential impacts of climate variability and change on health impacts of extreme weather events in the United States. *Environmental Health Perspectives* 109(Suppl. 2): 191–198.

Hu, Q., A. Weiss, S. Feng, and S. Baenziger. 2006. Earlier winter wheat heading dates and warmer spring in the Great Plains of the United States. *Agricultural and Forest Meteorology* 135: 284–290.

IPCC. 2007. Contributions of working group I to the fourth assessment report of the intergovernmental panel on climate change. In *Climate Change 2007: The Physcial Science Basis*, eds. S. Solomon, D. Qin, M. Manning, Z. Chen, M. Marquis, K.B. Averyt, M. Tognor, and H.L. Miller. New York: Cambridge University Press, p. 996.

Karl, T.R. and R.W. Knight. 1997. The 1995 Chicago heat wave: How likely is a recurrence? *Bulletin of the American Meteorological Society* 78: 1107–1119.

Lee, L.E., V. Fonseca, K.M. Brett, J. Sanchez, R.C. Mullen, L.E. Quenemoen, S.L. Groseclose, and R.S. Hopkins. 1993. Active morbidity surveillance after Hurricane Andrew—Florida, 1992. *JAMA* 270(5): 591–594. Erratum in *JAMA* 1993; 270(19): 2302.

Liang, W.M. W.P. Liu, S.Y. Chou, and H.W. Kuo. 2008. Ambient temperature and emergency room admissions for acute coronary syndrome in Taiwan. *International Journal of Biometeorology* 52(3): 223–229.

Lin, G. and T.L. Zhang. 2012. Examining extreme weather effects on birth weight from the individual effect to spatiotemporal aggregation effects. *Journal of Agricultural Biological and Environmental Statistics* 17: 490–507.

Lin, S., M. Luo, R.J. Walker, X. Liu, S.A. Hwang, and R. Chinery. 2009. Extreme high temperatures and hospital admissions for respiratory and cardiovascular diseases. *Epidemiology* 20(5): 738–746.

Lombardo, J. 2003. A systems overview of the Electronic Surveillance System for Early Notification of Community-Based Epidemics (ESSENCE II). *Journal of Urban Health* 80(2 Suppl. 1): i32–i42.

McMichael, A.J., R.E. Woodruff, and S. Hales. 2006. Climate change and human health: Present and future risks. *Lancet* 367: 859–869.

Meehl, G.A., C. Tebaldi, G. Walton, D. Easterling, and L. McDaniel, 2009. Relative increase of record high maximum temperatures compared to record low minimum temperatures in the U.S. *Geophysical Research Letters* 36: L23701.

Mills, D.M. 2009. Climate change, extreme weather events, and US health impacts: What can we say? *Journal of Occupational and Environmental Medicine* 51(1): 26–32.

Morganti-Gonzalez, K., S. Bauhoff, J.C. Blanchard, M. Abir, N. Iyer, A. Smith, J.V. Vesely, E.N. Okeke, and A.L. Kellermann. 2013. The evolving role of emergency departments in the United States. RAND RR 280-ACEP. Santa Monica, CA: RAND Corp., May.

Neuzil, K.M., C. Hohlbein, and Y. Zhu. 2002. Illness among schoolchildren during influenza season: Effect on school absenteeism, parental absenteeism from work, and secondary illness in families. *Archives of Pediatrics and Adolescent Medicine* 156(10): 986–991.

Olson, D.R., R.T. Heffernan, M. Paladini, K. Konty, D. Weiss, and F. Mostashari. 2007. Monitoring the impact of influenza by age: Emergency department fever and respiratory complaint surveillance in New York City. *PLoS Med* 4(8): e247.

Petersen, W.F. 1934. *The Patient and the Weather*. Vol. IV, Part 2, Ann Arbor, MI: Edward Brothers.

Rydman, R.J., D.P. Rumoro, J.C. Silva, T.M. Hogan, and L.M. Kampe. 1999. The rate and risk of heat-related illness in hospital emergency departments during the 1995 Chicago heat disaster. *Journal of Medical Systems* 23(1): 41–56.

Sargent, F. 1982. *Hippocratic Heritage: A History of Ideas about Weather and Human Health*. New York: Pergamon Press.

Schull, M.J., M.M. Mamdani, and J. Fang. 2005. Influenza and emergency department utilization by elders. *Academic Emergency Medicine* 12(4): 338–344.

Serfling, R.E., I.L. Sherman, and W.J. Houseworth. 1967. Excess pneumonia influenza mortality by age and sex in three major influenza A2 epidemics, United States, 1957–58, 1960 and 1963. *American Journal of Epidemiology* 86: 433–441.

Tam, K., K. Yousey-Hindes, and J.L. Hadlera. 2014. Influenza-related hospitalization of adults associated with low census tract socioeconomic status and female sex in New Haven County, Connecticut, 2007–2011. *Influenza and Other Respiratory Viruses* 8(3): 274–281.

Thompson, W.W., D.K. Shay, E. Weintraub, L. Brammer, N. Cox, L.J. Anderson, and K. Fukuda. 2003. Mortality associated with influenza and respiratory syncytial virus in the United States. *JAMA* 289: 179–186.

Travers, D., S.W. Haas, A.E. Waller, T.A. Schwartz, J. Mostafa, N.C. Best, and J.Crouch. 2013. Implementation of emergency medical text classifier for syndromic surveillance. *AMIA Annual Symposium Proceedings* 2013: 1365–1374.

Viboud, C., M.I. Nelson, Y. Tan, and E.C. Holmes. 2013. Contrasting the epidemiological and evolutionary dynamics of influenza spatial transmission. *Philosophical Transactions of the Royal Society of London B* 368: 20120199.

Weiss, B.P., L. Mascola, and S.L. Fannin. 1988. Public health at the 1984 Summer Olympics: The Los Angeles County experience. *American Journal of Public Health* s78(6): 686–688.

Section III

Data Integrations and Their
Applications in Health
Disparity Assessments

13 Linking Cancer Registry Data to Hospital Discharge Data

13.1 INTRODUCTION

State cancer registries are available for all 50 U.S. states and the District of Columbia. Incidence and mortality information from state cancer registries has been routinely used for cancer surveillance. The Centers for Disease Control and Prevention (CDC) National Program of Cancer Registries (NPCR) regularly publishes federal statistics on cancer incidence in the *United States Cancer Statistics* report, using combined state registry data. While previous studies using cancer registry data have made significant contributions to cancer incidence and staging surveillance (Koh et al., 2005; Niu et al., 2010), treatment information in cancer registries has rarely been used to inform the public and policy makers about treatment disparity and treatment effectiveness. As the national priority has moved from disease surveillance to eliminating disparities in cancer care (Reuben et al., 2011), demands for cancer care surveillance have increased. In 2000, the Institute of Medicine recommended that investigators use existing data systems, such as the NPCR, to measure variations in the use of appropriate standards of cancer care and to assess care outcomes (Hewitt and Simone, 2000). At the bedside, physicians use cancer sites, cancer stages, tumor sizes, and patient comorbidity profiles to decide which standard of care to follow. Standard cancer care is best practice accepted by medical experts as a proper treatment for a certain type of cancer. Major treatments include surgery, chemotherapy (CT), and radiation therapy (RT), and most of them are available in hospital records. Since each physician only accesses his or her patient profile and treatment information, a cancer registry is a natural home for comparing treatment outcomes at the population level.

However, cancer registry data do not usually have complete treatment information, making it difficult to investigate cancer care disparities using only registry data. To overcome registry data limitations, previous studies have used three approaches to examine cancer treatment disparities. One approach uses hospital data to examine treatment disparities (Hershman et al., 2005; Koscuiszka et al., 2011), but samples from selected hospitals tend to be small, and the results are not generalizable. The second approach uses an enlarged sample, linking Surveillance, Epidemiology, and End Results (SEER) and Medicare datasets. Studies using SEER–Medicare linked datasets are able to identify detailed CT and RT treatment among those aged 65 years or older (Brooks et al., 2000; Warren et al., 2002; Du et al., 2008; White et al., 2010; Gross et al., 2008), but those not eligible for Medicare are not included. The third approach uses surveys to supplement cancer registry information (Polednak, 2004). However, this method is costly, which makes it economically unfeasible for surveillance purposes that require updated CT or RT information for all cancer sites. In addition, survey data may be less complete than registry data, as many physicians would not respond to a survey.

Previous studies have linked a statewide cancer registry with hospital discharge data (HDD) for cancer case ascertainment (Penberthy et al., 2003) and for uncovering hard-to-find racial categories (Johnson et al., 2009; Jim et al., 2008). Polednak (2001) linked Connecticut (a SEER state) cancer registry data to HDD, and he expanded the linkage population to all ages. However, his purpose was to examine postmastectomy breast reconstructive surgery in the discharge data, rather than updating cancer treatment in the cancer registry. To our knowledge, a data linkage strategy has not been used to identify cancer treatment information, particularly outpatient treatment. Although the SEER database has been linked to Medicare claims data for gaining treatment and treatment cost information, there is a need to study treatment patterns for non-SEER states and non-Medicare patients.

In this study, we conducted cancer treatment surveillance by linking the Nebraska Cancer Registry (NCR) with Nebraska HDD. The intent was to develop a protocol for linking cancer registry data and HDD and assess its feasibility and utility. In Nebraska, more than 90% of RT is hospital based, and an overwhelming majority of CT is administered in outpatient settings. Similar to linked SEER–Medicare data, linked NCR–Nebraska HDD data will result in a population-based data source that will include both Medicare and non-Medicare patients. In addition, the linked dataset will be able to provide information on non-cancer-related diagnoses, treatments, and costs of care, and can therefore be used for treatment surveillance, clinical epidemiology, and health services research. We use the Charlson Comorbidity Index to demonstrate the potential application of this linked dataset.

13.2 METHOD

13.2.1 NCR DATA

Since 1995, the NCR has continuously received the North American Association of Central Cancer Registries' gold standard awards for quality, completeness, and timeliness. The NCR includes data on cancer patients who reside in Nebraska or who are diagnosed or treated for cancer in Nebraska, or both; patient-identifiable information, such as patient ID, name, social security number, and street address; and standard data items, such as patient demographics, tumor site, first-course treatment, and treatment facility ID. Although the majority of cancer cases are registered shortly after diagnosis, case finding is an ongoing process, and it takes up to 2 years after diagnosis to get all cancer cases into the system. For this reason, we selected data from 2005 to 2009, which assured complete case ascertainment.

13.2.2 NEBRASKA HDD

Nebraska HDD are maintained by the Nebraska Hospital Association (NHA) and are based on the standard UB-04 form. For the 2005–2009 study period, we had more than 13 million hospital records. Nebraska HDD include patient-identifiable information, disposition information, demographic variables, diagnostic codes, and procedure codes for all inpatient and outpatient visits to Nebraska's 87 nonmilitary hospitals. The identifiable information in the Nebraska HDD file includes patient ID, patient name, address, and hospital ID. Hospital visits include outpatient visits for such things as CT, RT, emergency room care, and rehabilitation care. Current procedural terminology (CPT) codes used for insurance billing purposes can be retrieved to identify different treatment categories. For the purpose of augmenting outpatient treatment information, we requested CPT codes that directly indicated RT or CT. For CT procedures, the CPT codes include 96400–96549, C8953–C9415, and G0355/G0359; for RT procedures, the CPT codes include 70010–79999 (Abraham et al., 2010).

Inpatient procedures with an International Classification of Diseases, Ninth Revision (ICD-9) code indicating cancer treatment were requested, but specific RT or CT codes for billing were not requested due to bundled diagnosis-related group (DRG) codes that cannot uniquely differentiate CT and RT. For this reason, we only searched the ICD-9 procedure codes up to the sixth procedure, and coded CT and RT procedures as much as we could. Specifically, we used ICD-9 procedure codes in 99.25, 17.70, 94.25, 94.22, and 99.28 to capture CT-related procedures, and we used the codes from 99.21 to 99.29 to capture RT-related procedures (CDC, 2012).

13.2.3 Linkage Strategy

We considered deterministic and probabilistic linkage methods and chose the probabilistic method due to lack of a unique identifier (e.g., social security number) between the two datasets. We designed the linkage process according to Newcombe's four steps (Newcombe, 1988)—(1) data preparation, (2) matching and merging, (3) manual review, and (4) verification—and used Link Plus 2.0 for data linkage (CDC, 2006). In addition, all these steps were conducted on site at the Nebraska Hospital Association. The resultant file for the analysis had no name, address, or zip code identifiers. The University of Nebraska Medical Center Institutional Review Board (IRB) approved this study design with the exempt status.

13.2.3.1 Data Preparation

Data preparation included checking data quality, deduplication, and parsing, and standardization of the linkage variables common in both datasets. The process started with the NCR data for January 1, 2005 to December 31, 2009, with 54,990 records (Figure 13.1, top circle), which included both in-state and out-of-state patients. The deduplication process was based on all linkage variables: first name, last name, date of birth, sex, resident county, resident zip code, and primary cancer site. Less than 0.1% of registry cases had missing linkage variables. This step resulted in 52,027 unique patients (Figure 13.1, diamond box).

Since the 5-year Nebraska HDD file was extremely large, we divided it into cancer-related and non-cancer-related datasets to increase computational efficiency (Figure 13.1, left circle). For each record, we searched ICD-9-CM diagnostic codes for up to the 10th diagnostic (Penberthy et al., 2003). If any of them were in the range of 140–208 or equaled 2386, we classified them as cancer related (CDC, 2012).

13.2.3.2 Matching

The cancer-related and non-cancer-related datasets were the basis for partitioning the Nebraska HDD file into mutually exclusive blocks and comparing only records within each block during the linkage process. When a linkage is complete, Link Plus calculates a score (total linkage weight) for all comparisons of a case in the NCR file with a case in the Nebraska HDD file, and generates a default cutoff weight above which values are considered potential matches (CDC, 2006; Blakely and Salmond, 2002).

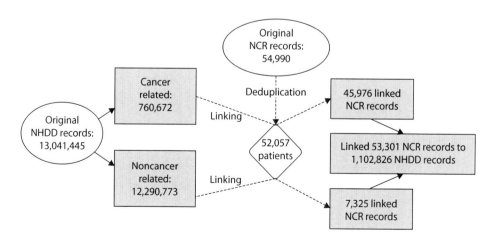

FIGURE 13.1 Flowchart for linking NCR and Nebraska HDD records.

13.2.3.3 Manual Review

All potential matched pairs above the cutoff weight were reviewed manually after each linkage run. We also considered a pair as a true match if there was a minor discrepancy, such as an obvious typo in the matching variables; transposed first and last names; either the first name, the last name, or the middle initial matched with another unmatched name; variants of a first name that the Soundex did not catch; a hyphenated first name or last name; transposed birth month and day; and closeness of the date of cancer diagnosis and the date of hospital admission. After manual review, we had 53,301 NCR records that corresponded to 1,102,826 Nebraska HDD records (Figure 13.1, leftmost box).

13.2.3.4 Validation

To valid the potential match of 54,990 records, we randomly selected 400 linkage pairs for an independent manual review, because a random sample of 381 records would be sufficient at the 95% confidence level.

13.2.4 Treatment Augmentation Strategy

We focus on RT treatment updates for colorectal cancer (CRC) and breast cancer. We chose all CRC records as a general case study because substantial rural and urban variation was reported (Kinney et al., 2006). We selected American Joint Committee on Cancer (AJCC) stage II and III breast cancer patients as a specific case study, because the National Comprehensive Cancer Network recommends adjuvant RT for all those patients (National Comprehensive Cancer Network, 2011).

In order to update RT for a cancer patient, it is necessary to meet two conditions: (1) the RT is for the diagnosed cancer site, and (2) the linked hospital stays or visits are after the site-specific diagnostic date. For condition 1, we selected patients with the primary site being either breast cancer or CRC only. For condition 2, we used date of diagnosis to exclude all outpatient visits prior to the diagnosis date. Although there were no missing service dates from the Nebraska HDD file, about 2.5% of the diagnostic months and days were missing from the NCR file. To preserve usable records with missing months or dates, we designed an algorithm to hierarchically delete missing records. When diagnosis and treatment years were the same, records with missing months were deleted, when diagnosis and treatment months and years were the same, records with missing dates were deleted.

13.3 RESULTS

13.3.1 Linkage Quality

Of the 54,990 NCR records, 53,301 were linked to Nebraska HDD records, with a crude linkage rate of 97%. The linkage rate varied by reporting sources (Table 13.1). The highest linkage rate, 99%, was from the NHA reporting hospitals. Because the quality of both datasets was high and because

TABLE 13.1

Linkage Assessment by Reporting Facility: 2005–2009

Reporting Facility	Linked	Unlinked	PPV (%)	95% CI
NHA hospitals	30,032	338	99	99, 99
Non-NHA hospitals	19,021	914	95	95, 95
Labs, physicians' offices, etc.	3,624	316	92	91, 93
Autopsy and death records only	624	121	83	81, 86

PPV, positive predictive value; CI, confidence interval.

we adhered to the protocol, the linkage rate was 97% overall. Only NCR records from autopsy or death certificates had a low linkage rate (83%), which was expected, as some individuals who died had not gone to a hospital. Manual reviews were conducted by three judges, with 12 variables for 400 randomly selected pairs. The most false positives were 3, followed by 1 and 0, resulting in an average false positive rate of 0.3%. These reviews, as well as the results presented in Table 13.1, suggested that the linkage yielded very high-quality data.

13.3.2   POTENTIAL TO AID CASE FINDING

Although the main purpose of this study was to improve treatment information, it is always important for the NCR to augment potential missing cases. In the Nebraska HDD cancer-related dataset (i.e., with a cancer-related ICD-9 code), we found 19,907 person-specific records with the primary diagnosis of cancer that were not in the NCR. We put these records in the potential NCR cancer case file (PNCCF). Since person-specific records included multiple visits, primarily for outpatient care, we selected a unique record for each person in the PNCCF, and this process yielded 4270 unique patient-based records. We also deleted suspected out-of-state patients who might have sought continued outpatient care in Nebraska while traveling. We selected all out-of-state outpatients and deleted those with only one RT or CT visit, which resulted in 3593 not captured by the NCR 2005–2009 incidence file.

However, some of the 3593 NHA records could be patients prior to 2005 from the NCR file that were not included in the linkage process above. We quickly linked these records with the NCR file from 1990 to 2005 by the deterministic data linkage method, using first name, last name, sex, and birth date. This process yielded 934 patients from the early NCR file who continued their cancer-related RT after 2005. By removing these records, we had 2659 records to be further assessed for case finding.

13.3.3   POTENTIAL TO IMPROVE RADIATION TREATMENT INFORMATION IN THE NCR

As mentioned earlier, we used RT for CRC as a general case study that included all linked CRC patients. Based on the method that placed a set of RT visits after the corresponding CRC diagnostic date, we had 2150 CRC patients in the linked NCR–Nebraska HDD file. Of those, 750 records in the NCR file showed RT as the first course of treatment, and 90 records were identified from outpatient CPT codes indicating receipt of RT after cancer diagnosis. These findings represented an improvement of 12.5% over the information captured by the NCR file. To assess rural–urban difference in CRC treatment, we divided the 90 newly found cases into metropolitan and nonmetropolitan counties. Although 60% of Nebraskans live in metropolitan counties, 60% of the augmented RT patients were from nonmetropolitan areas. This preliminary result suggested that missing treatment records in the NCR were more likely to be for rural patients than for urban patients. Although 12.5% of additional records may not seem like a large number, those records raised the overall RT treatment rate in Nebraska from 34.9% to 39.1%.

13.3.4   POTENTIAL TO IMPROVE RADIATION TREATMENT INFORMATION FOR PUBLIC HEALTH

In our analysis of CRC treatment, we did not know how many cancer patients would need RT, and therefore we could not assess whether the 12.5% additional RT records were adequate for updating CRC RT records. To investigate further, a total of 6887 AJCC stages II and III breast cancer patients, who were expected to have RT, were identified (Table 13.2). Of the total, 2454 had RT confirmed by both the NCR and the Nebraska HDD. Of the total, 1738 patients were reported to have received RT by the NCR, but records for these patients were not found in the outpatient CPT codes. Of the 2650 patients not reported to have received RT by the NCR, 620 (23.13%) were found in the Nebraska HDD file. These 620 records represent a 14.8% improvement over the 4129 records initially shown to have received RT. A total of 69.9% of patients had verifiable RT.

TABLE 13.2

Radiology Treatment Reported by NCR and Nebraska HDD

From NCR	From Nebraska HDD Outpatient Records				
	Yes	Row %	No	Row %	Total
Yes	2454	(58.54)	1738	(41.46)	4192
No	620	(23.13)	2075	(77.62)	2695
Total	3074	(44.53)	3813	(44.34)	6887

TABLE 13.3

RT (%) by Poverty: NCR vs. Nebraska HDD

	Pvt1	Pvt2	Pvt3	Pvt4	Pvt5
Directly from NCR	45.31	46.46	43.02	41.06	37.28
Nebraska HDD updated	52.12	51.85	48.42	47.02	44.74
Total *N*	1558	2484	1799	789	912

Note: Pvt1, <5%; pvt2, 5%–10%; pvt3, 10%–15%; pvt4, 15%–20%; pvt5, >20%.

One potential application of using the updated RT information is to investigate treatment disparity. Table 13.3 compares RT derived from NCR and NCR + Nebraska HDD by poverty level. Here, the poverty levels are in five categories (pvt1, <5%; pvt2, 5%–10%; pvt3, 10%–15%; pvt4, 15%–20%; pvt5, >20). If we use NCR numbers, the socioeconomic status (SES) gradient along the poverty levels is not as apparent as using the Nebraska HDD updated RT information. The former has a hump in the poverty 2 category, and then declines, while the latter has a linear decline from 52.12% to 44.74%. This shows that the poorer the neighborhood, the less chance of receiving RT, while not accounting for any other breast cancer patient variables.

13.3.5 POTENTIAL USE OF COMORBIDITY INFORMATION

Even though the standard cancer registry data items include comorbidities, they are rarely collected or compiled. If one wants to study survival or treatment disparities, comorbidities should be controlled. For surveillance purposes, we calculated Charlson comorbidity scores by major cancer sites. The calculation of the Charlson Index includes the diagnosis of 17 comorbidities for both inpatient and outpatient care: myocardial infarction, heart failure, peripheral vascular disease, cerebrovascular disease, dementia, chronic pulmonary disease, connective tissue disease—rheumatic disease, peptic ulcer disease, mild liver disease, diabetes without complications, diabetes with complications, paraplegia and hemiplegia, renal disease, cancer, moderate or severe liver disease, metastatic carcinoma, and HIV/AIDS. The algorithm works as follows:

1. Use cancer diagnosis date to set the cutoff date for calculating the index to make sure that the index is based on comorbidities prior to cancer diagnosis.
2. Count the number of distinct comorbidities for each patient by searching up to 10 diagnosis codes from inpatient visits until the diagnosis date.
3. Count the number of distinct comorbidities for each patient by searching up to 10 diagnosis codes from outpatient visits until the diagnosis date.

4. Retain the record with the maximum number of comorbidities from either in- or outpatient data. If the inpatient and outpatient comorbidities are equal, use those from the inpatient.
5. Calculate the score for each selected cancer site, and summarize all.

Table 13.4 lists the results for up to four comorbidities. This simply shows the potential from the Charlson Index; it does not represent the actual number of comorbidities. Even though we found that each cancer patient would have 1.2 inpatient visits and 19 outpatient visits, many of them were after the cancer diagnosis, and therefore had to be ignored. This is due to our data request restriction. We paired 2005–2009 cancer incidence data to 2005–2009 hospital data. If we had included 2004 or earlier HDD, the percent listed under the column heading "0" would be much smaller.

The application of the Charlson Index can be based on either the total number of comorbidities or discrete categories. In Figure 13.2, we use the latter to show the importance of controlling for comorbidities when investigating colon cancer survivals. The top line has 1 or 0 comorbidities, and it has the gentlest survival curve. On the other hand, the bottom curve is associated with four or more comorbidities, and has a relatively steepest survival curve.

TABLE 13.4

Charlson Comorbidity Scores by Cancer Sites: 2005–2009

Cancer Sites	0	1	2	3	4	Total
Oral cavity and pharynx	36.24	39.78	17.04	5.47	1.46	1297
Esophagus	21.49	41.17	25.87	9.47	2	549
Stomach	22.81	45.44	21.58	8.25	1.93	570
Colon and rectum	27.54	39.43	22.5	8.1	2.43	5876
Liver and intrahepatic duct	15.6	30.83	35.6	15.05	2.94	545
Pancreas	16.77	41.37	28.39	11.69	1.77	1240
Larynx	27.06	40.83	22.94	7.8	1.38	436
Lung and bronchus	13.17	35.59	33.04	14.14	4.06	6356
Skin melanoma	60.45	25.93	9.4	3.21	1.01	2584
Breast	51.06	34.26	10.52	3.4	0.76	8319
Cervix uteri	56.25	29.83	12.22	1.42	0.28	352
Corpus uteri	49.22	33.76	12.46	3.54	1.02	1469
Ovary	30.03	45.41	19.23	4.59	0.74	676
Prostate	51.98	30.51	11.52	4.48	1.5	7049
Testis	64.16	33.45	2.05	0.34	0	293
Urinary bladder	34.67	32.69	20.68	9.23	2.72	2166
Brain and other central nervous system (CNS)	52.63	29.49	12.39	4.27	1.22	1638
Thyroid gland	56.74	32.61	8.07	2.18	0.4	1239
Hodgkin's disease	66.06	23.76	7.83	2.09	0.26	383
Non-Hodgkin's lymphoma	50.22	29.66	14.09	4.67	1.36	2505
Multiple myeloma	39.05	35.29	17.48	6.21	1.96	612
Leukemia	45.75	31.93	15.27	5.39	1.66	1447
Kidney and renal pelvis	34.87	33.79	19.18	9.7	2.45	1752
Other sites	31.97	36.7	20.71	7.8	2.82	4510
Total	39.27	34.29	17.78	6.75	1.91	53863

Note: The cores were calculated from hospital in- and outpatient visits.

Product-limit survival estimates

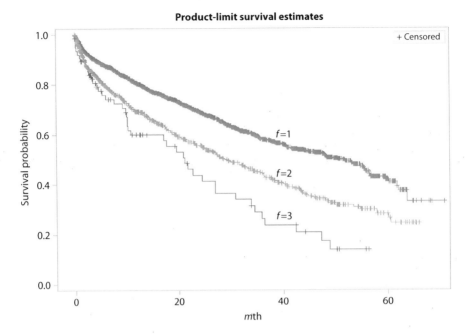

FIGURE 13.2   Age-, sex-, and stage-adjusted colon cancer survival curves by the number of comorbidities from the Charlson Index. *f*, comorbidity category; 1, one or no comorbidities; 2, two or three comorbidities; 3, four or more comorbidities.

13.4   CHAPTER SUMMARY

In this study, we linked NCR and Nebraska HDD data for 2005–2009. Our linkage was population based and included all age groups and all types of cancer. We used Link Plus 2.0 and followed an established procedure of data preparation, matching, manual review, and independent validation. Because the quality of both datasets was high and because we adhered to the protocol, the linkage rate was 97% overall. Only NCR records from autopsy or death certificates had a low linkage rate (83%), which was expected, as some individuals who died had not gone to a hospital.

Linking HDD with registry data has emerged as a major source of gaining diagnosis and treatment procedure information (Van Walraven et al., 2011; Chantry et al., 2011). This study piloted three applications: population-based case finding, cancer site-specific treatment information enhancement, and site- and stage-specific treatment information enhancement. Our experience suggests that population-based case finding via a potential false negative (i.e., those records with a cancer diagnosis in the Nebraska HDD file but not in the NCR) will require further study regarding a new definition of cancer cases. We identified 2659 Nebraska patients in the Nebraska HDD who had some indication of cancer treatment but who were not in the NCR file. They represented about 5% of the total cases from 2005 to 2009. One area that needs further clarification is outpatient care of out-of-state patients. If these patients' primary diagnosis and treatment took place outside the state of Nebraska, and their postsurgery CT or RT outpatient care took place in Nebraska, hospital staff may be less diligent about reporting their care in the NCR than they are about reporting the care of Nebraska citizens.

Our effort to identify additional RT cases for CRC and breast cancer patients resulted in modest 12.5% and 15% improvements. Since all breast cancer stage II and III patients were expected to have RT, we focused on their assessment. After the update, the RT rate increased from 61% to 70%, which still left about 30% of the records unverified for RT. Although the increase in treatment rate seems modest, it is consistent with a previous study that compared the cancer registry with a chart review for the patients of PacifiCare of California and showed that the RT rate increased from 30.6 to 41.1 (Malin et al., 2002). In addition, even though the 70% RT rate seems low, it was just about 8% below a chart review–based study from the AVON Center for Breast Care in Atlanta, where all breast patients were under age 72 and had breast-conserving surgery (Iyengar et al., 2010). Since our study had no age and surgery restrictions, the lower percentage was expected. Finally, the period in the current study covers the adoption of hospital-based quality indicator standards for stages I–III breast cancer, which was endorsed by the National Quality Forum (NQF) in 2007. As shown in a recent study, prior to the 2005–2006 period, the RT rate was not very high (76%), and the rate increased to 96% after 2007 (Rizzo et al., 2011).

There are several additional reasons for missing RT data in this linkage project. First, in order to ensure that outpatient RT care was cancer specific, we placed outpatient service dates for RT after cancer diagnosis. However, some dates are missing in the NCR, and we might have missed a small number of cancer treatments due to date discrepancies. Second, some RT may have been coded with other outpatient procedures that we were unable to straightforwardly select. Third, we assumed that hospital-based RT treatment was captured primarily by the NCR or from the inpatient data file, as it is primarily a hospital-based registry. However, some first-course RT might have been missed by the NCR, and the bundled DRG codes from the Nebraska HDD cannot uniquely identify RT. Although it is possible to uncover specific procedure codes, for both inpatient and outpatient care, doing so requires different protocols to govern the data request. Since our linkage protocol was designed to capture primarily outpatient cancer care, future studies should design a protocol that includes both inpatient and outpatient cancer treatment codes. In addition, some patients who had surgery in Nebraska hospitals may have received CT and RT out of state, which would not have been captured within the Nebraska HDD. Finally, we did not have a validation sample, and therefore were unable to tell whether the Nebraska HDD linked file captured all RT cases. We selected all stage II and III breast cancer patients, and all of them were supposed to have had RT. However, some patients may not have followed the prescribed treatment. Although we expected that more than 70% of these patients would have received RT, expecting a 100% RT rate might be unrealistic.

The limitations described above provide avenues for future studies. First, our study provided an empirical basis for further improvement of linkage protocols that may broaden the inclusion of CPT and other bundled procedure codes. Second, multiple and repeated procedures reinforced the need for improvement in diagnosis month and date completeness in the NCR. As information networks are connected throughout the state, it will be possible to develop a real-time treatment update information system, so that missed information can be checked in real time or close to real time.

In addition to the above cancer registry–specific data quality enhancement, the linked NCR–Nebraska HDD database provides access to additional information about comorbidity and service charges. For example, we can use primary to tertiary diagnosis codes to develop comorbidity indices. Since non-cancer-related hospital visits are also linked, postsurgery health service needs can be examined. Moreover, linked cancer treatment and other information will likely become a valuable asset for public health practitioners and researchers (Xia et al., 2010). Examples of possible uses of such linked data include surveillance of treatment disparity, health outcome comparison of different treatment regimens, and population-based health policy development.

REFERENCES

Abraham, M., J.T. Ahlman, A.J. Boudreau, J.L. Connelly, and D.D. Evans. 2010. *CPT 2011: Standard Edition.* Chicago, IL: American Medical Association Press.

Bernard, S.M., J.M. Samet, A. Grambsch, K.L. Ebi, and I. Romieu 2001. The potential impacts of climate variability and change on air pollution-related health effects in the United States. *Environmental Health Perspectives* 109(Suppl. 2): 199–209.

Blakely, T. and C. Salmond. 2002. Probabilistic record linkage and a method to calculate the positive predictive value. *International Journal of Epidemiology* 31: 1246–1252.

Brooks, J., E. Chrischilles, S. Scott, J. Ritho, and S. Chen-Hardee. 2000. Information gained from linking SEER cancer registry data to state-level hospital discharge abstracts. *Medical Care* 38(11): 1131–1140.

CDC (Centers for Disease Control and Prevention). 2006. Link Plus version 2.10 [probabilistic record linkage software]. Atlanta, GA: CDC.

CDC (Centers for Disease Control and Prevention). 2012. International Classification of Diseases, Ninth Revision, Clinical Modification (ICD-9-CM). Atlanta, GA: CDC. http://www.cdc.gov/nchs/icd/icd9cm.htm (accessed May 17, 2012).

Chantry, A.A., C. Deneux-Tharaux, C. Cans, A. Ego, C. Quantin, and M.H. Bouvier-Colle; GRACE Study Group. 2011. Hospital discharge data can be used for monitoring procedures and intensive care related to severe maternal morbidity. *Journal of Clinical Epidemiology* 64(9): 1014–1022.

Drake, D. 1850. *A Systematic Treatise, Historical, Etiological, and Practical, on the Principal Diseases of the Interior Valley of North America.* Cincinnati, OH: W. B. Smith, p. 447.

Du, X.L., S.Y. Fang, and T.E. Meyer. 2008. Impact of treatment and socioeconomic status on racial disparities in survival among older women with breast cancer. *American Journal of Clinical Oncology* 31: 125–132.

Gross, C.P., B.D. Smith, E. Wolf, and M. Andersen. 2008. Racial disparities in cancer therapy: Did the gap narrow between 1992 and 2002? *Cancer* 112: 900–908.

Hershman, D.L., R. McBride, J.S. Jacobson, L. Lamerato, K. Roberts, V.R. Grann, and Al. Neugut. 2005. Racial disparities in treatment and survival among women with early-stage breast cancer. *Journal of Clinical Oncology* 23: 6639–6646.

Hewitt, M. and J.V. Simone. 2000. *Enhancing Data Systems to Improve the Quality of Cancer Care.* Washington, DC: National Academy Press.

Iyengar, R., M.J. Lund, P. Lamson, L. Holmes, M. Rizzo, H. Bumpers, J. Okoli, D. Senior-Crosby, R. O'Regan, and S.G. Gabram. 2010. Using National Quality Forum breast cancer indicators to measure quality of care for patients in an AVON comprehensive breast center. *Breast Journal* 16(3): 240–244.

Jim, M.A., D.G. Perdue, L.C. Richardson, D.K. Espey, J.T. Redd, H.J. Martin, S.L. Kwong, J.J. Kelly, J.A. Henderson, and F. Ahmed. 2008. Primary liver cancer incidence among American Indians and Alaska Natives, US, 1999–2004. *Cancer* 113(5 Suppl.): 1244–1255.

Johnson, J., A.S. Soliman, D. Tadgerson, G.E. Copeland, D.A. Seefeld, N.L. Pingatore, R. Haverkate, M. Banerjee, and M.A. Roubidoux. 2009. Tribal linkage and race data quality for American Indians in a state cancer registry. *American Journal of Preventive Medicine* 36(6): 549–554.

Kinney, A.Y., J. Harrell, M. Slattery, C. Martin, and R.S. Sandler. 2006. Rural-urban differences in colon cancer risk in blacks and whites: The North Carolina Colon Cancer Study. *Journal of Rural Health* 22(2): 124–130.

Koh, H.K., C.M. Judge, B. Ferrer, and S.T. Gershman. 2005. Using public health data systems to understand and eliminate cancer disparities. *Cancer Causes Control* 16(1): 15–26.

Koscuiszka, M., D. Hatcher, P.J. Christos, A.E. Rose, H.S. Greenwald, Y.-L. Chiu, S.S. Taneja, M. Mazumdar, P. Lee, and I. Osman. 2011. Impact of race on survival in patients with clinically nonmetastatic prostate cancer who deferred primary treatment. *Cancer* 118(12): 3145–3152. doi: 10.1002/cncr.26619.

Malin, J.L., K.L. Kahn, J. Adams, L. Kwan, M. Laouri, and P.A. Ganz. 2002. Validity of cancer registry data for measuring the quality of breast cancer care. *Journal of the National Cancer Institute* 94(11): 835–844.

National Comprehensive Cancer Network. 2011. Clinical practice guidelines in oncology: Breast cancer. V. 2.2011. http://www.nccn.org/professionals/physician_gls/PDF/breast.pdf. (accessed February 2011).

Newcombe, H.B. 1988. *Handbook of Record Linkage: Methods for Health and Statistical Studies, Administration, and Business.* Oxford: Oxford University Press.

Niu, X., K.S. Pawlish, and L.M. Roche. 2010. Cancer survival disparities by race/ethnicity and socioeconomic status in New Jersey. *Journal of Health Care for the Poor and Underserved* 21(1): 144–160.

Penberthy, L., D. McClish, A. Pugh, W. Smith, C. Manning, and S. Retchin. 2003. Using hospital discharge files to enhance cancer surveillance. *American Journal of Epidemiology* 158: 27–34.

Polednak, A.P. 2001. How frequent is postmastectomy breast reconstructive surgery? A study linking two statewide databases. *Plastic and Reconstructive Surgery* 108: 73–77.

Polednak, A.P. 2004. Chemotherapy of non-elderly breast cancer patients by poverty-rate of area of residence in Connecticut. *Breast Cancer Research and Treatment* 83(3): 245–248.

Reuben, S.H., E.L. Milliken, and L.J. Paradis. 2011. America's demographic and cultural transformation: Implications for cancer. President's Cancer Panel • 2009–2010 Annual Report. http://deainfo.nci.nih.gov/advisory/pcp/annualReports/pcp09-10rpt/pcp09-10rpt.pdf.

Rizzo, M., H. Bumpers, J. Okoli, D. Senior-Crosby, R. O'Regan, A. Zelnak, L. Pan, M. Mosunjac, S.G. Patterson, and S.G. Gabram. 2011. Improving on national quality indicators of breast cancer care in a large public hospital as a means to decrease disparities for African American women. *Annals of Surgical Oncology* 18(1): 34–39.

Van Walraven, C., C. Bennett, and A.J. Forster. 2011. Administrative database research infrequently used validated diagnostic or procedural codes. *Journal of Clinical Epidemiology* 64(10): 1054–1059.

Warren, J.L., C.N. Klabunde, D. Schrag, P.B. Bach, and G.F. Riley. 2002. Overview of the SEER-Medicare data: Content, research applications, and generalizability to the United States elderly population. *Medical Care* 40(8 Suppl.): IV-3-18.

White, A., S.W. Vernon, L. Franzini, and X.L. Du. 2010. Racial disparities in colorectal cancer survival: To what extent are racial disparities explained by differences in treatment, tumor characteristics, or hospital characteristics? *Cancer* 116(19): 4622–4631.

Xia, Q., J.L. Westenhouse, A.F. Schultz, A. Nonoyama, and W. Elms. 2010. Matching AIDS and tuberculosis registry data to identify AIDS/tuberculosis comorbidity cases in California. *Health Informatics Journal* 17(1): 41–50.

14 Mother Index and Its Applications

14.1 BIRTH CERTIFICATE DATA LINKAGE: A BRIEF REVIEW

Data linkage based on birth certificate data is commonly used for interprogram collaborations among various public health programs at the state level. Birth records were routinely linked to death records to study perinatal, neonatal, and infant mortality. The Centers for Disease Control and Prevention (CDC)-funded Pregnancy Risk Assessment Monitoring System (PRAMS) is based on state surveys of women who had a live birth, in which birth certificate information is linked to mothers' attitudes and experiences before, during, and shortly after pregnancy. Medicaid claims are often linked to birth certificates to monitor birth outcomes for Medicaid program participants in comparison to nonparticipants. Birth defects in a typical state birth defect registry are linked not only to birth records, but also to birthing mothers. In addition, birth certificate data are often used in infant hearing screening; the Special Supplemental Nutrition Program for Women, Infants, and Children (WIC); social welfare; and foster care programs. In this chapter, we first review major birth record linkage application areas, and then present Nebraska's experience of building a birth index file. In this section, we review major birth record linkage projects.

14.1.1 LBW Paradox

Birth weight is a strong predictor of infant mortality. Birth weight became a standard item on the national birth certificate system in 1950, and it is one of the most commonly used health indicators. For most states, some forms of birth certificate data can be obtained from 1950 onward, but the use of linked birth and death records at the national level started 30 years later. Allen Wilcox (2001) synthesized his previous works with his collogues, in which a normalized z-score was used to show that birth weight among different segments of the population followed almost identical normal curves with a left tail. In particular, their comparisons were infants at high altitude versus low altitude, Blacks versus Whites, and infants exposed versus unexposed to their mothers' smoking. In all examples, the reference groups, or the ones with a lower risk, had higher infant mortality. This is the so-called low-birth-weight (LBW) paradox, in that population groups with a higher proportion of LBW babies often have higher infant mortality, yet LBW babies in high-risk populations tended to have lower mortality than LBW babies in the comparison groups (Adams et al., 1997). Wilcox (2001, p. 1039) concluded that "the mortality difference must be due either to a difference in small preterm births or to differences in weight-specific mortality that are independent of birthweight." Wilcox focused on the importance of preterm delivery in infant mortality, and implied a universal and biological process where interventions can tweak but not change the shape of birth weight curves (David, 2001; Hertz-Picciotto, 2001).

Even though some of Wilcox's conclusions are debatable, the derivation of normalized birth weight curves for different populations and the emphasis of the left tails of these curves generated many further inquiries that require sophisticated record linkage. For example, the work on the outcomes of second births by Skjaerven et al. (1988) required indexing mothers to identify mothers of multiple births. Traditional studies of birth outcomes are based on the individual pregnancies. Based on rich data from Norway from 1967 to 1984, Skjaerven et al. were among the first to examine infant mortality by conditioning on first births. They found that although birth weight and perinatal mortality had a strong and positive association, this association is negatively correlated with the birth weight of first births, suggesting that it was relative birth weight to the first births, rather than absolute birth weight, that predicted perinatal mortality. Using the long period of the same dataset, Skjærven et al. linked mother's birth weight to the survival of her infants. It was found that there is a positive and strong correlation between mother and baby birth weights. In particular, when a mother's birth weight was below a threshold of 2000 grams, the maternal birth weight was associated with the perinatal survival.

14.1.2 INTERGENERATIONAL DATA FILES

In the United States, there has been a number of intergenerational data linkage projects based on birth certificate data. The earliest one was for mothers born in Tennessee in 1959–1966 and their children in 1979–1984 (Klebanoff and Yip, 1987). The study showed a strong association between maternal and infants' birth weights. About 10 years later, the Illinois transgenerational birth file (TGBF) was constructed by Richard David's group (David et al., 2010), who linked infants' birth certificates in 1989–1991 to mothers' birth certificates from 1956 to 1976 in Illinois. In addition, paternal birth certificates were also linked to obtain parental birth weights. Later on, both parental and infant files were geocoded at the census tract and other community levels. An early study by Coutinho et al. (1997) using the dataset showed a stronger correlation between infant and maternal birth weights than between infant and paternal birth weights. The lowest-birth-weight mothers, however, did not necessarily have smallest infants. The TGBF has generated more than a score of journal articles, mainly led by Collins and David. About the same time the TGBF was constructed, Emanuel and his research team (1999) constructed an international file from birth and hospital discharge data in Washington State. The file was based on obstetric and neonatal admissions between 1987 and 1995, which were linked to the birth certificates of mothers. These data were also linked to vital records for mortality information. Besides confirming findings from Klebanoff and Yip (1987) and Coutinho et al. (1997), most of its applications were related to clinically specific conditions, such as maternal weight versus cesarean delivery, the risk of gestational diabetes mellitus, and respiratory distress (Li et al., 2003; Shy et al., 2000; Strandjord et al., 2000; Williams et al., 1999).

More recently, the Maternal and Child Health Bureau of the Health Resources and Services Administration funded a number of projects to promote a life-course health development perspective. Chapman and Gray (2014) reported the process of building a maternally linked birth dataset for life-course studies with the state health department. They called it the Virginia Intergenerational Birth File, which was based on the 2005–2009 birth cohort with 170,624 births to 136,021 mothers aged 11–48 years within the state. The study also demonstrated the utility of the linked file by showing a strong correlation between maternal LBW and infant LBW, and the overall improvement of birth weight over the generation.

The Georgia longitudinal linked file had the largest birth cohort from 1994 to 2007, or 14 years of birth records, which is essentially a birth record file indexed by mothers. In addition, the project team developed a neighborhood deprivation index to examine neighborhood exposure during multiple pregnancies (Kramer et al., 2014). It was found that mothers with multiple deliveries had upward socioeconomic mobility as they got older. Black women with (1) a history of preterm birth (PTB) and (2) a high cumulative neighborhood deprivation index had double doses of risk for preterm and LBW. Coincidentally, there was a similar longitudinal file built for a Georgian birth cohort of 1980–1992, which did not include the geocoding component (Adams et al., 1997). However, it is not known if this file could be combined with the later file to form a 1980–2007 birth cohort. In any case, the most recent longitudinal dataset from Georgia has generated some novel research studies, such as linking mothers' behavior risk factors and infant birth weight and PTB status to school performance (Feng et al., 2013; Williams et al., 2013). The dataset was also used to follow a cohort of women as they moved from housing projects to seek causal inference of housing transition with birth outcomes (Kramer et al., 2012).

14.1.3 BIRTH RECORD LINKAGE TO MEDICAID CLAIMS DATA

Linking birth records with Medicaid claims and other social services has been ongoing since at least 1970 for program eligibility and other mandatory requirements. Linking Medicaid claims data to birth records for health outcome assessments is fairly recent. Early works on Medicaid–birth record linkage were mostly methodologically oriented (Gyllstrom et al., 2002). Piper et al. (1990) evaluated a linked birth–Medicaid enrollment file that was also linked to fetal death and death certificate files. The authors cautioned the users about representation issues, stating that fully linked files tended to underestimate important outcomes, such as fetal, perinatal, and neonatal mortality rates. Other studies also reported similar underestimation and inconsistency issues (Grisso et al., 1997; Buescher, 1999). A later study echoed some of the concerns raised, but it also pointed out that underestimation biases could mostly be avoided by teasing out different data selection processes during data linkage (see also Bronstein et al., 2009). Qayad and Zhang (2009), while evaluating linkage quality, reported that 48.4% of newborns had Medicaid-paid delivery claims. This number is quite astonishing, and it would not be surprising if some states had more than 50% of Medicaid-paid deliveries. Given the overwhelming numbers of newborns covered by Medicaid, CDC and Centers for Medicare and Medicaid Services (CMS) have joined forces to monitor health outcomes among participants and nonparticipants. A website was also created for linkage training (http://academyhealth.org/datalinkageproject). Several states reported in their training materials that Medicaid-paid deliveries tended to be preterm and LBW.

14.1.4 OTHER DATA LINKAGES BASED ON BIRTH CERTIFICATE DATA

Birth certificate data have been linked to many other data. A notable set of data is birth defects. Most states have a birth defect registry. Although birth defects have the longest history of linking to birth and death records, the linkage part is trivial. Most birth defect registries use the birth certificate number as the ID, and most infant deaths are reviewed and linked back to birth records. A study in Michigan reported that 33.7% of infant deaths could be found in the state birth defect registry (Copeland and Kirby, 2007). Such a proportion is widely reported in birth defect surveillance. Variable reporting is an important issue in birth defect registries, as most states do not have an active registry, relying instead on voluntary reporting by birthing hospitals and attending physicians. A study in New York State showed that the misleading spatial clusters could be attributed to late reporting or nonreporting by some hospitals (Forand et al., 2002).

Birth certificate data are also comprehensively linked to other datasets. The Massachusetts's Pregnancy to Early Life Longitudinal (PELL) system is a statewide longitudinal database partnered with both public and private sectors. The core of the system is based on birth and fetal death files that were linked to hospital discharge records of the mother's delivery and child's birth. In addition, various public health program data, vital statistics datasets, clinical datasets, and census data at a neighborhood level were integrated into the core. Examples include birth defects, childhood mortality, maternal mortality, WIC, newborn screening, substance abuse services, and assisted reproductive technology (Kotelchuck et al., 2014). Like the TGBF in Illinois, the PELL system garnered much research interest with dozens of publications. In addition, it was also used by the Massachusetts's Department of Public Health.

Finally, there are some specialized linkage projects for a particular program. Washington State WIC program data were linked to birth records to assess its protective effect on preterm delivery (El-Bastawissi et al., 2007). The Michigan infant screening and birth record linkage is a web-based program to ensure screening compliance (Korzeniewski et al., 2010). The Ohio home vesting program for high-risk women provides a resource file that links birth records to hospital discharge records, and neighborhood data, so that home visiting staff can be well prepared with personalized materials (Hall et al., 2014). An Oregon linkage project connects children with parents in electronic health records and the state's public health insurance data (Angier et al., 2014). In California, data from Child Protective Services are linked to birth records to study birth outcomes among teen or adolescent mothers who had been involved in Child Protective Services because of maltreatment, neglect, or being foster care children (Putnam-Hornstein et al., 2013, 2015).

14.2 NMI AND ITS APPLICATIONS

14.2.1 Setting Up NMI

All the intergenerational linkage files reviewed above had major federal funding, and most of them were not intended for public health practice. Except studies based on the Illinois TGBF, most birth linkage studies did not attempt to disentangle race, socioeconomic status (SES), and other environmental stressors that pass through generations (Collins et al., 2009). We built the Nebraska mother index (NMI) for data infrastructure enhancement rather than for a research project or for a specific funding requirement. We felt that by building the index now, it would expand our vision for life-course intervention. Although birth records were also linked to hospital discharge data and other data files, we only describe the mother index and its application in this chapter.

Prior to 1989, the birth certificate only identified the parent's name and, in most cases, the father. In theory, we could index mothers by tracing our cohort to 1989, because a new birth certificate form was introduced that explicitly included the mother's identification information. However, the digital data with the new form only became available from 1995 onward. For this reason, we used data from 1995 to 2011 to build a mother index, and updated it using 2012 data when this book was written.

Mother's identification variables included first name, last name, maiden name, birth date, social security number, and residential address. However, since the social security number was not allowed for building the mother index, we relied on a deduplication process in Link Plus 2.0. In addition, we used a mother ID generated from the social security number to conduct quality control and use the vital records query system to manually check potential inconsistencies. The whole process was straightforward, and it took an experienced person close to a week of full-time equivalent (FTE) time to build the index for nearly a half-million birth records with more than 280,000 unique mother records. However, the whole process took more than a month due to getting feedback from the vital records office and a health statistician for questionable or inconsistent records. This quality control process identified 464 mother records that may or may not be the true link. Examples of questionable pairs include some bad data, such as very short birth intervals (e.g., <15 weeks). We simply did not index these mothers, so that their births would be treated as if they had single birth records.

14.2.2 Construction of Longitudinal Datasets for Descriptive Analysis

The NMI file can be used to investigate birth outcomes for mothers with multiple births. In addition, since birth certificate numbers are preserved in the indexed birth file, any program data that had birth certificate numbers can be linked to the mother index. Examples include birth defects, fetal death, infant mortality, and PRAMS. Most of these linked dimensions have been investigated by others. Since both parents were linked to hospital discharge data from 2005 to 2011, the indexed mothers could also be found in hospital discharge data. This linkage file can be used to answer comorbidity and other hospital-related questions. Finally, most birth and death records after 2005 were geocoded with an acceptable quality; some analysis on the impact of neighborhood environments on birth outcomes could also be carried out. In the following section, we demonstrate how the mother index might be used for public health programs within a state. We selected singleton births from the mother index and restrict our analysis to mothers who had at least two singleton births from 1995 to 2012.

Skjaerven et al. (1988) divided the birth weights of first children into eight categories (i.e., <1500, 1500–1999, 2000–2499, 2500–2999, 3000–3499, 3500–3999, 4000–4499, and ≥4500) and then plotted the distribution of birth weight of second children conditional on the birth weight of the first-born children. We attempted to replicate this analysis to see if a cohort in the United States would follow similar patterns. However, if we chose first live birth followed by the second birth, we would have deleted many mothers who either moved from other states with first births elsewhere or had their first live births prior to 1995. In order to preserve the number of records, we used the first-birth record from each mother as the "first birth." We cross-tabulated the first births by second and third births according to the eight birth weight categories.

The column headings of Table 14.1 were for the first births, and the rows for subsequent birth weight categories. The upper panel is based on mothers with two or more births in the database; the lower panel includes mothers with three or more children. If there was a very strong trend of giving births within a birth weight category, then birth weight frequencies would be dominated by diagonals. In fact, only the middle birth weight categories were dominated by diagonals. Even though it is not easy to inspect all distributions, the trends are consistent with Skjaerven et al.'s (1988) findings. If the first births were from a lower end of birth distributions, it is more likely that the second births were also from relatively lower birth weight categories. Likewise, mothers having first births in higher birth weight groups were more likely to find their second births in the higher birth weight categories. These relationships were persistent for third children, but they were much weaker and less apparent.

While Table 14.1 is informative to birth weight trends by birth order, it is not informative in terms of LBW. Table 14.2 displays the same data in two categories: LBW or not in first births versus the same categories in second births, where LBW is defined as having <2500 grams in birth weight among infants. For those mothers with LBW first births, the chance of having second births in the same category was 20%, as opposed to 3.11% when the first births were not LBW. Similar but weaker relationships were passed on to third births.

We found similar patterns for PTB, where PTB is for a gestation age of <37 weeks (Table 14.3). When the first births were full term, the chances of PTB for second births were 5.3%. When the first births were preterm, the chances of PTB for second births were 23.5%. The impact of first PTB also permeates to third births, with 20% of them being preterm when the first births were preterm.

14.2.3 CONSTRUCTION OF LONGITUDINAL DATASETS FOR MULTIVARIATE ANALYSIS

Given the strong predictive powers of LBW and PTB from the first births, we wanted to conduct a multivariate analysis to examine if infants from racial minority mothers and smoking mothers would suffer extra deficits in LBW and PTB independent of the predictive powers of first births. We used three racial categories: White, Black, and Other races. We used three education levels—high school graduates, some college, and having a college degree—to control for SES. Preliminary results showed no difference between less than high school and high school, so we treated high school graduates or less as a single category. Although it is possible that a mother gained additional education between the first birth and subsequent births, preliminary results show no effect, so we did not include this variable. In addition, both teen births (mother of age 19 or younger) and rapid repeat births (RRBs) (birth interval within 24 months) are important program concerns that require intervention, so we included indicator variables for both.

We constructed three smoking indicator variables against the nonsmoker from the birth certificate variable: if the mother smoked during pregnancy at birth 1 (time 1) and birth 2 (time 2). A persistent smoker is defined as the mother smoking during the first and second pregnancies if the outcomes are in reference to second births. When the outcomes are in reference to third births, persistent smokers are defined as the mother either having smoked during the first or second pregnancy or still smoking during the third pregnancy. A new smoker is defined as the mother not smoking during the first pregnancy but smoking during the second. When the outcomes are in reference to third births, new smokers are defined as the mother not smoking during the first two pregnancies but smoking during the third. Finally, previous smokers are defined as mothers who smoked during the first pregnancy but not during the second or third pregnancy.

TABLE 14.1
Cross-Tabulations of First Child's Birth Weight by Second and Third Children's Birth Weights

First Birth	<1,500	1,500–1,999	2,000–2,499	2,500–2,999	3,000–3,499	3,500–3,999	4,000–4,499	≥4,500	Total
By Second Birth									
<1,500	8.33	7.03	11.89	23.52	33.07	12.59	3.04	0.52	1,152
1,500–1,999	4.47	4.56	16.62	28.95	30.03	12.96	2.14	0.27	1,119
2,000–2,499	1.96	2.92	11.62	33.55	35.53	12.08	2.03	0.31	4,182
2,500–2,999	0.91	1.12	5.82	27.77	45.68	16.39	2.16	0.15	18,775
3,000–3,499	0.41	0.46	1.96	14.43	46.07	30.9	5.23	0.54	47,351
3,500–3,999	0.35	0.24	1.03	6.05	32.7	43.48	14.29	1.85	35,756
4,000–4,499	0.33	0.21	0.39	2.88	19.67	44.06	25.74	6.73	9,069
≥4,500	0.31	0.23	0.61	1.77	12.91	34.59	33.21	16.37	1,301
Total	751	790	3,243	16,494	46,240	38,496	10,901	1,790	118,705
By Third Birth									
<1,500	6.71	3.8	8.95	23.71	31.54	22.37	2.46	0.45	447
1,500–1,999	5.37	5.08	7.91	28.25	33.33	16.67	3.39	0	354
2,000–2,499	0.85	2.98	10.43	31.42	36.6	14.89	2.48	0.35	1,410
2,500–2,999	0.94	1.22	6.4	26.57	43.77	18.09	2.81	0.2	6,395
3,000–3,499	0.51	0.59	2.47	14.71	44.09	31.04	5.9	0.68	16,032
3,500–3,999	0.37	0.32	1.11	6.62	33	42.43	14.03	2.14	12,033
4,000–4,499	0.51	0.31	1.05	2.82	21.11	43.58	24.17	6.46	2,942
≥4,500	0	0.77	1.28	4.35	13.81	36.32	30.18	13.3	391
Total	262	300	1,189	5,602	15,289	13,033	3,701	628	40,004

TABLE 14.2
LBW Cross-Tabulation by Birth Order

First Child	No	Yes	Total
Second Birth			
No	96.89	3.11	112,252
Yes	79.99	20.01	6,453
Third Birth			
No	96.3	3.7	37,793
Yes	84.03	15.97	2,211

Note: Second child is for mother with more than one child. Third child is for mother with more than two children.

TABLE 14.3
PTB Cross-Tabulation by Birth Order

First Child	No	Yes	Total
Second Birth			
No	94.7	5.3	110,344
Yes	76.47	23.53	8,361
Third Birth			
No	93.89	6.11	37,260
Yes	79.96	20.04	2,744

Note: Second child is for mother with more than one child. Third child is for mother with more than two children.

We first ran a logistic regression with the dependent variable being LBW. The first two columns of Table 14.4 show that if the first birth is LBW, the odds of being LBW for the second birth is close to six times higher than those whose first births were not LBW. Teenage births were about 26% more likely to be LBW. Blacks were twice as likely to have LBW than Whites. Both associate and college degrees reduced the odds of having LBW by about 16% and 36%, respectively. All indicators listed under smoking status were significant predictors of LBW infants. The odds ratio was strongest for persistent smokers and weaker for new and previous smokers. Finally, RRBs were 17% more likely to be LBW than the comparison group.

All the directional effects for the third-birth model (last two columns of Table 14.4) were consistent with those from the second-birth model. The point estimate to the LBW effect on third births was weaker than that for second births. It was a coincidence that point estimates for persistent smokers and new smokers were identical. Note, however, that there was a minor difference in persistent smoker definition from the second-birth model to the third-birth model. The latter targets mothers who either smoked during the first or second pregnancy or still smoked during the third pregnancy (Table 14.5).

We ran two PTB models that correspond to LBW models. Almost all the directional effects were retained, although point estimates tend to be weaker than LBW models. The only exception was

TABLE 14.4
Predictors of LBW for Second and Third Births

	Second Birth		Third Birth	
	OR	95% CI	OR	95% CI
First-birth LBW	6.733	6.27–7.231	4.186	3.68–4.762
Teen birth	1.263	1.087–1.468	1.376	0.857–2.208
Race (ref = White)				
Black	2.36	2.15–2.589	2.158	1.87–2.491
Other race	1.349	1.219–1.494	1.278	1.076–1.518
Education (ref ≤ High School)				
Some college	0.842	0.781–0.907	0.748	0.657–0.852
College or high	0.638	0.584–0.696	0.53	0.453–0.62
Smoking Status (ref = Nonsmoker)				
Persistent smoker	2.116	1.95–2.297	2.032	1.795–2.301
New smoker	1.598	1.414–1.806	2.032	1.62–2.548
Previous smoker	1.432	1.252–1.639	1.157	0.95–1.409
RRB	1.172	1.099–1.249	1.144	1.032–1.268

Note: Sample is based on NMI for mothers with multiple births during 1995 and 2012. OR, odds ratio; CI, confidence interval.

TABLE 14.5
Predictors of PTB for Second and Third Births

	Second Birth		Third Birth	
	OR	95% CI	OR	95% CI
First-birth LBW	4.702	4.411–5.012	3.726	3.359–4.133
Teen birth	1.322	1.164–1.501	1.648	1.119–2.426
Race (ref = White)				
Black	1.44	1.321–1.569	1.557	1.369–1.771
Other race	1.146	1.056–1.245	1.154	1.003–1.327
Education (ref ≤ High School)				
Some college	0.944	0.891–1.002	0.814	0.736–0.901
College or high	0.785	0.736–0.837	0.612	0.546–0.686
Smoking Status (ref = Nonsmoker)				
Persistent smoker	1.371	1.276–1.473	1.389	1.247–1.546
New smoker	1.215	1.093–1.351	1.294	1.048–1.598
Previous smoker	1.156	1.033–1.294	1.207	1.037–1.405
RRB	1.184	1.126–1.245	1.22	1.124–1.324

Note: Sample is based on NMI for mother with multiple births during 1995 and 2012.

some college education for the second PTB model, which was not significant. Most point estimates tended to be weaker than corresponding estimates for LBW models.

14.2.4 Using NMI to Construct Longitudinal Datasets for PRAMS

PRAMS is part of the CDC initiative to reduce infant mortality and LBW, and it covered more than 35 states in 2014. PRAMS is a mixed-mode surveillance system consisting of a mail survey with telephone follow-up for nonrespondents. The system is designed to provide state-specific information on maternal behaviors and experiences during pregnancy and a child's early infancy. In Nebraska, PRAMS samples about 10%–11% of in-state mothers each year based on information from birth certificates. The sample is stratified by race and regions with an annual sample size of more than 2300 respondents based on a response rate of close to 80%.

PRAMS has been used for assessing (1) maternal and infant health from pregnancy through the early postpartum period, (2) behavior health issues (e.g., smoking, prenatal care visits, partner violence, postpartum depression, and breast feeding), and (3) disparities by race ethnicity, income, and neighborhood income distribution. In a review article, Kotelchuck (2006) pointed to future use of PRAMS. He emphasized the data integration approach of linking PRAMS to program data such as WIC program data and birth defects. He also promoted the multilevel approach of linking to census tract data and the longitudinal approach of following PRAMS subjects. It was stated that "PRAMS needs to explore and facilitate other database linkages more aggressively." Since Kotelchuck's articles, there have been many more articles that used census tract information. However, few have constructed longitudinal datasets using PRAMS or between PRAMS and birth certificate data. Since we already had the mother index, it was straightforward to build a longitudinal dataset either for PRAMS or between PRAMS and the mother index. However, before letting a program use it, it is necessary to evaluate its feasibility.

First, a longitudinal sample based solely on multiple PRAMS data may not have sufficient statistical power. For example, (1) we had an average of 2297 respondents each year from 2004 to 2010, (2) 56.8% of new mothers would have new babies in the next 3 years according to the 18-year average from the mother index, and (3) the PRAMS sample in each year is largely random and independent, with about 10% new birth mothers. According to (2), we expected 1305 mothers to have live births within 3 years. If these mothers were to be randomly selected according to (3), regardless of which year, 133 of them would be in the next few years of PRAMS samples. In the end, we expect 800–917 (7×131) mothers with second births, depending on if we want to count the 2010 PRAMS. Note that the above calculations do not need data linkage, and one can simply use the result to guide the feasibility. Now let's have a reality check using the mother index that had a link to PRAMS. During 2004–2010, Nebraska PRAMS had 16,081 new mothers; among them, 986 were found in the next few waves of PRAMS for having second live births, and scores of them had third births. Our rough estimate was not too far off given that we did not know exactly the sample probability. In addition, we suspected that the sample had more minority racial and ethnic groups, which tend to have a high fertility. Depending on outcome measures, the power may not be sufficient. Suppose that we want to study the rural–urban difference in LBW by some aspects of PRAMS variables, such as LBW, late recognition of pregnancy, and change in smoking status; the number of positive outcomes (e.g., yes for LBW) would be less than 100 respondents. When they are further stratified by rural–urban, race, and behavioral variables, the sample size for a particular cell gets smaller and smaller. Note also that we have not considered item response rate yet, which could further reduce sample size. We concluded that a construction of a longitudinal dataset based solely on three or four waves of PRAMS samples would have limited power for most maternal–child health analyses.

The second concern is sample bias. When a longitudinal PRAMS sample is not large enough, one can use the mother index to follow PRAMS mothers for their second births. Here, the PRAMS sample would be wave 1 in the longitudinal dataset, and mothers of second births and associated birth certificate information would be wave 2. In this case, how to use the sample weight of wave 1

TABLE 14.6

Sample Comparisons among Those Having Second Births: 2004–2010

	Mother Index	Unweighted PRAMS	Weighted PRAMS
Total *N*	69,376	10,535	125,056
Age Group			
<16	1.19	0.57	0.28
16–19	13.90	9.87	6.2
20–29	65.57	58.38	57.99
30–39	19.03	29.82	34.03
≥40	0.30	1.36	1.5
Race of Mother			
White	82.42	52.74	83.12
Black	6.08	21.4	6.62
Other	11.51	25.86	10.26
Ethnicity			
Non-Hispanic	87.61	83.73	88.40
Hispanic	12.39	16.27	11.60
Mother's Education			
Less than high school	17.67	21.62	13.01
High school	23.37	24.62	20.58
Some college	28.00	29.52	32.99
College degree and beyond	30.96	24.23	33.42

Note: Mother index, mothers having two or more children, or second-birth mother index; unweighted PRAMS, PRAMS linked to second-birth mother index without wave 1 weight; weighted PRAMS, PRAMS linked to second-birth mother index with wave 1 weight.

is a concern. Table 14.6 lists sociodemographic characteristics between the second-birth full sample and second-birth PRAMS–mother index longitudinal sample. The latter is also compared with non-weighted and weighted samples. The mother index sample represents all in-state mothers having second births during 2004–2010, in which the proportion of mothers aged 19 or younger at the first birth is very high (15.09%). The corresponding proportion in the unweighted PRAMS sample was 10.44%. The proportion in the weighted sample was even lower, suggesting that teenage pregnancy or RRBs for teenagers would be biased toward underestimates. In addition, the unweighted PRAMS had much fewer White mothers, while the weighted PRAMS had rates of White mothers that were comparable to those in the mother index. There were also some inconsistencies in ethnicity and educational level that we are not going to describe in detail. Evidently, there were some significant differences in sociodemographic variables that could lead to biased estimation in univariate and multivariate analyses.

The third evaluation was for potential outcome biases. Table 14.7 shows some cross-tabulation results between time 1 (first birth) and time 2 (second birth) for smoking status, PTB, and LBW. Under the mother index, 30.37% of time 1 smokers were nonsmokers during their second pregnancies. The corresponding percentages for unweighted and weighted PRAMS samples were 18.82% and 15.06%, respectively. The outcome measures for both PTB and LBW were very close between the mother index and weighted PRAMS samples. Again, if one uses smoking as an outcome measure, the results could be biased. Depending on the study design, if we want to examine behavior change without comparing at the population level, smoking as an outcome measure can still be used. For example, in the PRAMS sample, 19% of mothers quit smoking during time 2 if they had

TABLE 14.7

Consistencies among Outcome Measures by Different Samples

	Mother Index	PRAMS Unweighted	PRAMS Weighted
	Smoking Status		
	T2 nonsmoker	T2 nonsmoker	T2 nonsmoker
T1 nonsmoker	94.45	96.86	97.8
T1 smoker	30.37	18.82	15.06
	PTB		
	T2 non-PTB	T2 non-PTB	T2 non-PTB
T1 non-PTB	94.52	94.05	94.58
T1 PTB	76.13	73.48	76.55
	LBW		
	T2 non-LBW	T2 non-LBW	T2 non-LBW
T1 non-LBW	96.8	95.94	96.83
T1 LBW	80.26	77.28	80.36

Note: All numbers are in percentage. T1, time 1; T2, time 2.

no smoking cessation treatment, but for those who had the treatment, the quitting rate was 33.3% (not shown in the table). In Nebraska, only a small fraction of people are eligible for smoking cessation treatment, and the effectiveness of the treatment has significant policy implications, and such information from time 1 PRAMS data is certainly helpful.

Finally, we stress the importance of careful selection and use of time 1 and time 2 variables from a combined PRAMS–mother index longitudinal dataset. As one might notice, all the measures in Tables 14.6 and 14.7 are from birth certificate data. When the PRAMS sample is followed for second births, there is a smaller sample with PRAMS questions about second birth, and the full second-birth sample with birth certificate data. Although studies have suggested high concordance among many measures between birth certificate and PRAMS data, this limitation has to be recognized. For consistency, one should incorporate PRAMS time 1 variables into a study design, but use consistent outcome measures between time 1 and time 2 based on birth certificate data.

14.3   USING NMI AND GEOCODED DATA TO CONSTRUCT RESIDENTIAL MOBILITY INFORMATION

14.3.1   Residential SES Mobility

Insofar, we have not touched neighborhood SES via census tract data. Since most birth certificate data have census tract information, they can be easily linked to census tract poverty data from the American Community Survey (ACS). There have been many studies using census tract poverty as a proxy for neighborhood SES in birth outcome analysis. Very few studies take a longitudinal view, as it would be taxing without the mother index or a longitudinal birth file. Since we already have the mother index, we can tract mothers when they moved within the state.

Table 14.8 cross-tabulated mothers whose first and second births were during 2005–2011 by their residential SES or poverty levels according to the first and second births, in percentage. The row and column totals were poverty levels according to times 1 and 2, respectively. The diagonals show the percentages that did not contribute to residential SES mobility. Off-diagonals show movers who crossed at least one poverty group in their residential mobility. Cells above the diagonals show propensities of moving into a neighborhood with a higher percentage of poverty; cells below

TABLE 14.8
Neighborhood Mobility among Poverty Groups between First and Second Births

Pvt: Birth 1/Birth 2	<5	5–9.99	10–14.99	15–19.99	>20	Row N
<5	81 (12.98)	7.94	6.3	2.04	2.72	7,365
5–9.99	8.31	73.09 (10.15)	8.83	3.94	5.83	10,482
10–14.99	7.29	11.34	69.48 (10.4)	4.73	7.16	8,202
15–19.99	5.63	10.49	10.54	62 (5.53)	11.34	3,871
>20	5.16	10.53	9.4	6.56	68.36 (26.72)	7,990
Column N	8,065	10,423	8,248	3,875	7,299	37,910

Note: Sample, mothers who had no previous birth and had their first and second births during 2005–2011; pvt, percent of population under the federal poverty line. Cross-tabulations are in row percent, and diagonals are stayers and movers who moved within each poverty level between two births.

the diagonals indicate propensities of moving into a neighborhood with a lower poverty rate. Both are in reference to their previous neighborhoods. The main diagonals without parentheses include both nonmovers and movers, so that the sum of row percentages equals 100%. Intrapoverty movers (shown in parentheses) are those who moved into a neighborhood with an identical poverty category according to time 1 births. Intrapoverty group movers include but are not necessarily intraneighborhood movers, who move within the same census tract.

Table 14.8 shows that there was a fair amount of mobility even though the diagonals were dominant. First, there were 42% movers (by adding all the off-diagonals and numbers in parentheses), but at least 62% of mothers were either stayers or intrapoverty group movers. Second, movers in the 15%–19.99% poverty group were the least likely to be intrapoverty group movers, as only 5.53% stayed in the same poverty group, while more than 10% moved to adjacent poverty groups in each direction. Third, among movers, mothers from the most impoverished neighborhoods were the most likely to be intrapoverty movers (26.72%). Finally, there was a propensity of moving to a less impoverished neighborhood.

14.3.2 RAPID REPEAT BIRTH

In the following, we demonstrate how we might use poverty and residential mobility information to investigate short-interval births. We included birth intervals less than 18 and 24 months. The former was found to be most likely associated with adverse birth outcomes in a metadata analysis (Conde-Agudelo et al., 2006). The latter, when including repeat teen births, is loosely defined as RRB. RRBs are associated with many adverse maternal and neonatal outcomes, such as PTB, LBW, congenital anomalies, increased perinatal mortality, cerebral palsy, maternal mortality, and anemia. As parents move beyond perinatal care, RRB is a predictor of parenting stress and child maltreatment (El-Kamary et al., 2004).

We were interested in the effects of neighborhood environment and moving between neighborhoods on RRB. We already know the importance of neighborhood poverty status. We added neighborhood mobility because young mothers are the most active segment of movers. We set up a model by controlling for age, sex, and educational level. We included some behavioral risk factors: a lack of prenatal care (fewer than six prenatal care visits) and smoking during the first pregnancy reflect less awareness of the harmful effects on infants. For smoking, we broke it down into persistent smokers (smoked during the first and second pregnancies), new smokers (did not smoke during the first but did smoke during the second pregnancy), and those who quit (smoked during the first but not the second pregnancy).

We used five poverty levels: <5%, 5%–9.99%, 10%–14.99%, 15%–19.99%, and >20% of the population within a census tract who lived below the poverty line. This design is slightly different from that in most studies using neighborhood SES in maternal and child health, in that we used it to refer to first births instead of current births. In addition, we included residential mobility in terms of moving up or down among different poverty levels. If a mother resided in poverty group 1 (5%) and moved to poverty group 5 (>20%), she moved up in poverty rate, or to a poorer neighborhood, by more than two levels. If she moved to poverty group 2, she moved up one level. Conversely, people could move from a well-off neighborhood to a poorer neighborhood. We chose a six-category geographic mobility coding with nonmovers being the referent, and compared them with those who moved within a poverty level, up one level, up two levels, down one level, and down two levels. Note that we were not looking for the most parsimonious model or a concept-driven model.

Table 14.9 shows the results. Let's focus on RRB on the far right, which includes birth interval less than 24 months or repeat birth by those ages 19 or younger. As it has been identified through the literature, the RRB is the highest among age groups less than 17 at the time of first birth, who could either give birth up to 36 months or until age 19. The next age group that had the strongest effect was 40 or older, who presumably hurried up fertility to beat the odds of fertility decline and other adverse birth outcomes. The third age group is age 17–19. Hence, if we used 19 or younger as an age break point, we would have a group with the greatest likelihood of having RRB, but that would require age coding according to the second birth instead of the first birth. Compared to Whites, RRBs were 45% and 19% more likely among Blacks and "other race," respectively. Hispanics were barely less likely to be associated with RRB (p-value = 0.0508).

Having some college education or a college degree is protective against having an RRB. Having fewer than six prenatal visits during the first pregnancy is associated with a 26% higher chance than the referent of having an RRB. Both persistent smokers and those who smoked in the first but not the second pregnancy were 33% and 16% more likely than nonsmokers to have RRBs. Note that we included all combinations of smoking status between time 1 and time 2; the reference group was nonsmokers. New smokers, who did not smoke during the first pregnancy but smoked during the second pregnancy, were somehow likely to have RRBs.

As expected, those residing in poor neighborhoods in time 1 tended to be associated with RRBs. The odds ratios were around 1.07 or 1.08 for poverty groups 3 and 4, and increased to 1.25 in group 5, suggesting some type of linear relationship. If we used a multilevel model with the same setup, but using a continuous poverty variable, the odds ratio would be 1.008 (not shown). Examining odds ratios for residential mobility suggested that it is really just mover categories versus nonmovers, because all the odds ratios ranged from 0.5 to 0.67 against nonmovers. Moving, in general, is a sign of temporary fertility control or having births in longer rather than shorter intervals.

In the 18-month birth interval model (Table 14.9, first two columns), the most obvious differences were those odds ratios associated with age. Both age 17–19 and 40 and older groups had an odds ratio greater than 1, suggesting an increased likelihood of having 18-month-interval births. In addition, new smokers became nonsignificant. Only poverty groups 4 and 5 had elevated odds of having 18-month-interval births. Again, all the odds ratios pertaining to residential mobility had a protective effect on 18-month-interval birth, and grouping them together might be better in terms of model fitting.

14.4 CHAPTER SUMMARY

In this chapter, we first reviewed major linkage projects based on birth certificate data. One common characteristic is external funding, which is a double-edged sword. On the one hand, major funding sources provide opportunities for state and cities to initiate data linkage projects. These projects provided data infrastructure for state and local governments moving forward with other data linkage

TABLE 14.9
Factors Associated with RRBs: 2005–2011

	Birth Interval < 18 Months		RRB	
	OR	95% CI	OR	95% CI
Age at the First Birth				
Age 17–19 vs. age < 17	1.348	1.084–1.676	0.242	0.194–0.302
Age 20–29 vs. age < 17	1.073	0.868–1.325	0.179	0.144–0.223
Age 30–39 vs. age < 17	1.105	0.887–1.377	0.197	0.158–0.246
Age ≥ 40 vs. age < 17	1.669	1.045–2.665	0.337	0.222–0.51
Mother's Race/Ethnicity				
Black vs. White	1.546	1.391–1.717	1.445	1.318–1.583
Other race vs. White	1.31	1.176–1.46	1.188	1.086–1.3
Hispanic vs. non-Hispanic	0.95	0.851–1.059	0.914	0.835–1
Mother's Education (ref = High School or Less)				
Some college	0.788	0.734–0.845	0.864	0.815–0.915
College degree or higher	0.48	0.44–0.524	0.689	0.644–0.736
Behavioral Risk Factors				
PNC < 6 first birth (yes vs. no)	1.429	1.289–1.585	1.264	1.154–1.385
Persistent smoker (yes vs. no)	1.48	1.361–1.609	1.328	1.236–1.428
New smoker (yes vs. no)	0.92	0.788–1.075	0.851	0.75–0.965
Previous smoker (yes vs. no)	1.436	1.27–1.623	1.161	1.044–1.291
Time 1 Poverty (ref = pvt1 < 5%)				
Pvt2: 5–9.99%	0.985	0.905–1.073	0.997	0.936–1.063
Pvt3: 10–14.99%	1.017	0.93–1.114	1.078	1.006–1.154
Pvt4: 15–19.99%	1.117	1–1.246	1.09	1–1.189
Pvt5: >20%	1.164	1.053–1.287	1.254	1.159–1.356
Poverty Mobility (ref = Nonmover)				
Moved within a poverty group	0.659	0.605–0.717	0.641	0.599–0.685
Moved one poverty group up	0.836	0.747–0.936	0.735	0.671–0.805
Moved two poverty groups up	0.813	0.726–0.91	0.73	0.666–0.8
Moved one poverty group down	0.611	0.545–0.685	0.609	0.559–0.664
Moved two poverty groups down	0.536	0.479–0.6	0.544	0.5–0.593

Note: Sample is based on those who had no previous birth and had their first and second births during 2005–2011. Persistent smoker, smoked during first and second pregnancies; new smoker, did not smoke during the first pregnancy but smoked during the second pregnancy; previous smoker, smoked during the first pregnancy but did not smoke during the second. Poverty mobility, stayers are those who either did not move or moved within their own poverty groups between first and second births. Pvt, poverty. Time 1, first birth; time 2, second birth. Moved 1 poverty group up, time 2 pvt – time 1 pvt = 1; moved 1 poverty group up, time 2 pvt – time 1 pvt ≥ 2; moved 1 poverty group down, time 2 pvt – time 1 pvt = –1; moved 1 poverty group down, time 2 pvt – time 1 pvt ≤ –2. PNC, prenatal care visits.

projects. One the other hand, funded projects tend to have a short funding period. As a project ends, the linked datasets are often left without caretakers to maintain, update, and make further use of them.

Building a simple and low-cost mother index from birth certificate data is an easy task. Although building and maintaining the mother index requires some effort, extra to regular job duties typically found within a state's vital records or health statistics office, such as crosscutting, are well worth it. To demonstrate the utility of the mother index, we provided three applications.

First, we investigated the relationship between first and later birth outcomes in terms of LBW and PTB. Cross-tabulation based on first and second births showed that 20% of mothers would have LBW infants when their previous births were LBW. Likewise, 23% of mothers who had PTBs would have PTBs next time. To a lesser degree, these trends also repeat for third births. It seems natural to devote some intervention effort to those mothers who had LBW infants and who intend to have a second child. Home visiting programs sponsored by the CDC served some, but their coverage is not wide enough. We also ran multivariate analyses of LBW and PTB for both second and third births based on predictors from first births. The results were generally consistent with previous literature in that smoking has a harmful effect independent of LBW of the first births. Our simple analysis also demonstrated that we did not need to have a large-scope or funded project to conduct longitudinal studies using a longitudinally constructed mother index.

Second, we explored potential applications for administratively constructed PRAMS longitudinal datasets. Since all the mothers in birth records within the state were indexed, it was really a straightforward deterministic linkage of PRAMS mothers back to the birth certificate system. We found that more than 50% of PRAMS mothers gave birth again after they entered into PRAMS. Due to some methodological concerns, we did not proceed with analytical assessment of public health issues. We instead looked at some of the potential biases from a longitudinally constructed PRAMS mother who had given birth again. We used the PRAMS–mother index sample and the full mother index sample for comparisons and found some differences between the two, especially the unweighted PRAMS. We concluded that without some proper adjustments, caution needs to be exercised when using the longitudinal data constructed from the PRAMS–mother index.

Finally, we demonstrated the utility of geocoded mother indices by constructing residential mobility between first and second births. In addition, we used neighborhood SES to predict RRB, which would be quite taxing without the mother index. With the availability of the mother index or NMI, it would be a simple query to select mothers of multiple births and short birth intervals. With additional geographically attached information, it would be easy to classify residential locations by poverty levels. With some caution, a retrospective cohort study design can be constructed with ease.

REFERENCES

Adams, M.M., H.G. Wilson, D.L. Casto, C.J. Berg, J.M. McDermott, J.A. Gaudino, and B.J. McCarthy. 1997. Constructing reproductive histories by linking vital records. *American Journal of Epidemiology* 145: 339–348.

Angier, H., R. Gold, C. Crawford, J.P. O'Malley, C.J. Tillotson, M. Marino, and J.E. DeVoe. 2014. Linkage methods for connecting children with parents in electronic health record and state public health insurance data. *Maternal and Child Health Journal* 18(9): 2025–2033.

Bronstein, J.M., C.T. Lomatsch, D. Fletcher, T. Wooten, T.M. Lin, R. Nugent, and C.L Lowery. 2009. Issues and biases in matching Medicaid pregnancy episodes to vital records data: The Arkansas experience. *Maternal and Child Health Journal* 13: 250–259.

Buescher, P.A. 1999. Method of linking Medicaid records to birth certificates may affect infant outcome statistics. *American Journal of Public Health* 89: 564–566.

Chapman, D.A. and G. Gray. 2014. Developing a maternally linked birth dataset to study the generational recurrence of low birthweight in Virginia. *Maternal and Child Health Journal* 18: 488–496.

Collins, J.W. Jr., R.J. David, K.M. Rankin, and J.R. Desireddi. 2009. Transgenerational effect of neighborhood poverty on low birth weight among African Americans in Cook County, Illinois. *American Journal of Epidemiology* 169(6): 7.

Conde-Agudelo, A., A. Rosas-Bermudez, and A.C. Kafury-Goeta. 2006. Birth spacing and risk of adverse perinatal outcomes: A meta-analysis. *The Journal of the American Medical Association* 295: 1809–1823.

Copeland, G.E. and R.S. Kirby. 2007. Using birth defects registry data to evaluate infant and childhood mortality associated with birth defects: An alternative to traditional mortality assessment using underlying cause of death statistics. *Birth Defects Research Part A: Clinical and Molecular Teratology* 79(11): 792–797.

Coutinho, R., R.J. David, and J.W. Collins Jr. 1997. Relation of parental birth weights to infant birth weight among African Americans and whites in Illinois: A transgenerational study. *American Journal of Epidemiology* 146: 804–809.

David, R. 2001. Commentary: Birthweights and bell curves. *International Journal of Epidemiology* 30: 1241–1243.

David, R. K. Rankin, K. Lee, N. Prachand, C. Love, and J. Collins Jr. 2010. The Illinois transgenerational birth file: Life-course analysis of birth outcomes using vital records and census data over decades. *Maternal and Child Health Journal* 14(1): 121–132.

El-Bastawissi, A.Y., R. Peters, K. Sasseen, T. Bell, and R. Manolopoulos. 2007. Effect of the Washington Special Supplemental Nutrition Program for Women, Infants and Children (WIC) on pregnancy outcomes. *Maternal and Child Health Journal* 11(6): 611–621.

El-Kamary, S.S., S.M. Higman, L. Fuddy, E. McFarlane, C. Sia, and A.K. Duggan. 2004. Hawaii's healthy start home visiting program: Determinants and impact of rapid repeat birth. *Pediatrics* 114(3): e317–326.

Emanuel, I., W. Leisenring, M.A. Williams, C. Kimpo, S. Estee, W. O'Brien, and C.B. Hale. 1999. The Washington State Intergenerational Study of Birth Outcomes: Methodology and some comparisons of maternal birthweight and infant birthweight and gestation in four ethnic groups. *Paediatric and Perinatal Epidemiology* 13: 352–369.

Feng, J., M.R. Kramer, B.V. Dever, A.L. Dunlop, B. Williams, and L. Jain. 2013. Maternal smoking during pregnancy and failure of the Georgia first grade criterion-referenced competency test. *Paediatric and Perinatal Epidemiology* 27(3): 275–282.

Forand, S.P., T.O. Talbot, C. Druschel, and P.K. Cross. 2002. Data quality and the spatial analysis of disease rates: Congenital malformations in New York State. *Health Place* 8(3): 191–199.

Grisso, J.A., J.L. Carson, H.I. Feldman, I. Cosmatos, M. Shaw, and B. Strom. 1997. Epidemiologic pitfalls using Medicaid data in reproductive health research. *Journal of Maternal-Fetal Medicine* 6: 230–236.

Gyllstrom, M.E., J.L. Jensen, J.N. Vaughan, S.E. Castellano, and J.W. Oswald. 2002. Linking birth certificates with Medicaid data to enhance population health assessment: Methodological issues addressed. *Journal of Public Health Management and Practice* 8: 38–44.

Hall, E.S., N.K. Goyal, R.T. Ammerman, M.M. Miller, D.E. Jones, J.A. Short, and J.B. Van Ginkel. 2014. Development of a linked perinatal data resource from state administrative and community-based program data. *Maternal and Child Health Journal* 18(1): 316–325.

Hertz-Picciotto, I. 2001. Commentary: When brilliant insights lead astray. *International Journal of Epidemiology* 30(6): 1243–1244.

Klebanoff, M.A. and R. Yip. 1987. Influence of maternal birth weight on rate of fetal growth and duration of gestation. *Journal of Pediatrics* 111(2): 287–292.

Korzeniewski, S.J., V. Grigorescu, G. Copeland, G. Gu, K.K. Thoburn, J.D. Rogers, and W.I. Young. 2010. Methodological innovations in data gathering: Newborn screening linkage with live births records, Michigan, 1/2007–3/2008. *Maternal and Child Health Journal* 14(3): 360–364.

Kotelchuck, M. 2006. Pregnancy risk assessment monitoring system (PRAMS): Possible new roles for a national MCH data system. *Public Health Reports* 121(1): 6–10.

Kotelchuck, M., L. Hoang, J.E. Stern, H. Diop, C. Belanoff, and E. Declercq. 2014. The MOSART database: Linking the SART CORS clinical database to the population-based Massachusetts PELL reproductive public health data system. *Maternal and Child Health Journal* 18(9): 2167–2178.

Kramer, M.R., A.L. Dunlop, and C.J. Hogue. 2014. Measuring women's cumulative neighborhood deprivation exposure using longitudinally linked vital records: A method for life course MCH research. *Maternal and Child Health Journal* 18(2): 478–487.

Kramer, M.R., L.A. Waller, A.L. Dunlop, and C.R. Hogue. 2012. Housing transitions and low birth weight among low-income women: Longitudinal study of the perinatal consequences of changing public housing policy. *American Journal of Public Health* 102(12): 2255–2261.

Li, C., J.R. Daling, and I. Emanuel. 2003. Birth weight and risk of overall and cause-specific childhood mortality. *Paediatric and Perinatal Epidemiology* 17: 164–170.

Piper, J.M., W.A. Ray, M.R. Griffin, R. Fought, J.R. Daughtery, and E. Mitchel Jr. 1990. Methodological issues in evaluating expanded Medicaid coverage for pregnant women. *American Journal of Epidemiology* 132: 561–571.

Putnam-Hornstein, E., J.A. Cederbaum, B. King, A. Lane, and P. Trickett. 2015. A population based, longitudinal examination of intergenerational maltreatment among teen mothers. *American Journal of Epidemiology* 181: 496–503.

Putnam-Hornstein, E., B. Needell, B. King, and M. Johnson-Motoyama. 2013. Racial and ethnic disparities: A population-based examination of risk factors for involvement with Child Protective Services. *Child Abuse and Neglect* 37(1): 33–46.

Qayad, M.G. and H. Zhang. 2009. Accuracy of public health data linkages. *Maternal and Child Health Journal* 13(4): 531–538.

Shy, K., C. Kimpo, I. Emanuel, W. Leisenring, and M.A. Williams. 2000. Maternal birthweight and cesarean delivery in four race-ethnic groups. *American Journal of Obstetrics and Gynecology* 182: 1363–1370.

Skjærven, R., A.J. Wilcox, N. Øyen, and P. Magnus. 1997. Mother's birth weight and survival of their off-spring: Population based study. *British Medical Journal* 314: 1376–1380.

Skjaerven, R., A.J. Wilcox, and D. Russell. 1988. Birthweight and perinatal mortality of second births conditional on weight of the first. *International Journal of Epidemiology* 17: 830–838.

Strandjord, T.P., I. Emanuel, M.A. Williams, W.M. Leisenring, and C. Kimpo. 2000. Respiratory distress syndrome and maternal birthweight effects. *Obstetrics and Gynecology* 95: 174–179.

Wilcox, A.J. 2001. A review: On the importance—and the unimportance—of birthweight. *International Journal of Epidemiology* 30: 1233–1241.

Williams, B.L., A.L. Dunlop, M. Kramer, B.V. Dever, C. Hogue, and L. Jain. 2013. Perinatal origins of first-grade academic failure: Role of prematurity and maternal factors. *Pediatrics* 131(4): 693–700.

Williams, M.A., I. Emanuel, C. Kimpo, W.M. Leisenring, and C.B. Hale. 1999. A population-based cohort study of the relation between maternal birthweight and risk of gestational diabetes mellitus in four racial/ethnic groups. *Paediatric and Perinatal Epidemiology* 13: 452–465.

15 Assessing and Managing Geocoding of Cancer Registry Data

15.1 INTRODUCTION

In cancer registry practice, states are required to submit latitude and longitude data based on patient residential address at diagnosis and to assign addresses to corresponding census tract units. However, the quality of the data that the states submit to the North American Association of Central Cancer Registries (NAACCR) is not regularly assessed. Except for states with stable and sufficient funding sources (e.g., those with National Cancer Institute Surveillance, Epidemiology, and End Results [SEER] sites), most states in the United States have very limited resources for geocoding services. In order to establish a standard and improve geocoding among state and provincial cancer registries in North America, the NAACCR surveyed state cancer registries and recently published a best practices guide (Goldberg, 2008). Based on the practice guide, a geocoding service has been established for NAACCR member states. However, most states, including Nebraska, still use the existing geocoding protocols to guide geocoding practices. In this chapter, we review some of early geocoding practices, establish some empirical benchmarks for geocoding quality, and provide some management tips for geocoding projects.

Geocoding is a process of matching residential addresses to geographically identifiable locations, often by latitude and longitude. The quality of geocoding in a cancer registry ranges from the "gold standard" of global positioning system location to less exact centroids of geographic areas, such as zip codes, cities, and counties. Poor geocoding can result in erroneous conclusions (Oliver et al., 2005; Zimmerman et al., 2007). For example, if multiple unmatched addresses are all represented by the centroid of a zip code, the census tract that is assigned to that zip code in the cancer registry would have an artificially inflated relative risk (Boscoe, 2008). Some analysts would simply delete those unmatched addresses by treating them as a random sample, but the unmatched addresses tend to be unevenly distributed—they are more likely to occur in rural areas and newly developed suburban areas, and less likely to occur in inner city areas. In many spatial cancer epidemiological studies, there is no other choice but to improve matching rate and location specificity for unmatched records.

A state cancer registry provides a common verifiable data source for all data users. If the geocoding quality of a cancer registry needs to be improved, state cancer registries should logically take responsibility for initiating the process. When a spatial cancer cluster is detected, a common data source with the same data quality will help to independently replicate and verify study findings. In addition, the most efficient way to address poor geocoding quality of past registry data is to update registry data within the cancer registry operation so that data requesters do not have to duplicate geocoding efforts. It is therefore important for a state registry to review its past geocoding practices and improve its future geocoding procedures.

A number of studies provide geocoding guidelines or discuss lessons learned, for example, the notable geocoding tips from the New York State Cancer Registry (Boscoe, 2008) and a best practices guide published by NAACCR (Goldberg, 2008). In general, the first step in geocoding a cancer registry is to clean and standardize the patient address. Although private vendors can provide address standardization services, limited budgets often prevent smaller states from using these services.

Second, the geodatabase for geocoding addresses should be updated regularly. Geodatabases are often updated using either a point feature, such as a parcel center point file, or a line feature, such as a TIGER file. The U.S. Postal Service zip + 4 is another data source for geocoding at a point location, usually the center of a building or a building complex. A drawback is that TIGER and its variants available from geographic information system (GIS) software vendors are not updated regularly. Third, one may want to experiment with different geocoding engines and specifications. The former may point to different software packages or platforms, and the latter is to set parameters within a geocoder. Both may affect geocoding quality in terms of match rate and location or position accuracy. Finally, since post office box (PO box) addresses are not available from traditional parcel or street centerline databases, it is often necessary to convert PO box and delivery route addresses to physical location addresses. However, how to process the conversion remains an issue (Hurley et al., 2003).

15.2 GEOCODING ASSESSMENTS

In an early assessment in 2010, we ran a query to check the availability of GIS coordinates in the registry data and found that the most recently geocoded data were from 2005. Based on this information, we included 15 years of geocoded data from 1991 to 2005 in the query and found that 66.4% of records had a pair of GIS coordinates coded at the street level. There was a metro–nonmetro difference, with 82.8% of urban addresses being coded at the street level, in contrast to 51.0% of nonmetro addresses. Since we know that African Americans are more likely to live in urban areas than in rural areas, we also cross-tabulated the quality variable with race. We found that addresses for African American patients were more likely to be coded at the street level (84.4%) than were addresses for Whites and other races (66%). These results suggest that cancer epidemiological studies conducted for metro areas or for the African American population would require much less additional geocoding. However, since only 48.4% of Nebraska cancer patients resided in metropolitan counties in this period, and only 12% of Nebraska Cancer Registry (NCR) patients were African American, the chance of working with only these populations was at most 50%.

Based on the assessment above, the NCR identified a contractor in 2012 to geocode registry data up to 2010. In preparation for this chapter, we again evaluated geocoded data from the NCR based on an October 2013 file. There were a total of 249,357 address records from the October 2013 cancer registry file; among them, 89,168 addresses were not geocoded. It is determined that at least 54,203 records were geocoded before because they could be found in the recently geocoded addresses. But the remaining records could not be determined, as the identifiers returned by the contractor were not useful due to lack of documentation for an NCR-generated ID. Upon further inquiry, we learned that in order to protect patient confidentiality, a temporary ID was created for the contractor, together with addresses from 1995 to 2010. However, due to poor bookkeeping, the bridge between the temporary ID and the permanent record ID was lost. We revealed this issue because state cancer registry staff often face frequent turnover. It is especially a concern when a registry-generated ID is not documented well enough for the new staff member to link back to the central NCR data system.

Our assessments identified two issues. One is how to improve geocoding coverage, and the other is how to manage geocoding workflows. Below, we first identify remedies to improve geocoding coverage and develop a protocol that ensures an acceptable quality of geocoding. By acceptable quality, we mean that more than 90% of physical addresses are geocoded at the street level, while the remaining are placed at the centroid of a city, zip code, or other administrative unit. An early study by McElroy et al. (2003) provided a step-by-step geocoding process based on a Wisconsin breast cancer case-control study that had a street-level match rate of >95%. However, the steps in the process were not evaluated empirically. The geocoding guide developed by Goldberg (2008) for the NAACCR is a good reference to developing a state-specific geocoding protocol. In this study, we followed the principle used by Goldberg and the detailed steps by McElroy et al. In assessing geocoding quality, we accepted the location generated from a geocoder in ArcGIS, whether it was an exact match to the address location or an interpolated one within a matched address range. Location uncertainties in terms of street centerline and matched address range interpolation were not evaluated, as many have done so (Oliver et al., 2005; Wu et al., 2005).

In order to properly manage geocoding, we wanted to assess the whole geocoding coding process for geocoding not only recent addresses, but also addresses from more than 10 years ago. In addition, we have experimented with the idea of the master address index by (1) identifying duplicated addresses in the NCR database and (2) geocoding all unique addresses, so that all addresses would be coded only once, similar to the master patient index (MPI) for data linkage. Once we develop this practice, we can then move forward to addresses from participating programs and provide latitudes and longitudes for a master address index (MAI). In this way, if the geocoding quality of a particular address needs to be improved, the informatics unit would initiate the process of editing the coordinates based on new updates. When this process is finished, the updated information will be available for all participating programs. Interprogram sharing of high-quality geocoding can enhance program collaboration. If a spatial cancer cluster is detected, we could use the same cluster to examine if other programs' participants had elevated risks in a clustered area.

15.3 GEOCODING WORKFLOW DEVELOPMENT

This part was partially based on an early study (Lin et al., 2010). We sampled 3000 records from 2005 to 2008 NCR data, of which 2706 were within Nebraska, and attempted to use them to develop an in-house geocoding process. We used the TeleAtlast 2008 street map as the basis, as it was close to the sampling period. We start with ArcGIS-based geocoding with address standardization and address sensitivity analysis, and then move to identify PO box locations and unmatched addresses.

15.3.1 GEOCODING WITHIN THE ARCGIS ENVIRONMENT

Although one can get street files from many places, the TIGER/Line® shapefiles and their variation in ArcGIS and other mapping databases are the most frequently used. For geocoding, one should download he Address Range Feature (ADDREFEAT) shapefiles, which contain all address range–feature name and address range–edge database relationships at the county level. One can easily merge county files to create a state file.

To create an address locator for geocoding within ArcGIS, one needs to open ArcToolbox and select "Create Address Locator" from the geocoding tools. Since the street is the centerline that has left- and right-side addresses, one needs to set the Address Locator Style to "US Address—Dual Ranges." In the next selection box, put the downloaded Address Range Feature shapefile as the reference data. In the Field Map section, map the five required fields as follows: From Left—LFROMHN, To Left—LTOHN, From Right—RFROMHN, To Right—RTOHN, Street Name—FULLNAME. After the creation of an address locator, one can select Geocode Addresses in the geocoding tools to perform geocoding.

In real-world geocoding, one would start with address cleaning and standardization. Since we wanted to use the software function to recognize some spelling errors and other issues for self-adjustment, we started with unstandardized addresses. Using the default setup, we were able to match 2020 out of 2706 records with a match score of 80 or higher.

A matched address is given a match score based on the number of address fields and exact wording matched. If more than three fields from the U.S. address style (address, city, state, zip code) are not matched perfectly, the match score is generally below 50, or unmatched. The reported match scores are inversely related to spelling sensitivity set by the ArcGIS. A spelling sensitivity level is the percentage of characters of a street name that must be matched exactly. If 1 out of 20 characters in an address field is misspelled, the sensitivity score will likely be 95. Spelling sensitivity is implicitly set above 80 in the ArcGIS address locator.

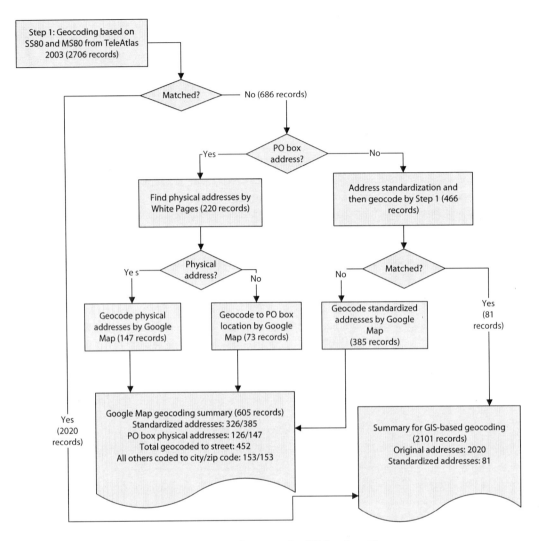

FIGURE 15.1   Geocoding process assessment for a sample of Nebraska addresses.

Note that 2020 out of 2706 matches were done without any address standardization. Among the 2020 matched addresses, 389 addresses had a match score between 80 and 100. When we used the U.S. Postal Service address validation service to verify those 389 addresses, we were able to correct 346 for spelling errors and 43 for zip code errors. After this address standardization, all 2020 addresses had a 100 match score. However, we still had 686 unmatched addresses (Figure 15.1). When we repeated the address validation process for 466 unmatched addresses that excluded post office boxes and delivery routes, an additional 81 addresses were matched. Upon further investigation, we found that an additional 165 addresses could be matched because they were standard addresses out of range of the street base map that we used. These results suggest that before geocoding, one should always find the newest street base map and perform address standardization.

15.3.2   Geocoding Addresses from Other Sources

Since we had 385 standardized addresses (including 165 out-of-range addresses) and 220 PO box and rural delivery route addresses that could not be coded by ArcGIS, we tried to stitch together a complete geocoding process by adding White Pages and other Internet search engines in the PO box location referencing, and Google Map geocoding. Since Internet geocoding cannot be solely for patients due to confidentiality, we mixed the patient addresses with other addresses and then took out the patient addresses after Google Map geocoding.

We treated PO box and rural delivery route addresses separately. We used both the White Pages and Google reverse geocoding to assess the availability of physical addresses for those PO box and delivery route addresses. If the returns from both the White Pages and Google pointed to the same address, the physical address was assured; if only one of them returned a physical address, we geocoded it with less certainty. If neither returned an address, it could not be coded at the street level.

We chose to include only those addresses for which Google Map returned exactly the same street numbers, together with other address fields. We used the Google Map geocoder to code the remaining 385 unmatched addresses and the 147 PO box addresses for which the White Pages returned a physical address. The remaining addresses that could not be coded to street address locations and PO boxes for which the White Pages did not return a physical address were coded to the centroids of zip codes or city centers, whichever was smaller, in a geographic area. In other words, if a zip code was in a rural area that covers two or three small cities, we coded the location according to the city center. If a zip code was in a large urban area, we coded it to the zip code center.

The Google Map geocoding results are shown in Figure 15.1, at the bottom left corner. The process of coding 385 standardized addresses and 147 PO box street addresses yielded 326 and 126 locations at the street level, respectively. These two processes left 80 addresses unfound. In addition to seven addresses with an empty street field, most of those unfound addresses at the street level were in rural areas. If rural (county or state) highways were standardized and uniquely identifiable, most of these 80 street addresses would be found. However, for the purpose of the current evaluation, we combined them with all other PO box addresses and coded them to city or zip code centroids. Together, they represented low-quality geocoding, or 5.7% of the total records.

Since Google Map and ArcGIS used different geocoding base maps, we also made some comparisons by selecting 1500 high-quality addresses to geocode them with both methods. The results showed that the median distance between each pair of 1500 coded locations was 36.6 meters, with an average of 92.8 meters. In the exploratory analysis, we found that 22 rural addresses could be geocoded with a quite different uncertainty due to rural route box location and exact farmhouse location. If we deleted those from the summary statistics, the average distance would be 52.5 meters.

15.3.3 EXTRACT CENSUS TRACT NUMBER FOR EACH GEOCODED LOCATION

After geocoding, we can load the geocoded shapefile to ArcGIS and perform a spatial join with the census tract shapefile that can be downloaded from the census TIGER file website. In the ArcToolbox, click Overlay from Analysis Tool, and then select Spatial Join. Select the geocoded address shapefile for the Target Features and the census tract shapefile for the Join Features. Under Output Feature Class, choose a folder and name for the joined file. Check the box for Keep All Target Features, and for Match Option, select HAVE_THEIR_CENTER_IN and then OK. In this way, the attributes from the area shapefile will be joined to the geocoded point file. Note that an 11-digit census tract number normally named GeoID2 should be attached. This field can then be used to link to other census tract–level variables from the census and American Community Survey (ACS) data files.

15.4 OTHER SECURED INTERNET DATA SOURCES FOR GEOCODING

Since we developed the above geocoding protocol for cancer and other geocoding within the Nebraska Department of Health and Human Services (DHHS) in 2009, other welcome developments have emerged. Recent developments in Web 2.0 and volunteer geographic information (VGI) provide various low-cost geographic information gathering modes. However, the public health agency cannot rely on crowdsourcing from the general public due to quality, time, and confidentiality concerns.

However, if an address is not found within a database, we can use databases within local governments that update street data more frequently. We experimented with a collaborative model between state and local governments. With the assistance of local governments in Douglas County (Omaha) and Lancaster County (Lincoln), and coordination by the Nebraska GIS Council, the Nebraska DHHS piloted a hybrid and multimodal geocoding system that includes (1) in-house geocoding within the Nebraska DHHS, (2) the state centerline geocoding service from the NebraskaMap, and (3) several geocoding services from major cities in the state. In addition, we use Google API to identify some addresses that either have bad components or are too new to be included in any databases. Addresses found by Google Map provide feedback to (1) the Nebraska DHHS for corrected address components and 2) local governments for potential checking and updating of new street segments. The project not only demonstrates a number of advantages of leveraging existing web services from local and state governments, but also serves as a network connecting geospatial information specialists for promoting efficient GIS workflows among various levels of governments. Using the same 2706 addresses above, we found an additional 228 address records that could be coded by GIS servers within local government GIS servers.

More recently, Texas A&M University provided a geocoding service to all NAACCR members. In general, it received a very positive review. Our review based on its 2013 web version was also positive, with a few glitches. We experimented with 50,000 records that had already been coded by the ArcGIS system to identify some potential issues. We first wanted to see if the system would accept some common formats, such as a SAS dataset. Only three data formats—access, comma-separated value (CSV), and text—were accepted. However, if one uses SAS to generate access data with an extension of .mdb, the system might not accept it. We found that CSV was the most user-friendly at that point. However, one needs to watch some special characters, such as ½ and the comma (,). When uploading a large dataset, one needs to watch the number of observations loaded. Otherwise, it was generally manageable once we learned how to fix special character issues.

We next compared geocoding quality. We randomly selected one-third of the addresses, or 16,464 records, to perform real geocoding. Among them, 14,819 could be geocoded by both the GIS and geocoder at the address range level. Most of the unmatched and matched addresses at the zip code level by the geocoding service system at Texas A&M were unmatchable at the address range level. Among the matched, 91.13%, or 13,505 records, had identical census tract codes, and about 8.51% (1261) had different census tract codes; only 0.36%, or 53 records, had different county codes. Since different treatments by different software may come with different geocoding outcomes, it was hard to say which one is more accurate at this point without ground truth. In general, the geocoding service provided by Texas A&M to NAACCR member states seems to be of high quality.

15.5 CONCLUDING REMARKS

Geocoding is a process that should include (1) address cleaning and standardization, which includes reverse address finders from White Pages; (2) multiple in-house street centerline files; (3) the most recent local government data to find addresses for new construction and so forth; and (4) use of an Internet geocoding service or Google Map or Earth to geocode the remaining addresses. Note that using Internet services requires some measure of mixing up real patient addresses with other addresses, so that patient addresses alone are not revealed outside the agency. If we strictly follow these major steps, we can achieve a 90% or even a 95% match rate. With the availability of the geocoding service by NAACCR, there could be simply a two-step process, one for address standardization and the other for NAACCR-sponsored geocoding.

Our assessment of unmatched addresses from ArcGIS suggests that a fair amount of address cleaning, reverse address finding, and regeocoding of Internet-based geocoding would be required. In anticipation of the increased role of Internet-based geocoding, we evaluated a clean address dataset. We found that the quality of Google Map geocoding is likely to be acceptable in most cases. Address reversion is labor-intensive. To reduce the manual or contract workload for Internet-based geocoding, using the zip + 4 file could be a viable option. Most standardized addresses would return a zip + 4 number as a by-product of address standardization. Pilot studies are needed to assess the quality and efficiency of including the zip + 4 number and using the zip + 4 file for geocoding problem addresses.

In order to perform meaningful spatial analysis of registry and other public health data, we have to improve the spatial specificity of cases. Many public health data have not been geocoded. For legacy data that were geocoded in the early years, it is common that more than 30% of cases are not coded at residential or street locations, which are not appropriate for population-based data analysis at the subcounty level. Due to different training, most data managers and occasion users do not have proper knowledge to check geocoding quality. Since geocoded data are normally processed within a GIS, a common myth is that checking geocoding quality also requires GIS knowledge. While this is true for sophisticated visualization, checking geocoded data quality does not necessarily require GIS. For example, by using a frequency table, one could easily see how many geocoded addresses are missing coordinates. In addition, for cancer, hospital discharge, birth, and death, it would be rare for a dozen people to have the same coordinates within a particular year if they were not from an institution (prison, nursing home, etc.). For this reason, one can run a Structured Query Language (SQL) by checking to see the distribution of identical coordinates from a geocoded product. If a dozen or even a half-dozen coordinates come from the same location, a flag should be raised for quality checking.

In addition, a data manager should prepare a gold standard geocoding dataset that has been geocoded to the highest standard. The dataset should have between 300 and 400 addresses with a variety of difficulties, such as few PO box and rural delivery route addresses. These addresses should be coded by a professional who can then manually check the data quality, and then be attached with some needed geographic units, such as census tracts. These gold standard addresses should be included periodically with a set of addresses to be sent for geocoding. In this way, geocoding quality can be checked for a subsample of the gold standard addresses.

Finally, a data manager can also play a proactive role by developing address geocoding quality insurance measures. A common practice for a data manager is to simply pass an address plus a unique ID to a contractor or in-house GISer. If this ID is temporarily assigned, at least two persons within the program should know that, so that a sudden staff turnover would not jeopardize the geocoding work. If an ID field is too long, one should include two sets of IDs whenever possible, because many GIS-based geocoders do not handle long IDs, and it can be truncated without warning. Another measure that may enhance geocoding quality is to include an address indicator to differentiate residential from other types of addresses. On the one hand, many nursing homes may come with a street address that does not look like an institutionalized facility. On the other hand, many individuals only list institution names without addresses. If a geocoding contract has both types of information, it would be a lot easier to cross-reference between the two types of addresses for better coding. Sometimes, it also helps if data managers can provide some education to address entry persons or data form receptionists to obtain better addresses. But the best way is to install address validation software that would standardize and valid an address at the patient encounter.

REFERENCES

Boscoe, F.P. 2008. The science and art of geocoding: Tips for improving match rates and handling unmatched cases in analysis. In *Geocoding Health Data: The Use of Geographic Codes in Cancer Prevention and Control, Research and Practice*, ed. G. Rushton, M.P. Armstrong, J. Gittler, B.R. Greene, C.E. Pavlik, M.M. West, and D.L. Zimmerman. Boca Raton, FL: CRC Press, pp. 95–110.

Goldberg, D.W. 2008. *A Geocoding Best Practices Guide.* Springfield, IL: North American Association of Central Cancer Registries.

Hurley, S.E., T.M. Saunders, R. Nivas, A. Hertz, and P. Reynolds. 2003. Post office box addresses: A challenge for geographic information system-based studies. *Epidemiology* 14(4): 386–391.

Lin, G., J. Gray, and M. Qu. 2010. Improving geocoding outcomes for the Nebraska Cancer Registry: Learning from proven practices. *Journal of Registry Management* 37(2): 49–56.

McElroy, J.A., P.L. Remington, A. Trentham-Dietz, S.A. Robert, and P.A. Newcomb. 2003. Geocoding addresses from a large population based study: Lessons learned. *Epidemiology* 14(4): 399–407.

Oliver, M.N., K.A. Matthews, M. Siadaty, F.R. Hauck, and L.W. Pickle. 2005. Geographic bias related to geocoding in epidemiologic studies. *International Journal of Health Geographics* 4: 29. doi: 10.1186/1476-072X-4-29.

Wu, J., T.H. Funk, F.W. Lurmann, and A.M. Winer. 2005. Improving spatial accuracy of roadway networks and geocoded addresses. *Transactions in GIS* 9(4): 585–601.

Zimmerman, D.L., X. Fang, S. Mazumdar, and G. Rushton. 2007. Modeling the probability distribution of positional errors incurred by residential address geocoding. *International Journal of Health Geographics* 6: 1.

16 Sex Difference in Stroke Mortality

16.1 INTRODUCTION

In the United States, stroke contributes significantly to cardiovascular disease (CVD) death and was the third highest cause of mortality in 2007, with more than 130,000 deaths. Approximately 3% of Americans have had a stroke, with a direct and indirect cost of $73.7 billion (Lloyd-Jones et al., 2010). Since 1990, cardiovascular mortality, including stroke mortality, has declined, but the decline is greater for men than for women (Lloyd-Jones et al., 2010). Since women live longer and have considerably lower mortality from cardiovascular disease than men, the slow decline in stroke mortality for women coupled with disproportionately more women aged 60 years and older will likely create a greater need for stroke care among women than among men.

Past research has shown little consistency in sex differential in stroke case fatality. Some early studies suggest a female survival advantage shortly after a hospital stay (Andersen et al., 2005; Olsen et al., 2007, 2009), while others suggest the opposite advantage for ischemic stroke patients (Niewada et al., 2005). An overwhelming majority of studies in both the United States and other countries show higher female short-term (28 days) case fatality (Thorvaldsen et al., 1995; Appelros et al., 2009). A notable exception is from a study among the Get With the Guidelines—Stroke population in the United States (Reeves et al., 2008), which shows no sex difference in case fatality after controlling for other factors.

The apparent sex difference inconsistency in hospital-based case fatality may be due to the interplay between hospital-based and community-based findings. It is known that women, in general, have lower stroke mortality than men. This fact, coupled with no sex difference in case fatality within the hospital-based stroke care system, suggests that men are more likely to die in the community, without having been hospitalized for stroke care. On the other hand, if women do not have a survival advantage in the short term (4–6 weeks after hospitalization) but are disadvantaged in stroke case fatality, then they would be more likely to die from stroke in the community, while men would be more likely to die from other causes. In addition, there are more women than men aged 60 years and older, and an age-specific adjustment may not fully account for the overall sex difference. In any case, evidence from hospital-based data cannot be straightforwardly added up with cause-specific mortality that includes both community- and hospital-based data.

Sex differences in stroke mortality can also be examined from community-based studies. First, difference in treatment in hospitals affects both in- and out-of-hospital survival prognoses. Treatment for stroke is dependent on time, and a late response can be deadly (Lloyd-Jones et al., 2010; Pedigo and Odoi, 2010). Gargano et al. (2009) suggests that women receive treatment later than men because older women tend to live alone and cannot call for help at the onset of stroke. Second, informal care provided by a spouse may differ, and the difference usually favors men due to the traditional role of women (Persky et al., 2010). Third, community resources, follow-up care, and public health interventions may affect people differently. Women are more likely to live alone and have less social support than men (Persky et al., 2010). Those who receive hospital-based rehabilitation services are more likely to survive than those who receive no rehabilitation services or other types of rehabilitation services (Thorsen et al., 2005). In addition, both treatment effectiveness and after-hospital survival depend on follow-up care at the community level. Since rural areas tend to have fewer community-based resources and are less likely to provide timely access to emergency stroke care (Pedigo and Odoi, 2010), rural residents tend to have poorer survival outcomes than urban residents (Sergeev, 2011). Fourth, survivors of stroke who are functionally dependent have a higher risk of dying than functionally independent survivors (Slot et al., 2009), and women are more likely to be functionally dependent, with a greater number of comorbidities. Finally, women are also more likely to be depressed after their stroke, which can impact recovery (Persky et al., 2010). Since most of these factors favor male stroke patients, female stroke patients may have more survival disadvantages both in and out of the hospital.

In this chapter, we take an integrated approach to sex differences in stroke case fatality, connecting hospital and community settings. We include patients who received hospital-based stroke care and patients who died of stroke outside of a hospital. We linked 5-year Nebraska hospital discharge data from 2005 to 2009 to the Nebraska vital statistics record system in 2010 to obtain 30-day case information for both fatality and survival. The results also provide stroke prognosis factors for Nebraska, which is important for stroke care providers and public health practitioners.

16.2 METHODS

16.2.1 DATA

Two data sources were used. Hospital discharge data for all Nebraska residents who were hospitalized for stroke and discharged between January 2005 and December 2009 were retrieved from the Nebraska Hospital Association. Nebraska death records from January 2005 to September 2010 were obtained from the Nebraska Department of Health and Human Services Office of Vital Records. The hospital discharge data records have inpatient and outpatient (including emergency department and rehabilitation) discharge information from all hospitals in Nebraska. The system is based on the standard UB-04 form, which includes patient demographic information, diagnostic codes, procedure codes, and disposition. The UB-04 also includes International Classification of Diseases, Ninth Revision (ICD-9) codes for primary, secondary, and other diagnoses. CVD is the leading cause of hospitalization in Nebraska, and it accounted for about 8,000 emergency department visits and more than 30,000 hospitalizations each year. The current study is restricted to Nebraska residents only. Out-of-state patients who sought care in Nebraska hospitals were excluded, and records for Nebraska patients who sought care in other states were not available.

To obtain survival information, we linked hospitalization data to death certificate data. A probabilistic linkage strategy was used, with major linkage variables being patient name, date of birth, sex, and residence zip code. The linkage was processed using Link Plus at the Nebraska Hospital Association website, which hosts the hospital discharge data. After linkage, the data were deidentified by removing name and address information. Our check of linkage quality showed that 97.4% of hospital discharge records with an indication of "expired" (24,368 records) among all CVD patients (ICD-9-CM 390–459) matched death certificate data. The remaining 2.6% of unmatched records might be due to transportation of patients who died to adjacent states for a funeral or for other unknown reasons. Since the linkage is patient based, readmissions or other hospital encounters are also available, and we had to decide which admission to use as the baseline. If a patient had multiple hospitalizations, we selected the first hospitalization, which should be similar to the "first-ever" stroke used by Roquer et al. (2003). Stroke is defined by the primary diagnosis (ICD-9 430–434, 436–438), where transient ischemic attacks were excluded.

Control variables included age group and urban core county or not (counties with 100,000 or more population). In addition, we included a number of comorbidities identified in previous literature (Niewada et al., 2005; Roquer et al., 2003). Diabetes mellitus is a potential factor in stroke survival and has been shown to be more prevalent in men. Atrial fibrillation is a known risk factor for stroke and has been shown to be more prevalent in women. Heart failure is a potential factor in stroke survival. Anemia is a risk factor for stroke and has been shown to significantly increase the likelihood of death 1 year after stroke hospitalization. Chronic kidney disease is an independent risk factor for mortality and also leads to complications such as anemia. If a diagnosis indicated one or more of these comorbidity factors, it was coded 1; otherwise, it was coded 0. We also controlled for stroke subtype—ischemic and hemorrhagic—because hemorrhagic stroke is much more deadly and difficult to treat and is more common in women (Kapral et al., 2005). Finally, we included the use of hospital-based rehabilitation, which has been shown to increase the likelihood of survival in stroke patients (Paolucci et al., 2006).

16.2.2 STATISTICAL ANALYSIS

Statistical analyses were performed using SAS version 9.2. Crude and standard stroke mortality were calculated based on the 2005–2009 underlying cause of death, with the at-risk population of 2007 multiplied by 5 for an annualized rate. The U.S. 2000 standard population was used. In the descriptive analysis, all variables were in proportions, and sex difference was tested using Pearson's χ^2 test. In the multivariate analysis, a logistic regression was used to predict 30-day fatality, and the Cox proportional hazard model was used to predict overall survival. The results from the logistic and Cox regressions are presented as odds ratios (ORs) and hazard ratios (HRs), respectively, with corresponding 95% confidence intervals (CIs). Since both regressions included age and sex, the corresponding ORs and HRs are age and sex adjusted. To assist the interpretation and discussion, the Kaplan–Meier curves for males and females were plotted by number of comorbidities.

16.3 RESULTS

16.3.1 DESCRIPTIVE STATISTICS

Between 2005 and 2009, there were 4509 deaths in Nebraska with the underlying cause of stroke. Based on the at-risk populations by sex in 2007, the crude stroke mortality rates were 40.5 and 61.2 per 100,000 for males and females, respectively (Table 16.1). This difference implies a substantial sex disparity in stroke care burden, as there were 2311 deaths due to stroke among females aged 75 years and older, almost double the number for males (1214) aged 75 years and older. Even though the crude rates for males and females are quite different, the standardized mortality rates are very close (44.07 vs. 45.58 per 100,000 for males and females, respectively).

Among the 15,806 stroke patients identified in the 2005–2009 Nebraska hospital discharge data, 7,656 (48%) were males and 8,150 (52%) were females. The mean age was higher for females than for males (74.1 years for females vs. 70.3 years for males, $p < 0.001$). Mean hospital stays, not including rehabilitation, were significantly shorter for males than for females (3.8 days for males vs. 4.2 days for females, $p < 0.01$). Table 16.2 lists frequencies and crude survival rates for selected

TABLE 16.1

At-Risk Population (2007), Stroke Deaths (2005–2009), and Mortality

Age Group	Males			Females		
	Population	Deaths	Mortality	Population	Deaths	Mortality
0–14	189,055	5	0.53	180,429	3	0.33
15–29	200,003	7	0.70	189,662	5	0.53
30–44	169,438	30	3.54	163,512	24	2.94
45–59	179,511	165	18.38	181,849	106	11.66
60–74	92,416	353	76.39	101,637	286	56.28
75+	46,350	1,214	523.84	76,050	2,311	607.76
Total	876,773	1,774	40.47	893,139	2,735	61.24
Age-standardized rate			44.07			45.58

Note: Mortality is annualized per 100,000.

TABLE 16.2

Descriptive Statistics among Hospitalized Stroke Patients by Sex

			Crude Survival Rate, %	
Total	7656	8150	69.5	64.2
Age, years	**Male**	**Female**	**Male**	**Female**
<40	200 (2.6)	216 (2.7)	87.0	93.1
40–59	1396 (18.2)	1115 (13.7)	83.8	85.8
60–75	2840 (37.1)	2297 (28.2)	76.8	76.7
76–90	2985 (39)	3775 (46.3)	58.4	56.0
>90	235 (3.1)	747 (9.2)	31.5	30.4
	Stroke Subtype			
Ischemic stroke	5679 (74.2)	5927 (72.7)**	73.2	66.5
Hemorrhagic stroke	1381 (18)	1427 (17.5)	55.0	52.1
	Comorbidities			
Atrial fibrillation	1083 (14.2)	1429 (17.5)**	55.7	44.1
Anemia	454 (5.9)	681 (8.4)**	56.2	56.7
Diabetes	1880 (24.6)	1868 (22.9)*	69.8	65.4
Chronic kidney disease	587 (7.7)	514 (6.3)**	53.7	52.1
Heart failure	573 (7.5)	732 (9)**	49.7	36.6
	Rehabilitation			
Had rehab	1240 (16.2)	976 (12)	83.6	86.8
Did not have rehab	6416 (83.8)	7174 (88)**	67.1	61.5
Urban core counties	3361 (43.9)	3669 (45)	72.8	66.3
	Expired Status			
30 days	930 (12.1)	1314 (16.1)**		
30+ days	1384 (18.1)	1575 (19.3)*		

Note: Percentages as male and female totals are in parentheses.
* $p < 0.05$; ** $p < 0.01$.

variables by sex. There was a greater proportion of males than females in the 40–59 years and 60–75 years age groups, but the age-specific survival rates were comparable (Table 16.2, last two columns). Males were slightly more likely than females to have ischemic stroke, while there was no sex difference for the more deadly hemorrhagic stroke. All the comorbidity conditions were significant. With the exception of chronic kidney disease, females were more likely than males to have other comorbidities, such as atrial fibrillation, anemia, and heart failure. Males were more likely than females to receive rehabilitation services after stroke, but there was no rural–urban difference between males and females, suggesting that a lack of access to rehabilitation services in rural areas may not be a factor contributing to more males seeking rehabilitation services. Finally, although females were more likely than males to die both in and out of the hospital, the increased likelihood of out-of-hospital death was significant.

Among all stroke patients, females had a lower unadjusted survival rate (64.2% for females vs. 69.5% for males). This result is not surprising because males tended to be younger than females and more likely to receive rehabilitation services. However, females were more likely than men to survive if they were less than 40 years of age or if they received rehabilitation services after stroke. Interestingly, while females were less likely to have ischemic stroke, they were more likely to die from it afterwards. Since ischemic stroke is less life threatening than hemorrhagic stroke, this result suggests that comorbidities, which are more prevalent among females, made them more likely to die in nonhospital settings. This implication was further reinforced by the fact (not shown) that females were much less likely to be discharged to home (50.39% for males vs. 38.87% for females, $p < 0.01$), requiring more skilled nursing and other prescribed care after discharge.

16.3.2 MULTIVARIATE ANALYSIS

In the multivariate analysis, we examined the degree to which the findings from the descriptive analysis were attenuated by other variables. Table 16.3 lists results from a logistic regression on 30-day case fatality using the same set of variables from Table 16.2. Age effects were all significant, but the effects for the 40–59 years and 60–75 years age groups were similar to the effects seen in the descriptive analysis. Stroke patients who had chronic kidney disease, atrial fibrillation, or heart failure were significantly more likely to die within 30 days of onset. Having received rehabilitation services reduced mortality. Both ischemic and hemorrhagic strokes increased the likelihood of 30-day fatality. Females were still more likely than males to die within 30 days, even after controlling for the above factors.

For the survival analysis, we analyzed the overall sample (Table 16.4, first two columns). In addition, we ran an out-of-hospital survival model using the same set of variables as in Table 16.2 (Table 16.4, last two columns). The age effects followed a gradient: the older the patient, the greater the HR. Those aged 91 years and older had an HR of 11.195 compared to those aged 39 years or younger. With the exception of diabetes, comorbidity factors, such as anemia, chronic kidney disease, atrial fibrillation, and heart failure, were all independent predictors of mortality, with HRs around 1.5. Receiving rehabilitation services reduced case fatality. As expected, hemorrhagic stroke increased case fatality (HR 2.33, 95% CI 2.08–2.619). After controlling for age, comorbidity, and stroke type, sex was not an independent predictor of mortality. This finding was true for the overall and out-of-hospital survival models. Finally, compared to the overall survival models, the HRs in the out-of-hospital survival model had either similar or stronger effects.

TABLE 16.3
Logistic Regression on 30-Day Stroke Case Fatality

	OR	95% CI
Ref (age < 40)		
Age 40–59	1.91	1.239–2.943
Age 60–75	2.314	1.517–3.528
Age 76–90	4.637	3.058–7.031
Age 91 and older	10.857	7.015–16.805
Females	1.149	1.04–1.269
Diabetes	0.988	0.877–1.113
Anemia	1.045	0.875–1.248
Chronic kidney disease	1.32	1.104–1.579
Atrial fibrillation	1.542	1.369–1.738
Heart failure	1.995	1.717–2.317
Had rehabilitation	0.843	0.763–0.931
In urban core counties	1.382	1.115–1.713
Stroke Type (ref = others)		
Ischemic stroke	6.528	5.219–8.164
Hemorrhagic stroke	0.035	0.019–0.061

Note: If a CI does not contain 1, then the OR is significant.

TABLE 16.4

Cox Survival Models for the Overall and Out-of-Hospital Stroke Patients

	Overall Survival		Out-of-Hospital Survival	
	HR	95% CI	HR	95% CI
Age 40–59	1.828	1.319–2.534	1.773	1.174–2.677
Age 60–75	3.145	2.292–4.316	3.505	2.353–5.222
Age 76–90	5.968	4.36–8.169	7.009	4.717–10.415
Age 91 and older	11.195	8.113–15.447	14.335	9.574–21.464
Diabetes	1.06	0.993–1.133	1.08	1.004–1.162
Anemia	1.151	1.046–1.266	1.248	1.124–1.385
Chronic kidney disease	1.513	1.377–1.662	1.63	1.47–1.807
Atrial fibrillation	1.419	1.327–1.517	1.424	1.321–1.536
Heart failure	1.65	1.521–1.791	1.696	1.547–1.858
Had rehabilitation	0.343	0.306–0.384	0.394	0.351–0.442
In urban counties area	0.9	0.85–0.953	0.856	0.803–0.913
Ischemic stroke	1.043	0.937–1.16	1.024	0.915–1.146
Hemorrhagic stroke	2.334	2.081–2.619	1.525	1.34–1.737
Female sex	1.024	0.968–1.084	1.023	0.96–1.09

Note: If a CI does not contain 1, then the HR is significant.

Because the descriptive and multivariate analyses results differed, we plotted the Kaplan–Meier curves by sex after hospitalization to gain insight into the timing and comorbidities of stroke (Figure 16.1). To aid the visual effect, censored points were not shown, and 250 weeks or more were truncated. Figure 16.1 shows that after an initial drop in survival, male and female stroke patients had similar survival profiles when they had no comorbidities or one comorbidity. Differences in sex survival stem mainly from those with two comorbidities: males with two comorbidities had a significant survival advantage within the first 10 weeks after the admission compared to males with

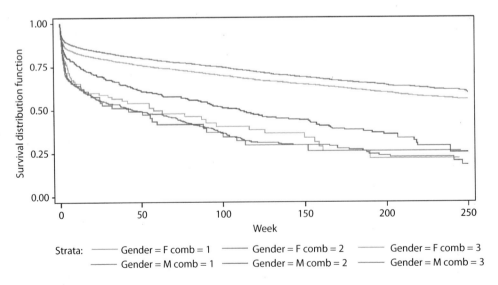

FIGURE 16.1   (See color insert) Kaplan–Meier curves by sex and comorbidities.

three or more comorbidities. Females with two comorbidities, on the other hand, were as likely to die as males and females with three or more comorbidities.

16.4 CHAPTER SUMMARY AND DISCUSSION

In this study, we integrated hospital-based stroke data with community-based vital statistics records to provide a fuller picture of sex differences in case fatality and mortality. Consistent with previous literature, we found that females in Nebraska have higher stroke case fatality than males. Adjusting baseline characteristics does not remove this disparity (OR 1.149, 95% CI 1.04–1.269). Females also have slightly higher 30+-day case fatality than males, but these differences are not significant after adjusting for baseline risk factors, such as age and comorbidities.

Other factors reduce female survival rate. Because females outnumber males in older age groups, females overall are more likely to have age-related comorbidities. These comorbidities have been shown to reduce survival and thus make females more likely to die than their male counterparts. Two or three morbidity conditions contribute similarly to the survival prognosis for females, while each additional morbidity condition progressively reduces males' chance of survival.

Several factors protect patients from stroke mortality. Similar to Carinci et al.'s study (2007), we found that patients who receive rehabilitation services after hospitalization are more likely to survive than those who do not receive rehabilitation services. In the context of our descriptive findings, it is difficult to know how rehabilitation services may contribute to or reduce sex difference in stroke case fatality. Although we did not have data on stroke severity for our study group, a Swedish national study suggests that when stroke conditions are comparable (Eriksson et al., 2009), females are less happy than males with rehabilitation consultation. These results suggest that rehabilitation programs should be more accessible and amenable to women, so that they would be more willing to participate in them. In addition, living in an urban environment also appears to increase survival because of greater access to rehabilitation and other stroke care resources. Although there was no significant difference between males and females in rural–urban living, women may have greater transportation needs to access different stroke care resources, such as emergency and home care, as many women stroke victims live alone.

Women outlive men by a large margin, which has significant and wide implications in stroke incidence, mortality, and case fatality. In the current study, females aged 75 years and older account for more than 60% of female stroke hospitalizations and 83% of female stroke deaths. The number of stroke deaths in this age group was almost twice as high for females as for males. Although the age-adjusted stroke mortality rates were similar by sex (44.07 per 100,000 for males vs. 45.58 for females), stroke places a large burden on those caring for elderly women, and the age-standardized rate may mask stroke care burden for women from both health care delivery and public health systems (Reeves et al., 2008). This possibility is supported in our analysis of hospitalization data, where we found no sex difference in survival after adjusting other risk factors, but found that females stay in the hospital longer and require more community-based services after discharge.

The study has several limitations. First, we did not have a severity index for stroke victims, and we did not have information about risk factors for the patients, such as whether they smoked or had chronic hypertension. Second, we had very limited prehospital and in-hospital treatment and timing information, which is critical for stroke patient survival. In addition, a Nebraska resident could have been hospitalized for stroke within the state but moved out of the state and died; we had such records only if the death occurred in a state with which Nebraska has a vital statistics exchange agreement. In addition, no data were available about Nebraska residents who may have been hospitalized for stroke in another state. Finally, although our check of the data linkage quality showed a high degree of accuracy, it is possible that some patients were missed during the linkage process.

REFERENCES

Andersen, M.N., K.K. Andersen, L.P. Kammersgaard, and T.S. Olsen. 2005. Sex differences in stroke survival: 10-year follow-up of the Copenhagen Stroke Study Cohort. *Journal of Stroke and Cerebroovascular Disease* 14: 215–220.

Appelros, P., B. Stegmayr, and A. Terént. 2009. Sex differences in stroke epidemiology: A systematic review. *Stroke* 40: 1082–1090.

Carinci, F., L. Roti, P. Francesconi, R. Gini, F. Tediosi, T. Di Iorio, S. Bartolacci, and E. Buiatti. 2007. The impact of different rehabilitation strategies after major events in the elderly: The case of stroke and hip fracture in the Tuscany region. *BMC Health Services Research* 7: 95.

Eriksson, M., E.L. Glader, B. Norrving, A. Terént, and B. Stegmayr. 2009. Sex differences in stroke care and outcome in the Swedish national quality register for stroke care. *Stroke* 40: 909–914.

Gargano, J.W., S. Wehner, and M. Reeves. 2009. Do presenting symptoms explain sex differences in emergency department delays among patients with acute stroke? *Stroke* 40: 1114–1120.

Kapral, M.K., J. Fang, M.D. Hill, F. Silver, J. Richards, C. Jaigobin, A.M. Cheung; Investigators of the Registry of the Canadian Stroke Network. 2005. Sex differences in stroke care and outcomes: Results from the Registry of the Canadian Stroke Network. *Stroke* 36: 809–814.

Lloyd-Jones, D., R.J. Adams, T.M. Brown, M. Carnethon, S. Dai, G. De Simone, T.B. Ferguson, et al. 2010. Heart disease and stroke statistics—2010 update: A report from the American Heart Association. *Circulation* 121: e46–e215.

Niewada, M., A. Kobayashi, P.A. Sandercock, B. Kamiński, and A. Członkowska. 2005. Influence of gender on baseline features and clinical outcomes among 17,370 patients with confirmed ischemic stroke in the international stroke trial. *Neuroepidemiology* 24: 123–128.

Olsen, T.S., C. Dehlendorff, and K.K. Andersen. 2007. Sex-related time-dependent variations in post-stroke survival—Evidence of a female stroke survival advantage. *Neuroepidemiology* 29: 218–225.

Olsen, T.S., C. Dehlendorff, and K.K. Andersen. 2009. The female stroke survival advantage: Relation to age. *Neuroepidemiology* 32: 47–52.

Paolucci, S., M. Bragoni, P. Coiro, D. De Angelis, F.R. Fusco, D. Morelli, V. Venturiero, and L. Pratesi. 2006. Is sex a prognostic factor in stroke rehabilitation? A matched comparison. *Stroke* 37: 2989–2994.

Pedigo, A.S. and A. Odoi. 2010. Investigation of disparities in geographic accessibility to emergency stroke and myocardial infarction care in East Tennessee using geographic information systems and network analysis. *Annals of Epidemiology* 20: 924–930.

Persky, R.W., L.C. Turtzo, and L.D. McCullough. 2010. Stroke in women: Disparities and outcomes. *Current Cardiology Reports* 12: 6–13.

Reeves, M.J., C.D. Bushnell, G. Howard, J.W. Gargano, P.W. Duncan, G. Lynch, A. Khatiwoda, and L. Lisabeth. 2008. Sex differences in stroke: Epidemiology, clinical presentation, medical care, and outcomes. *Lancet: Neurology* 7: 915–926.

Roquer, J., A.R. Campello, and M. Gomis. 2003. Sex differences in first-ever acute stroke. *Stroke* 34: 1581–1585.

Sergeev, A.V. 2011. Racial and rural-urban disparities in stroke mortality outside the stroke belt. *Ethnicity and Disease* 21: 307–313.

Slot, K., E. Berge, P. Sandercock, S.C. Lewis, P. Dorman, M. Dennis, on behalf of the Oxfordshire Community Stroke Project, the Lothian Stroke Register, and the International Stroke Trial. 2009. Causes of death by level of dependency at 6 months after ischemic stroke in 3 large cohorts. *Stroke* 40: 1585–1589.

Thorsen, A.M., L.W. Holmqvist, J. Pedro-Cuesta, and L. von Koch. 2005. A randomized controlled trial of early supported discharge and continued rehabilitation at home after stroke: Five-year follow-up of patient outcomes. *Stroke* 36: 297–303.

Thorvaldsen, P., K. Asplund, K. Kuulasmaa, A.M. Rajakangas, and M. Schroll. 1995. Stroke incidence, case fatality, and mortality in the WHO MONICA project. *Stroke* 26: 361–367.

17 Model Outcomes of Acute Myocardial Infarction (AMI) by Residence and Hospital Locations

17.1 INTRODUCTION

Acute myocardial infarction (AMI), or heart attack, is a major cause of death in the United States. AMI mortality and rehospitalization rates are standard quality of care indicators. To improve AMI patient care, the American Heart Association (AHA), the Centers for Medicare and Medicaid Services (CMS), and the American College of Cardiology (ACC) have introduced sets of quality improvement guidelines (Bernheim et al., 2010). Despite these efforts, disparities in health outcomes exist among different population groups and over different geographic areas (Xian et al., 2010). Previous studies, both before and after the introduction of the clinical improvement guidelines, have found significant differences in process measures for AMI care between rural and urban hospitals. Earlier evidence from the mid-1990s showed that rural hospitals provide poorer quality of inpatient care for AMI patients than do urban hospitals. Patients admitted to rural hospitals were less likely to receive recommended lifesaving treatments or key quality of care indicators, such as aspirin at discharge, beta-blockers, and thrombolytic therapy (Tsai et al., 2010). Despite overall improvement in adhering to recommended treatments for AMI patients, and closing gaps between large rural hospitals and urban hospitals (Baldwin et al., 2010), these treatment disparities persisted 5–10 years after the introduction of the guidelines (Barnato et al., 2005; O'Connor and Wellenius, 2012). A study showed rural hospitals had a higher 30-day mortality rate than urban hospitals (Baldwin et al., 2004), while another study from Iowa found no difference in AMI in-hospital mortality rate between rural and urban hospitals (James et al., 2007). Due to data limitation, almost all the studies used 30-day or in-hospital mortalities as the outcome. In this case study, we extend it to survival beyond 30 days after discharge.

From an AMI patient's perspective, Nebraska rural residents face a number of barriers to receiving optimal care after an AMI attack (Sheikh and Bullock, 2001; Nawal et al., 2007). First, the overwhelming majority of cardiology care services and facilities are in the two largest cities: Omaha and Lincoln. In 2010, there were 130 cardiologists according to the Nebraska Professional Tracking System, among whom 102 were in these two cities. Second, sometimes an AMI patient has to seek care in a rural hospital when travel time to a major urban hospital is greater than 30 to 40 minutes, or the so-called golden hour for AMI patients. However, only 7 out of 90 rural counties have cardiologists. As a result, some of the patients might end up seeking care from a generalist rather than a cardiologist. Third, even though an AMI patient can ultimately receive care at a major urban hospital, emergency medical services (EMS) travel time and treatment during transport may not be optimal. In this case study, we included physical access measures, such as travel distance and bypassing rural hospital information, to evaluate rural–urban differences in AMI readmission and survival.

Cardiac rehabilitation, such as physical exercise, is an important step for AMI patient recovery. Cardiac rehabilitation can improve survival and reduce hospital readmissions. The AHA and ACC guidelines recognized physical exercises for AMI patients as Class I (B) recommendations on secondary prevention (Smith et al., 2006): all patients with an AMI should be offered a cardiac rehabilitation program. However, cardiac rehabilitation is still seriously underutilized, because of the geographic distribution of available programs and the failure of physicians to refer patients (Ades, 2001). In addition, most studies using hospital discharge data do not include cardiac rehabilitation, because most rehabilitation programs are carried out in outpatient facilities. In this case study, we address this data gap by linking inpatient data to outpatient data so that all AMI patients who had cardiac rehabilitation are identified.

Finally, interactions among known risk factors are common. In previous chapters, we assessed readmission and mortality outcomes according to major diagnoses for hospitalizations. In this chapter, we include comorbidities and neighborhood socioeconomic status (SES) that may affect AMI outcomes. In addition, we tapped the location information from geocoding and added two important pieces of information to each record: distance to hospital and five poverty levels at the census tract level. We know from Chapter 8 that AMI events are more likely to be found in the poorest poverty category (>20%) than in the least poor category (<5%). We know from Chapter 10 (Table 10.2) that the same is true for urban areas, but the opposite for rural areas. However, we do not know if poverty had any impact on mortality outcomes, and if bypassers had a better survival prognosis.

17.2 METHOD

17.2.1 DATA LINKAGE

We used the 7-year (2005–2011) hospital discharge data and selected the first hospitalization due to AMI based on the primary diagnosis. This practice is different from that in Chapter 8, where we used the first two diagnoses to define an AMI admission. We excluded patients who resided outside of Nebraska, leaving 14,457 unique patient records in the current study. If we used the primary and secondary diagnoses, the number of patients would be about 19,000. In order to obtain information on (1) mortality beyond 30 days, (2) rural hospital bypassing, (3) cardiac rehabilitations, and (4) neighborhood SES, we needed to link four different datasets. Each individual linkage was described in previous chapters, and we summarize them here again.

We first linked hospital discharge data to vital records mortality data (2005–2012). The linkage was conducted at the Nebraska Hospital Association based on the probabilistic linkage methodology. The variables used for the linkage include patient name, date of birth, sex, and residence zip code. Residential address and driver's license information were used during the manual review process to ascertain true links. Second, all patients were already geocoded to their residential locations using the method described in Chapters 6 and 15. After geocoding, poverty information, such as percent of people living under the federal poverty line, was attached to each patient based on the ACS 2006–2010 data. Third, we linked inpatient data and outpatient data by using the master patient ID created by the Nebraska Hospital Association. This is the deterministic linkage that goes through more than 17 million outpatient records between 2005 and 2011. In addition, all hospitals were geocoded according to their addresses. Although there could be some small variation in terms of exact latitudes and longitudes, hospital addresses were 100% correct in terms of county locations to identify rural hospital bypassing events. Based on hospital location and patient residential location, Euclidean distance is calculated for each hospitalization event.

17.2.2 VARIABLES

We were interested in rural and urban differences, and used three urban counties (Douglas, Sarpy, and Lancaster) in contrast to the rest of the state as rural. Key outcome variables included 30-day all-cause readmission and 1-year mortality or survival after discharge. The 30-day readmission was in reference to the first AMI admission. The out-of-hospital mortality excluded those who died in hospitals, as we wanted to focus on those who died in communities. Likewise, the out-of-hospital survival did not use 30 days as the cutoff, it started from the discharge date for those who were discharged alive, and lasted until 2012 or 1 year to the last discharge date in the study period.

The control variables used in the study included patient age, sex, and neighborhood poverty level, and a list of comorbidities, transfer status, and location factors. As the presence of comorbidities can alter the effectiveness of a treatment, a number of important comorbidity conditions, such as chronic obstructive pulmonary disease (COPD), diabetes, renal disease, hypertension, cerebrovascular disease, and cancer, were included as control variables. The exploratory analysis suggested that there were very weak correlations among these comorbidity variables, with the strongest being 0.13 between diabetes and hypertension. If a health condition appeared in one of nine diagnosis fields, we coded it as 1; otherwise, it was 0. A patient having an outpatient cardiac rehabilitation session was coded 1; otherwise, the patient was coded 0. Transfer status was derived from the admission source variable. To measure geographic access, we created two variables: distance to a hospital and rural bypasser. Distance to a hospital is based on crow flying distance from a geocoded patient location to the corresponding hospital location. A rural bypasser is defined as one who bypasses rural counties, presumably a rural hospital, to seek care in an urban hospital. Finally, we also included an indicator variable of transfer admission. If an admission was a transfer from a hospital, then it was 1; otherwise, it was 0. As we shall see, this variable overlaps significantly with the rural bypasser variable.

17.2.3 STATISTICAL ANALYSIS

The study followed a retrospective cohort design, with the key exposure being rehabilitation status. Information such as cardiac rehabilitation, all-cause readmission, and mortality information was all after the initial hospitalization. The longitudinal components were generated through data linkage and master patient ID. Both descriptive and multivariate analyses were used. In the descriptive statistics, all the variables were in proportion, and Pearson chi-squared tests were used for the rural–urban difference. In multivariate regression analyses, AMI readmission and 1-year mortality and survival were examined, controlling for demographic, poverty, comorbidities, and other risk factors. Two distinct models were estimated. A place of residence model was to contrast rural and urban residents; a place of hospital model was to contrast rural and urban hospitals. Logistic regression was used in mortality analysis, and a Cox proportional hazard model was used in the survival analysis. The odds ratio (OR) and hazard ratio (HR) were calculated to estimate AMI death and survival rates, respectively, with 95% confidence intervals (CIs). In all statistical analyses, a backward-fitting process was used to remove variables that were not significant ($p \geq 0.05$).

17.3 RESULTS

Table 17.1 shows that between 2005 and 2011, there were 14,457 patients hospitalized with AMI in Nebraska, of which more than half of the patients (45.7%) resided in urban counties, with urban hospitals taking 69.1% of all AMI patients. While there was no rural–urban difference in sex, urban hospitals took significantly more male patients (62.6% vs. 57.9%). Rural areas had many more patients age 80 years or older than urban areas (30.73% vs. 22.21%), and so did the rural hospitals (37.31% vs. 22.15%). Rural residents were less likely to be in the poorest and well-off neighborhoods, and so were the rural hospitals. Except for cerebrovascular disease (mainly stroke), there was no significant rural–urban difference in comorbidities, while rural hospitals took more AMI patients with COPD and cancer than urban hospitals.

The crude readmission rate is lower among rural than urban hospitals (6.15% vs. 7.29%). In contrast, the percentage of patients who died within 1 year of AMI hospitalization was higher in rural hospitals than in urban hospitals (23.05% vs. 13.83%); the corresponding difference between rural and urban residents was smaller (18.23% vs. 14.83%). These rates, although broadly comparable to the national rates, should not be compared, as they are not age and sex standardized. Based on

TABLE 17.1

Descriptive Statistics by Rural–Urban Residents or Hospitals

	Rural Residents	Urban Residents	Rural Hospitals	Urban Hospitals
Total cases	7851	6606	4468	9989
Demographics				
Sex (male)	61.01	61.31	57.9	62.6*
Age < 40	1.34	2.33*	1.34	1.99*
Age 40–59	24.61	33.3	21.71	31.65
Age 60–80	43.32	42.16	39.64	44.2
Age > 80	30.73	22.21	37.31	22.15
Poverty Groups				
>5%	9.48	27.32*	6.92	22.42*
5%–10%	36.45	26.58	35.7	30.26
10%–15%	37.29	14.37	38.16	21.74
15%–20%	9.49	9.28	10.27	9
20+%	7.29	22.45	8.95	16.57
Comorbidities				
COPD	15.11	14.26	16.88	13.76*
Diabetes	21.76	22.25	22.74	21.64
Renal disease	9.09	8.2	6.98	9.45*
Hypertension	56.04	55.21	56.47	55.3
Cerebrovascular disease	4.99	4.04*	4.77	4.46
Cancer	2.53	2.51	2.98	2.32*
Other variables	0		0	
Having rehab	40.19	46.73*	35.16	46.76*
Rural hospital bypassers	43.42			34.13
Transfer admission	21.62	2.68*	10.94	13.87*
Crude readmission rate	6.84	7.05	6.15	7.29*
Died in hospital	6.27	5.6	7.39	5.33*
Died less than 30 days	10.23	8.28*	12.56	7.9*
Died less than 1 year	18.23	14.83*	23.05	13.83*

Note: All percentages should be added to 100% by each selected variable. 100% – male percent in urban = female percent in urban. Rural hospital bypassers are rural residents who went to urban hospitals.

* Significant at $p < 0.05$.

outpatient data only, we found that rural areas tend to have a lower rate (40.19%) of rehabilitation than urban areas (46.13%). In addition, 43.2% of rural residents bypassed a rural county to seek AMI care in urban hospitals. By definition, rural (hospital) bypassers exclude urban residents, but there were less than 0.60% urban hospital bypassers in the study sample. Another way to look at bypassers is through an admission that was transferred from another hospital. As we can see, 21.6% of rural patients had a transfer admission, in contrast to 2.7% for urban residents. In fact, transfer patients accounted for 13.87% of urban AMI admissions. Given the high proportion of transfer admissions, our definition of bypassers may not necessarily mean actively bypassing a rural hospital to seek care in an urban hospital, because many of them could be transferred patients.

Table 17.2 lists the results for readmission models by place of residence and place of care. All the variables above the readmission entry in Table 17.1 were entered into models. The only difference between the two models was the last dummy variable: the place of residence or hospital. Consequently, both models yielded similar results. Having rehabilitation was associated with 17%

TABLE 17.2
Predictors of 30-Day Readmission by Place of Care or Residence

Age Effects (ref = <40)	Place Model	Hospital Model
Age 40–49	0.754 (0.464, 1.226)	0.751 (0.462, 1.222)
Age 50–59	0.973 (0.603, 1.57)	0.97 (0.601, 1.566)
Age 60–80	1.229 (0.755, 1.999)	1.25 (0.768, 2.034)
Renal disease	1.449 (1.173, 1.789)	1.423 (1.152, 1.758)
Cancer	0.568 (0.324, 0.997)	(Not significant)
Rehab (yes vs. no)	1.204 (1.05, 1.38)	1.197 (1.044, 1.373)
Urban vs. rural	(—)	1.193 (1.026, 1.387)

Note: Results from the models are based on backward selections that removed poverty. Distance to hospital is crow flying distance. Place model: Urban vs. rural—urban residents vs. rural residents. Hospital model: Urban vs. rural—urban hospitals vs. rural hospitals.

more odds of readmission, while urban patients and urban hospitals had about 20% greater odds of readmission than their rural counterparts. Rural hospital bypassers had 27.5% greater odds of readmission. Finally, sex, poverty, and most comorbidity conditions were not significant except renal disease, which increased the readmission odds by about 37%.

The results from 1-year out-of-hospital mortality were quite different from those for readmissions (Table 17.3). We focus on the results from the place of residence (column 2), as both models had similar results. There was an obviously rising trend in odds ratio for each age group in both models, and older patients were much more likely to die than younger patients. In both models, patient gender was not significant, and it was dropped by the backward selection. Likewise, poverty

TABLE 17.3
Odds Ratio of 1-Year Mortality: Place- and Hospital-Based Models

Age Effects (ref = <40)	Place of Residence	Place of Care
Age 40–59	2.175 (0.68, 6.956)	1.974 (0.616, 6.321)
Age 60–80	7.156 (2.268, 22.582)	6.343 (2.008, 20.037)
Age > 80	22.242 (7.059, 70.083)	20.139 (6.385, 63.515)
Comorbidities		
COPD	1.502 (1.306, 1.727)	1.545 (1.332, 1.791)
Renal disease	1.996 (1.704, 2.338)	1.979 (1.667, 2.35)
Hypertension	0.59 (0.526, 0.661)	0.61 (0.54, 0.688)
Cerebrovascular disease	1.555 (1.249, 1.936)	1.553 (1.236, 1.95)
Cancer	4.539 (3.502, 5.882)	4.831 (3.691, 6.323)
Rehab (yes vs. no)	0.111 (0.091, 0.135)	0.113 (0.092, 0.139)
Rural bypassers	0.658 (0.545, 0.795)	
Distance to hospital	0.996 (0.994, 0.998)	0.91 (0.878, 0.943)
Urban vs. rural	0.73 (0.643, 0.83)	

Note: Confidence intervals are in parentheses. Results from the models are based on backward selections that removed poverty. Distance to hospital is crow flying distance. Bypassers are rural residents who went to urban hospitals for care. Place model: Urban vs. rural—urban residents vs. rural residents. Hospital model: Urban vs. rural—urban hospitals vs. rural hospitals.

level and transfer status were not significant and were dropped from the final models. In a preliminary analysis of an age- and sex-adjusted model, a high poverty rate was significantly associated with the likelihood of dying within 1 year of indexing admission, but its effect diminished after the introduction of rehab status. This result could be explained by a high percentage of rural patients who didn't have rehabilitation. Rural people had lower a proportion of the extremely poor, but they tended to be older. These two effects canceled some poverty effects, and when the rehabilitation variable was added, poverty lost its explanatory power.

With regards to comorbidity, COPD, renal disease, cerebrovascular disease, and cancer were all positively associated with 1-year mortality, while hypertension was associated with a 40% reduction in 1-year mortality. For the positive predictors, cancer and renal disease had the strongest effects. These effects were broadly comparable between the two models.

Finally, having at least one rehabilitation session in an outpatient setting significantly reduced the odds of 1-year mortality (OR 0.111, 95% CI 0.091–0.135). In fact, judging by the explanatory power or the reduction of deviance residuals per degree of freedom, it is the most important predictor of 1-year mortality. In the place of residence model, both the distance to a hospital and rural bypasser variables were significant. Bypassing rural hospitals reduced the odds of 1-year mortality by 34% (OR 0.658, 95% CI 0.545–0.795), and the shorter the distance, the less likely the patient was to die, all else being equal. Controlling all of the above effects, we found that people residing in urban areas had a 27% mortality benefit within a year (OR 0.73, 95% CI 0.643–0.83), while the hospital location was not significant, a result consistent with James et al.'s (2007) study in Iowa.

Table 17.4 displays the results from out-of-hospital survival models by hospital location and place of residence. Again, both models yielded similar results, and we focus on the place of residence model. The age effect followed a gradient similar to that of the logistic regression: older patients tended to have a greater HR. All the effects for comorbidities were significant, and they followed

TABLE 17.4

Hazard Ratios for Place of Residence and Place of Care Models

Age Effects (ref = <40)	Residence Model	Hospital Model
Age 40–49	1.647 (0.946, 2.867)	1.641 (0.943, 2.856)
Age 50–59	5.216 (3.02, 9.009)	5.187 (3.004, 8.959)
Age 60–80	12.502 (7.24, 21.588)	12.402 (7.183, 21.414)
Male vs. female	0.934 (0.877, 0.994)	0.933 (0.877, 0.993)
Comorbidities		
COPD	1.418 (1.316, 1.527)	1.419 (1.318, 1.528)
Diabetes	1.109 (1.031, 1.192)	1.11 (1.032, 1.193)
Renal disease	1.47 (1.344, 1.608)	1.467 (1.341, 1.604)
Hypertension	0.668 (0.628, 0.71)	0.668 (0.628, 0.71)
Cerebrovascular disease	1.364 (1.215, 1.53)	1.36 (1.212, 1.526)
Cancer	2.116 (1.854, 2.415)	2.118 (1.855, 2.417)
Rehab (yes vs. no)	0.25 (0.229, 0.273)	0.25 (0.229, 0.272)
Bypassers (yes vs. no)	0.808 (0.728, 0.896)	
Distance to hospital	0.95 (0.929, 0.972)	0.94 (0.923, 0.957)
Urban vs. rural	0.876 (0.818, 0.937)	0.856 (0.803, 0.912)

Note: Results from the models are based on backward selections that removed poverty. Distance to hospital is crow flying distance. Bypassers are rural residents who went to urban hospitals for care. Place model: Urban vs. rural—urban residents vs. rural residents. Hospital model: Urban vs. rural—urban hospitals vs. rural hospitals.

the same directional effects of the logistic regression, and in this case, diabetes became significant for both survival models. Those with COPD, diabetes, renal disease, cerebrovascular disease, and cancer had much less survival time with an HR significantly greater than 1. Likewise, having hypertension had a protective effect for survival time.

After controlling for age and comorbidity effects, geographic access measures exerted a significant effect. Those who bypassed rural hospitals had a better survival prognosis. The HR for rural bypassers was 0.855 (95% CI 0.802–0.912). However, the distance effect was reversed; the greater the distance, the longer the survival days. Controlling for these access measures, among others, we found that rehabilitation significantly increased survival time (HR 0.25, 95% CI 0.229–0.273). In addition, the rural–urban effect was still significant. Urban patients had a longer survival time, although the effect for the place of residence model (HR 0.876, 95% CI 0.818–0.937) was not as strong as that for the place of care model (HR 0.856, 95% CI 0.803–0.912).

It should be noted that one consistent finding from the readmission and 1-year mortality and survival models was that poverty levels were not significant. This result was inconsistent with findings from Chapters 8 and 10. Since both chapters used the first two diagnoses to define AMI, we also ran the three models above using this definition. In general, neighborhood poverty was not a significant predictor. The only difference was the contrast between poverty level 5 (>20%) and poverty level 1 (<5%), with the odds ratio of 1.832 (95% CI 1.201– 2.794). We therefore concluded that neighborhood poverty level, by and large, was not a significant risk factor when other clinical risk factors were included.

17.4   CONCLUDING REMARKS

This is a data-driven study based on integrated hospital discharge data that were linked to (1) community-based vital records, (2) outpatient records typically not included in hospital discharge data, (3) neighborhood poverty information, and (4) hospital and patient locations. In the published literature, linking one of those dimensions is not common, let alone linking all of them statewide. In an integrated public health data system, such a task is not difficult.

One seemingly inconsistent finding is that rural patients or those treated at rural hospitals were less likely to have 30-day all-cause readmission, yet they were more likely to die within a year of discharge. A lower readmission rate suggested better quality of care, while a higher 1-year mortality rate suggested a lower quality of care. However, we do not know to what degree all other conditions are equal. For instance, we cannot control for disease severity and complications. On the other hand, indicators such as bypassing and rehabilitation status suggested higher readmission rates could be generously related to lower 1-year mortality in the context of rural–urban health disparities. Instead of being data driven, an in-depth study should be driven by a substantive issue, such as why rural patients had a lower readmission rate.

We found that SES was no longer significant when clinical risk factors were introduced. In the surveillance chapters (Chapters 8 through 11), neighborhood SES is the most consistent risk factor, where poorer neighborhoods were associated with worse outcomes in terms of incidence, readmission, 1-year mortality, and so forth. While we believe that incidence results would still stand because no other control variables could be included, the results from readmission and 1-year mortality all became nonsignificant. While these results need to be confirmed with more elaborate study designs to tease out overlapping effects between poverty and comorbidity and rehabilitation, they do suggest a strong and direct pathophysiological association between AMI and comorbidity (Kapur and Palma, 2007). Once a patient was admitted into a hospital, clinical variables, such as comorbidities or rehabilitation, became more salient, and the neighborhood poverty status consistently dropped out in various models when it competed with clinical variables for predictive power.

Comorbidities were important risk factors for mortality, but they played limited roles in readmission. Previous history of COPD increased mortality but it had little effect on readmission. Renal disease, cerebrovascular disease, diabetes, and cancer were all strong predictors of mortality and survival. Only renal disease increased readmission. Interestingly, we found a negative association between hypertension and AMI mortality and survival. Although the literature is inconclusive, hypertension is generally either inconsequential (Ali et al., 2004) or a risk factor for AMI patient survival (Hansen et al., 2007). It seems that more information about hypertensive patients is needed, as we do not know their behavior risk factors, and we did not distinguish types of hypertension.

Both mortality and survival outcomes suggest that rural AMI patients received a lower quality of care. If all rural AMI patients ended up in rural hospitals, while all urban AMI patients ended up in urban hospitals, the disparity in survival would be much larger, because bypassers had better outcomes. The rural deficit in survival is expected; as the rural physicians are trained to treat a wider range of medical conditions, they tend to take a more general approach in their practice pattern than the urban physicians. In addition, the lack of cardiologists and emergency services in rural areas forces generalists to care for AMI patients. If the patients could not obtain timely care from specialists, they would seek care from generalists, who might not order the necessary tests or cardiology referrals. Furthermore, cardiac rehabilitation was associated with reduced mortality, a finding that has previously not been reported at the population level, and rural patients were more likely to participate in cardiac rehabilitation than urban patients. Finally, the care received in the prehospital settings in rural areas may not be adequate, because long-distance EMS transport and most ambulance systems in rural Nebraska are staffed by volunteers who are unable to provide advanced cardiac life support interventions compared to the facilities in urban areas.

However, we are puzzled by distance effects. Previous studies using rural and urban as indicator variables implicitly suggest travel distance is a barrier for timely care, as it may delay some critical treatment options within the golden hour. The current study showed that greater distance increased the odds of 1-year mortality, but it was positively associated with survival time. This inconsistency may help to generate hypotheses for future studies. For instance, if we plot neighborhood income according to distance to major hospitals, we may find that most hospitals are in the inner city areas near low-income neighborhoods. We could hypothesize that distance effects in inner city areas may behave like a donut hole, harmful within a short distance, protective in moderate distance, and hazardous again in long distance. However, this hypothesis may not hold in recent few years, as new hospitals have been built in suburban areas far from the central city.

There are several limitations to this study. First, we did not include in-hospital procedures that tend to predict survivals. Second, we did not account for case severity between the rural and urban patients, because hospital discharge data do not contain this information. Third, comorbidity information was simply derived from diagnosis codes, which could have some variation among patients and hospitals. In the detailed discharge data file, each diagnosis variable is accompanied by "present on admission," which indicates if the diagnosis is clinically determined or through other methods. This variable should be included in the study design in the substantive issue drive study. Fourth, we had no direct information on some of the important risk factors, such as smoking status, physical activity status, and body mass index. Fifth, outpatient cardiac rehabilitation may not cover all of the cardiac rehabilitation. Sixth, both rural hospital bypassing and distance-to-hospital measures could be improved by using a real-world road network.

REFERENCES

Ades, P.A. 2001. Cardiac rehabilitation and secondary prevention of coronary heart disease. *New England Journal of Medicine* 345(12): 892–902.

Ali, I., D. Akman, N.E. Bruun, L. Køber, B. Brendorp, M. Ottesen, J. Møller, and C. Torp-Pedersen. 2004. Importance of a history of hypertension for the prognosis after acute myocardial infarction—for the Bucindolol Evaluation in Acute Myocardial Infarction Trial (BEAT) study group. *Clinical Cardiology* 27: 265–269.

Baldwin, L.M., L. Chan, C.H. Andrilla, E.D. Huff, and L.G. Hart. 2010. Quality of care for myocardial infarction in rural and urban hospitals. *Journal of Rural Health* 26: 51–57.

Baldwin, L.M., R.F. MacLehose, L.G. Hart, S.K. Beaver, N. Every, and L. Chan. 2004. Quality of care for acute myocardial infarction in rural and urban US hospitals. *Journal of Rural Health* 20: 99–108.

Barnato, A.E., F.L. Lucas, D. Staiger, D.E. Wennberg, and A. Chandra. 2005. Hospital-level racial disparities in acute myocardial infarction treatment and outcomes. *Medical Care* 43: 308–319.

Bernheim, S.M., J.N. Grady, Z. Lin, Y. Wang, Y. Wang, S.V. Savage, K.R. Bhat, et al. 2010. National patterns of risk-standardized mortality and readmission for acute myocardial infarction and heart failure. Update on publicly reported outcomes measures based on the 2010 release. *Circulation: Cardiovascular Quality and Outcomes* 3: 459–467.

Hansen, E.B., C.A. Larsson, B. Gulberg, A. Melander, K. Bostorom, L. Råstam, and U. Lindblad. 2007. Predictors of acute myocardial infarction mortality in hypertensive patients treated in primary care: A population-based follow up study in the Skaraborg project. *Scandinavian Journal of Primary Health Care* 25: 237–243.

James, P.A., P. Li, and M.M. Ward. 2007. Myocardial infarction mortality in rural and urban hospitals: Rethinking measures of quality of care. *Annals of Family Medicine* 5: 105–111.

Kapur, A. and R.D. Palma. 2007. Mortality after myocardial infarction in patients with diabetes mellitus. *Heart* 93(12): 1504–1506.

Nawal Lutfiyya, M., D.K. Bhat, S.R. Gandhi, C. Nguyen, V.L. Weidenbacher-Hoper, and M.S. Lipsky. 2007. A comparison of quality of care indicators in urban acute care hospitals and rural critical access hospitals in the United States. *International Journal of Quality Health Care* 19: 141–149.

O'Connor, A. and G. Wellenius. 2012. Disparities in the prevalence of diabetes and coronary heart disease. *Public Health* 126: 813–820.

Sheikh, K. and C. Bullock. 2001. Urban-rural differences in the quality of care for Medicare patients with acute myocardial infarction. *Archives of Internal Medicine* 161: 737–743.

Smith, S.C., J. Allen, S.N. Blair, R.O. Bonow, L.M. Brass, G.C. Fonarow, S.M. Grundy, et al. 2006. AHA/ACC guidelines for secondary prevention for patients with coronary and other atherosclerotic vascular disease: 2006 update endorsed by the National Heart, Lung, and Blood Institute. *Journal of the American College of Cardiology* 47(10): 2130–2139.

Tsai, C.L., D.J. Magid, A.F. Sullivan, J.A. Gordon, R. Kaushal, H.P. Michael, P.N. Peterson, D. Blumenthal, and C.A. Camargo Jr. 2010. Quality of care for acute myocardial infarction in 58 U.S. emergency departments. *Academic Emergency Medicine* 17: 940–950.

Xian, Y., W. Pan, E.D. Peterson, P.A. Heidenreich, C.P. Cannon, A.F. Hernandez, B. Friedman, R.G. Holloway, G.C. Fonarow GC; GWTG Steering Committee and Hospitals. 2010. Are quality improvements associated with the Get With the Guidelines—Coronary Artery Disease (GWTG-CAD) program sustained over time? A longitudinal comparison of GWTG-CAD hospitals versus non-GWTG-CAD hospitals. *American Heart Journal* 159: 207–214.

18 Disparities in Motor Vehicle Crash Injuries
From Race to Neighborhood

18.1 INTRODUCTION

In the United States, injuries are the leading cause of death among persons ages 1–44 (CDC, 2012a). In 2010, 180,811 lives were lost due to unintentional injuries, 2,855,000 persons were admitted into hospitals because of injuries, and 29,757,000 persons were treated at emergency rooms for nonfatal injuries (CDC, 2010). Among all causes of injuries, motor vehicle crashes (MVCs) lead to the greatest number of deaths (CDC, 2012b). By themselves, MVCs were the leading cause of death in the 8–24 age group (National Highway Traffic Safety Administration [NHTSA], 2012a). In 2009, about 10.8 million MVCs were recorded across the country (U.S. Census Bureau, 2012a), claiming 33,883 lives and corresponding to a fatality rate of 1.15 deaths per 100 million vehicle miles traveled (NHTSA, 2012b). In the same year, it was estimated that 2.21 million people were injured in MVCs, which translated into 74 injuries per 100 million vehicle miles traveled (NHTSA, 2010). Although members of racial and ethnic minority groups are less likely to own vehicles (Baker et al., 1998), they are more likely than Whites to be involved in an MVC (Harper et al., 2000). However, previous studies on racial disparities in MVCs have focused on fatalities because of the wide availability of the Fatality Analysis Reporting System (FARS). Since FARS does not include information on nonfatal crashes (Briggs et al., 2005), and cohort data are costly and mostly from other countries (Whitlock et al., 2003; Hasselberg et al., 2005), the literature is skewed toward crash fatality disparities (Braver, 2003; Campos-Outcalt et al., 2003). Even with the availability of FARS, some popular statistics reports, such as the annual statistic abstract from the Census Bureau, do not include race in MVC statistics.

In Nebraska, Whites accounted for 91.6% of MVC fatalities (NHTSA, 2012c), compared to 88.5% of the White population (U.S. Census Bureau, 2011a), suggesting that Whites were slightly overrepresented in MVC fatalities compared to non-Whites. In addition, Whites in Nebraska had greater fatal crash incidences than the U.S. average: 12.1 per 100,000 in Nebraska versus 9.5 in the United States. These numbers could be explained by the fact that Whites disproportionately reside in rural areas in Nebraska and rural crashes are more likely to be fatal (NHTSA, 2008). The overrepresentation of MVC risk factors among Blacks and slight underrepresentation of fatal crash injury suggest that there might be overrepresentation of racial minorities in nonfatal injuries. However, current MVC injury reporting mechanisms from both police crash reports and medical records provide very limited information on race. As a result, it is very difficult to assess the racial profile of MVC injuries without conducting longitudinal transportation or travel surveys. The efforts of traffic safety advocates to develop prevention strategies for the minority population have therefore been hampered by the lack of race-specific characteristics.

One way to reduce nonfatal injuries is to reduce racial disparity in MVCs, because many MVC risk factors, such as safety awareness, blood alcohol content (BAC), driving skills and experiences, vehicle conditions, and safety features, are less favorable to African Americans (Briggs et al., 2005). For example, African American drivers are more likely than drivers of other races to drive under the influence of alcohol (Braver, 2003; Roudsari et al., 2009), and they are less likely to use seat belts (Vivoda et al., 2004) or drive a vehicle with more recent and effective safety features

(Cooper et al., 2010; Ryb et al., 2009). It is therefore reasonable to assume that African Americans are more likely to have nonfatal MVC injuries. By focusing on MVC injuries rather than fatalities, the sample size tends to be large enough for assessing race-specific characteristics, such as the place of residence.

In 2006, NHTSA conducted a study on fatal MVC disparity, and it called for further study of nonfatal MVC injuries by race and ethnicity (NHTSA, 2006). In collaboration with the Nebraska Office of Highway Safety, the Division of Public Health at the Nebraska Department of Health and Human Services piloted an MVC disparity surveillance project using crash and driver's license data from 2006 to 2010 that include race. The project was based on the Crash Outcome Data Evaluation System (CODES), which routinely links information from crash reports to driver's license, hospital, and death data. CODES was a program initially facilitated by NHTSA as a component of its state data program. CODES uniquely uses probabilistic methodology to link crash records to injury outcome records collected at the scene, en route by emergency medical services (EMS), by hospital personnel after arrival at the emergency department or admission as an inpatient, and at the time of death, if applicable, on the death certificate.

The first phase of this project was to conduct population-based surveillance for the injured drivers (Zhang and Lin, 2013). During the first phase, we extracted crash injury severity from police reports, commonly known as the KABCO scale (K = fatal injury, A = disabling injury, B = visible injury, C = possible injury, O = no injury, property damage only) (GHSA, 2012). However, the accuracy of police-reported injury severity has been long debated. Previous studies have suggested using injury severity derived from medical diagnosis from hospital records as a more accurate and reliable measure of injury severity (Farmer, 2003; McDonald et al., 2009). Therefore, in the second phase of the project, we analyzed drivers' records that were found in the hospital discharge data during the study period. Since race may interact with other socioeconomic status (SES) variables, we geocoded injured drivers by their residential locations and used census tract population by race and the number of vehicles as exposure to conduct further surveillance along the SES dimension in phase III. This chapter summarizes the three phases of the project and demonstrates (1) use of linked driver's license information for MVC injury disparity surveillance, (2) use of hospital-based MVC injury information, and (3) use of geocoded data from injured drivers.

18.2 PHASE I PROJECT: MVC DISPARITY BASED ON POLICE-REPORTED INJURY SEVERITY

18.2.1 Methods

Phase I study was based on injured MVC drivers and licensed drivers in Nebraska. Previous research on nonfatal MVC injuries has used driver's license information to determine residence location (Blatt and Furman, 1998) and SES (Whitlock et al., 2003), but not to identify race. We used information from crash reports to determine injury severity and from driver's licenses to determine race and other demographic information for all licensed drivers in Nebraska. Since the data field of driver's license number is available from both the crash reports and driver's license data, it was used as the primary key to link the two datasets. The driver's race and other demographic information were added to injury information recorded by the police. With the linked data, we were able to conduct MVC injury disparity surveillance for Nebraskan drivers.

18.2.1.1 At-Risk Population

We generated the at-risk population from the Nebraska driver's license data, collected by the Nebraska Department of Motor Vehicles. Demographic fields in this dataset include date of birth, sex, race, residence address, and driver's license number. Unfortunately, information on ethnicity was not available. We extracted from the driver's license database for all type of drivers with an age range between 15 and 85. Non-Nebraska drivers' records (drivers who hold a license from other states, but not Nebraska) and records with an unknown license type were excluded. Note that although non-Nebraska drivers account for a small percentage of the driver's license database (3.2%), they account for 11.3% of minor to fatal MVC injuries in Nebraska (NE-CODES, 2006). The final dataset had an average of 1,518,908 records each year between 2006 and 2010.

18.2.1.2 MVC Incidence Exclusion Criteria

Passengers were excluded, as their race information, together with their at-risk populations, was not available. No-injury records (injury severity recorded as "none" or blank) from crash data were excluded. Also excluded were fatal injuries, because they were not the focus of this study.

18.2.1.3 MVC Incidence Inclusion Criteria

Injured drivers' records in crash reports were extracted for those age 15–85. Injury categories included were severe ("disabling" in crash reports) injury and nonsevere ("visible" or "possible" in crash reports) injury. The final dataset contained 59,356 eligible records for the 5-year period (or 11,871 records each year, on average). Among them, 50,436, or 84.97%, of the records were successfully linked to the driver's license database by driver's license numbers.

We extracted MVC records for age, sex, place of injury, and injury severity and added race information through driver's license records. Injury rates were calculated for drivers injured in crashes per licensed driver by race. Due to small numbers, racial groups other than Black and White were grouped as "other races," and they include Asian and Pacific Islander, American Indian and Alaska Native, and all other races in the license database. To establish baseline rates and expected trends by race, we first graphically examined incidence rates for severe or nonsevere injuries by year. This approach was identical to the one used to identify disparities in MVCs among young drivers in North Carolina (Imai and Mansfield, 2008). In addition, we statistically evaluated three demographic variables (age, sex, and race) and place of residence, all of which were available from the driver's license database. Place of residence was defined as urban or rural according to the U.S. Census Bureau's (2011b) designation, where "urban area refers generally to urbanized areas of 50,000 or more population and urban clusters of at least 2,500 and less than 50,000 population."

We used Poisson regression and set up a series of four-way log-rate models. A general main effect log-rate model can be expressed as

$$\text{Log}(M)_{ijkl}^{ASRP} = C + \log\left(DLH_{ijkl}\right) + \text{Age}_i + \text{Sex}_j + \text{Race}_k + \text{Place}_l \tag{18.1}$$

where, on the left-hand side of Equation 18.1, M refers to the number of MVCs. The superscript *ASRP* indexes age, sex, race, and place, while the subscript indexes their respective categories for age group i, sex j, race group k, and place group l. On the right-hand side, C is the grand mean or the intercept, and $\log(LD_{ijkl})$ is the offset of the number of driver's license holders (i.e., at-risk population) in the corresponding cell indexed by $ijkl$ in the four-way table.

For the severe injury model, we modeled the expected rate of severe MVC injuries by age, sex, race, and place of residence, with corresponding counts of licensed drivers indexed by $ijkl$ as the offset. The nonsevere injury model is identical to the severe injury model, except that the dependent variable becomes the expected nonsevere injury rate. Just as in a linear regression model, two-, three-, and four-way interactions can be added. When all interactions are added, the model becomes saturated, with no degrees of freedom, and the model fully captures the expected rate indexed by $ijkl$. In our analysis, only statistically significant interaction terms were included in the final models of severe and nonsevere MVC injuries. Up to this point, we have only used three-way tables. Here, we expand to a four-way table.

18.2.2 Results

Let's first compare the demographic characteristics of Nebraska licensed drivers with a those of a comparable age-specific population in Nebraska from the American Community Survey (ACS) 2006–2010 (U.S. Census Bureau, 2012b). We restricted the ACS to the same age range and found that race distributions between the ACS and licensed driver populations were comparable (see Table 2.3 for those age 19+). About 4.3% of Black licensed drivers corresponded to 4.1% of the single-race Blacks in the ACS. The licensed driver population included about 87.9% Whites, slightly less than the 89.7% of single-race Whites in the ACS. The above numbers suggest that it is reasonable to use licensed drivers as an at-risk population for MVC injury surveillance.

From the ACS, we also found that Whites in Nebraska were much more likely than Blacks and those of other races to own cars or trucks. Among 647,158 White heads of household, 4.68% were without a car or truck; among 29,773 Black heads of household, 17.98% were without a car or truck; and among 34,838 other race (not Black or White) heads of household, 9.22% were without a car or truck. The average number of cars among White, Black, and other heads of household were 1.96, 1.37, and 1.72, respectively. These results suggest that if we could adjust the number of vehicles owned by a driver's household, we would be able to further adjust Black drivers downward as the at-risk population.

From 2006 to 2010, there were 5137 Nebraskan drivers severely injured in MVCs: 87% were White, 6% were Black, and 7% were other races. There were a total of 45,297 Nebraskan drivers receiving nonsevere MVC injuries over the 5 years. Among those, 38,020 were White, 3,492 were Black, and 3,785 were of other races. For every severely injured driver, there were more than 8 nonseverely injured drivers for Whites and more than 10 for non-Whites. Black drivers had the highest nonsevere injury incidence throughout the study period, but the difference of nonsevere injury incidence rate between Black and White drivers was much larger than that of severe injury incidence rate. On average, the nonsevere injury incidence rate was 87% higher for Black drivers than for White drivers.

Since racial differences in severe and nonsevere injuries might be explained by age and sex distribution, we pooled the 5-year injury data to examine interactions using a two-way tabulation and modeled them using a four-way tabulation (age, sex, race, and residence) for severe and nonsevere injuries.

The results for the two-way tabulation (Table 18.1) showed that Black drivers had higher severe and nonsevere injuries throughout almost all comparisons. The differences of severe injury rates between Black drivers and their two other counterparts were much more noteworthy for female (95 compared to 57 for Whites and 59 for other races) and urban (92 compared to 57 for Whites and 53 for other races). For the nonseverely injured, the rate differences were much wider for age groups 15–24 and 25–44, female (1329 vs. 658 for Whites and 812 for other races), and urban (1100 vs. 661 for Whites and 698 for other races) drivers. It is also worth noting that for both severe and nonsevere injuries, injury rates generally decline with the increase of age, and the decline for the nonseverely injured was much steeper for Whites than for the other two racial categories. In addition, female drivers had higher nonsevere injury rates than male drivers regardless of race; while for severe injuries, that was only true for Black female drivers and not for the other two racial groups. Higher nonsevere injury rates are found in urban areas; however, for severe injuries, the rate contrast was opposite among White drivers and drivers of other races. Since some of the rates were close, and some may interact with others, we further described them using the four-way log-rate model.

The results for the severe injury model showed that all the main effects were significant (Table 18.2, first two columns). All age groups younger than 65 years were more likely than drivers 65 years and older to have a severe MVC injury. Females were less likely than males to have a severe MVC injury. Black drivers were more likely than White drivers to have a severe MVC injury, while drivers of other races were less likely than White drivers to have a severe MVC injury. Rural residents were much more likely than urban residents to have a severe MVC injury. Besides these

TABLE 18.1

Annualized MVC Injury Rates and Incidences by Age, Sex, Place of Residence, and Race

	Severe Injury Rate			Nonsevere Injury Rate			No. of Licensed Drivers
	White	Black	Other	White	Black	Other	
			Age Group				
15–24	133	146	122	1,207	1,726	1,135	194,930
25–44	66	88	49	595	1,102	559	564,380
45–64	57	80	52	458	838	513	520,723
65+	43	50	36	290	531	400	238,875
			Sex				
Female	57	95	59	658	1,329	812	725,336
Male	77	88	63	487	860	513	793,572
			Place of Residence				
Rural	77	80	72	481	662	550	732,684
Urban	57	92	53	661	1,100	698	786,224
No. of injured drivers	895	60	73	7,604	698	757	
No. of licensed drivers	1,334,391	65,528	118,989	1,334,391	65,528	118,989	1,518,908

TABLE 18.2

Log-Rate Model Point Estimates on Severe and Nonsevere MVC Injuries

	OR	95% CI	OR	95% CI
		Age Group (reference = 65+)		
15–24	3.18	2.87–3.52	4.05	3.89–4.21
25–44	1.6	1.45–1.77	1.97	1.9–2.04
45–64	1.36	1.23–1.51	1.54	1.48–1.6
		Sex (reference = Male)		
Female	0.74	0.7–0.78	1.35	1.32–1.38
		Race (reference = White)		
Black	1.33	1.12–1.57	1.62	1.53–1.71
Other race	0.76	0.66–0.88	1.09	1–1.18
		Place of Residence (reference = Urban)		
Rural	1.38	1.31–1.46	0.76	0.74–0.77
		Interaction Terms		
Black female	1.41	1.11–1.79	1.11	1.03–1.19
Other race female	1.24	1–1.54	1.13	1.06–1.21
Rural Black	0.56	0.35–0.88	0.76	0.65–0.89
Black age 15–24			0.78	0.72–0.84
Other race age 25–44			0.8	0.73–0.88
Other race age 45–64			0.82	0.75–0.89

Note: Only significant interactions were reported.

main effects, three interaction terms were also significant. All else being equal, the interaction term between race and sex showed that female Black drivers were significantly more likely (41% higher odds) to have a severe injury, given that Black drivers were already worse off. While the main effects for both female and other race categories showed reduced likelihoods of severe injuries, females in the other race group were 1.24 times more likely than their respective main effects to have a severe MVC injury. For instance, although the main effect for the other race category had 0.76× reduced odds of MVC injuries compared to Whites, if they were females, the odds were increased to 0.94 (0.76*1.24). As for the interaction between race and place of residence, rural Black drivers only had about half of the odds (odds ratio [OR] of 0.56) to have a severe MVC injury, all else being equal. Although the tabulation result in Table 18.1 only showed a small rural–urban difference, when risk factors such as age and race were controlled, the observed higher crude rate by residence was much weaker than it appeared. Again, given that rural areas had greater odds of severe MVC injuries (1.38), the odds for Black drivers were reduced to 0.77 (1.38*0.56). This result also suggests that the greater likelihood of severe injuries among Black drivers was mainly contributed by MVCs in urban areas.

The results for the nonsevere injury model (Table 18.2, last two columns) were quite different from those for the severe injury model. While the age effects were similar, female drivers were more likely than male drivers to have nonsevere MVC injuries. Black drivers remained more likely than White drivers to have nonsevere MVC injuries, and other race drivers became more likely than Whites to have nonsevere MVC injuries. As expected, rural areas were less likely to observe nonsevere MVC injuries than urban areas. Among significant interaction effects, female Black drivers and female drivers of other races had a greater likelihood of nonsevere MVC injuries, while rural Black drivers were less likely to have nonsevere MVC injuries. If the driver was Black and aged 15–24 years, or was of other races and aged 25–64 years, his or her corresponding age-specific effects were reduced by about 20%.

18.2.3   Discussion

The driver sample has a number of advantages in MVC studies. First, the racial profile of licensed drivers is fairly consistent with the general racial profile for the comparable population segment by age. Thus, crash drivers and licensed drivers together are helpful to delineate racial disparities for nonfatal MVC injuries. Even though we showed that Blacks are less likely to have access to vehicles, it is difficult to adjust for race-specific car ownership and miles driven at the state level. Using licensed drivers for the at-risk population, we found that Black drivers had the highest severe and nonsevere injuries per 100,000 licensed drivers. Compared to White drivers, we found that Black drivers had 31.6% and 87% more severe and nonsevere injuries, respectively. These findings persist when controlling for age, sex, and place of residence, but they are not consistent with those reported from fatal crash reports in Nebraska. Our interpretations are that fatal crash reports normally use census population, instead of drivers' population, as the denominator, which might overlook two facts: (1) Whites dominate the rural population in Nebraska, where risks of fatal crashes are much higher than in urban areas, and (2) fatal crashes include both in-state and out-of-state drivers and passengers, while our study includes only drivers.

Second, focusing on drivers could directly lead to MVC injury prevention programs (Whitlock et al., 2003), but an intervention design should balance between statistical significance and population impact. Although Black drivers in this study had a greater likelihood of severe and nonsevere injuries, we would not recommend targeting them for prevention programs in rural areas in Nebraska, because their exposure there is low. However, we would recommend that prevention programs target female Black drivers, as they are more likely to be injured. At the program level, an intervention design includes factors such as driving behavior, car conditioning (Ryb et al., 2009), and awareness of neighborhood road conditions (Durkin et al., 1994; Abdalla et al., 1997), together with demographic and residence factors.

A third advantage of the driver sample is that it provides a rarely tapped administrative data source for MVC disparity surveillance. Age- and race-specific MVC and driver's license data, as shown in this report, provide rich information for modeling injury disparity that may lead to race-specific interventions for drivers. Reducing racial disparities in MVCs will reduce both fatal and nonfatal MVC injuries. Even though driver's license data may not be available for public health agencies, crash reports, together with the database, should be available from the state highway safety department, which can conduct similar surveillance in-house, in collaboration with a public health agency, as implemented in the current study.

Although we emphasized racial disparities, rural–urban disparity in severe MVC injuries should also be underscored. We found that drivers living in rural counties were more likely than their urban counterparts to have a severe injury crash, which is consistent with our knowledge about fatal MVC injuries by location. Previous studies have found that rural residents have more frequent ambulance transfers from a secondary to a tertiary hospital than urban residents, that EMS responders in rural areas are more likely than those in urban areas to be volunteer than paid, and that the EMS scene-to-facility distance is greater in rural areas than in urban areas (Grossman et al., 1997; Mueller, 1999). The large number of severe injuries puts added stress not only on rural residents in seeking quality and timely care, but also on the injury care system in rural areas (Blatt and Furman, 1998; Probst et al., 2007).

The phase I project has several limitations. First, MVCs incurred by non-Nebraska drivers on Nebraska roads were excluded. Second, the injury incidence rates by race might be slightly underestimated, because drivers who moved out of state permanently may not have changed their driver's license according to state law (i.e., within 30 days of moving to a permanent residence in another state). Third, the findings can only be applied to drivers. In addition, pedestrian injuries may also have disparities (Chakravarthy et al., 2010). Fourth, there are limited variable fields in the driver's license dataset that can be used for MVC data analysis. Finally, although police-assigned injury severity or KABCO is commonly used, the KABCO injury scale may not effectively capture clinically based injury severity, such as the Maximum Abbreviated Injury Scale (MAIS) (Farmer, 2003). For a more restricted sample that includes hospitalized patients who can be found in the police reports, future studies should use MAIS to assess racial disparities in nonfatal MVC injuries. We will deal with the last limitation in phase II of the project.

18.3 PHASE II PROJECT: USING MAIS FOR HOSPITAL-BASED SURVEILLANCE

18.3.1 METHOD

The phase II project is based on the same at-risk population and incidence data as in phase I, with one more exclusion criterion: injured drivers who could not be found in the hospital discharge data between 2006 and 2010 were dropped. Note that the major methodological aim of the project is to add MAIS information to each injured driver. Since the MAIS information is only available for hospitalized patients, the sample has be further restricted. Hence, the data flow is that we added race information to injured drivers through data linkage to driver's license data, and we further linked race-enhanced data to the hospital discharge data to obtain MAIS information. The first linkage was based on driver's license number, and it was a deterministic linkage. The second data linkage was probabilistic (McGlincy, 2004) between hospital discharge data and the newly expanded crash data. Variables that were considered for linkage included personal and event information (e.g., date of birth, sex, event date, event location, and injury type), and match probabilities were calculated to determine whether two records from crash data and hospital discharge data described the same person and the same event. Consequently, we were able to attach hospitalization information that includes medical diagnoses in International Classification of Diseases, Ninth Revision (ICD-9) codes. Among those records that indicate a transport to a hospital, the average linkage rate over the 5-year study period was 80%. This part of data linkage was routine work by the CODES program, and frequency tables were provided for the following analysis. After the probabilistic linkage in the CODES program, we have both police-reported injury scores (KABCO) and clinical-based MAIS.

The assessment of injury burden by race can then be carried out by either KABCO or MAIS. We used MAIS as the basis for this study, because it is based on ICD-9 (MacKenzie and Sacco, 1997) and is more reflective of hospital-based health care services than KABCO (Compton, 2005). Due to the nature of probabilistic linkage with missing data in CODES, we obtained five imputed sets for the linkage between crash and hospital discharge data. Analysis was done by taking the average over the five imputed sets (Heeringsa et al., 2010). Among all the drivers in crash data, 22,825 out of 217,945 drivers, or 10.5%, were matched to hospital discharge data. Among those matched, 1,145 were severely injured (MAIS of 3, 4, or 5) and 18,216 were nonseverely injured (MAIS of 1 or 2). When these numbers are used to calculate rates, they reflect hospital-based injury burden without further mentioning.

The analysis followed the same descriptive and multivariate approaches. In addition, we report our multivariable results based on KABCO, so that results from the two injury severity scales can be assessed for their consistencies and inconsistencies. In the KABCO analysis, among those rated as severe or nonsevere injuries according to MAIS, there were 3,090 and 14,579 records in the severe (A) and nonsevere (B, C) injury categories, respectively; 1,692 records were excluded, as they were in neither category.

18.3.2 Results

From 2006 to 2010, there were 1145 Nebraskan drivers severely injured in MVCs who were found in the Nebraska hospital discharge data. Among them, 89.6% were White, 4.9% were Black, and 5.5% were other races, corresponding to crude rates of 77, 86, and 53 injured drivers per 100,000, respectively. Of the 18,216 nonseverely injured drivers, 81.1% were White, 10.7% were Black, and 8.2% were other races, corresponding to crude rates of 1,107, 2,974, and 1,255 injured drivers per 100,000, respectively. These crude rates suggest that Black drivers were 12% and 169% more likely than White drivers to be severely and nonseverely injured, respectively (Table 18.3).

Table 18.3 further shows that for both severe and nonsevere injuries, injury rates generally decline as age increases, with a few exceptions. In the workforce age groups (25–44 and 45–64 years old), Black drivers had the highest severe injury rates among the three race groups. They also had the highest nonsevere injury rates across all age groups. The gaps between Black drivers and their two other counterparts were wider for the nonseverely injured than for the severely injured. While the severe injury rates among the three race groups were comparable for females, female Black drivers had a nonsevere injury rate that was almost triple those of the other two racial groups. In addition, male drivers had higher severe injury rates, while female drivers had higher nonsevere injury rates, regardless of race. Black drivers had higher severe and nonsevere injury rates in urban areas. However, the absolute rates pointed to different directions in rural areas. Part of the reason could be Whites dominate the rural population. The numbers of severe injuries among Blacks and other races were very small; therefore, injury rates were too unstable to be interpreted.

We further analyzed the results using the four-way log-rate model. The results for the severe injury model (Table 18.4, first two columns) showed that drivers younger than 25 years were more likely than drivers 65 years and older to have a severe MVC injury, while drivers aged 25–44 years were less likely to be found in the hospital discharge data among the severely injured. Females were less likely than males to have a severe MVC injury. Black drivers were more likely than White drivers to have a severe MVC injury, while drivers of other races were less likely than White drivers to have a severe MVC injury. Rural residents were much more likely than urban residents to have a severe MVC injury. The only interaction term that was marginally significant was rural Black drivers. While the odds ratio was 0.11, the confidence interval was too wide.

The results for the nonsevere injury model (Table 18.4, last two columns) were different in both main and interaction effects from those for the severe injury model. All three younger age

TABLE 18.3

Annual Severe and Nonsevere MVC Injury Rates by Race, Age, Sex, and Place of Residence

	Severe Injury Rate per 100,000 Licensed Drivers			Nonsevere Injury Rate per 100,000 Licensed Drivers			No. of Licensed Drivers
	White	Black	Other	White	Black	Other	
Age Group							
15–24	5.50	4.03	4.02	104.35	205.52	96.09	194,930
25–44	2.49	3.77	1.82	45.14	128.46	42.74	564,380
45–64	2.71	2.96	1.58	33.23	81.28	38.66	520,723
65+	3.25	2.12	1.80	22.10	47.02	31.30	238,875
Sex							
Female	2.11	1.81	1.84	49.48	151.75	62.94	725,336
Male	3.99	4.68	2.33	39.44	92.75	40.98	793,572
Place of Residence							
Rural	3.47	0	2.51	40.09	62.99	46.93	732,684
Urban	2.67	3.70	1.85	48.67	123.27	52.32	786,224
Summary rates	15.38	17.12	10.59	221.42	594.89	251.07	
No. of licensed drivers	1,334,391	65,528	118,989	1,334,391	65,528	118,989	1,518,908

TABLE 18.4

Odds Ratios for MAIS-Based Severe and Nonsevere MVC Injuries

	OR	95% CI	OR	95% CI
Age Group (reference = 65+)				
15–24	1.73	1.43–2.09	4.61	4.34–4.9
25–44	0.82	0.69–0.98	1.99	1.88–2.12
45–64	0.85	0.71–1.01	1.5	1.41–1.59
Sex (reference = Male)				
Female	0.53	0.47–0.6	1.25	1.21–1.29
Race (reference = White)				
Black	1.35	1.02–1.79	1.88	1.72–2.05
Other race	0.68	0.52–0.87	1.1	0.97–1.25
Place of Residence (reference = Urban)				
Rural	1.28	1.13–1.44	0.86	0.83–0.89
Interaction Terms				
Rural Black	0.11	0.01–1	0.55	0.43–0.69
Black age 25–44			1.34	1.22–1.47
Other race age 15–24			0.77	0.66–0.89
Other race age 25–44			0.8	0.7–0.92
Black female			1.25	1.13–1.37
Other race female			1.18	1.06–1.31

groups (15–24, 25–44, and 45–64) were more likely to have nonsevere injuries than drivers 65 years and older. Female drivers were more likely than male drivers to have nonsevere injuries. Black drivers were still at higher risks for nonsevere injuries than White drivers. Drivers in rural areas were less likely to have nonsevere injuries than urban areas. Among significant inter-action effects, rural Blacks remained significant, but five additional interaction terms became significant. If the driver was Black and aged 25–44 years, he or she was more likely to have a nonsevere injury. On the other hand, if the driver was of other races and aged 15–44 years, his or her corresponding age-specific effects would be reduced by about 20%, all else being equal. In addition, female Black drivers and female drivers of other races had greater chances of receiving nonsevere injuries.

Comparing the KABCO-based MVC injury findings in Table 18.2 to the MAIS-based results in Table 18.4, we found mostly consistent results. For nonsevere injury, the point estimates were quite similar between the two measurement approaches for almost all the effects. In addition, most inter-action terms were different, except a consistent finding for rural Blacks. Some inconsistencies were found among interaction terms. However, there was a major inconsistency for the severely injured model in terms of age main effects. The KABCO-based analysis found that the 25–44 age group was significantly more likely to have a severe injury than those 65+ (Table 18.A1), while the MAIS-based result was the opposite. We suspect that this inconsistency may due to different samples, a much larger sample for the KABCO-based than the MAIS-based result. For this reason, we ran an identical analysis with an identical sample to that in Table 18.4 by using the KABCO rather than MAIS score, and found the same inconsistency about the age group (25–44). The odds ratio for the KABCO-based effect was 1.49 (95% confidence interval [CI] 1.29–1.65), in contrast to the MAIS-based effect (OR 0.82, 95% CI 0.69–0.98), shown in Table 18.4. Other than those differences, the two injury schemes generate fairly consistent findings by either using different samples, as shown in Tables 12.4 and 18.2, or using identical samples.

Hospital-based MVC injury data by race have many potential applications, such as comorbid-ity, length of stay, and hospital-related cost. For instance, hospital discharge data have a discharge status that includes discharge to home, which together with injury severity can be used to assess community-based informal injury care needs. Here, we chose cost to demonstrate the benefit of linking to hospital discharge data. MVCs have tremendous costs for individuals and society, and reducing injury disparities has been a top policy agenda for many government agencies, such as CDC and NHTSA. Estimated fatal and nonfatal MVC injury costs in 2005 were $52 and $42 bil-lion, respectively (Naumann et al., 2010). Even though major racial and ethnicity disparities have been reported, associated race-specific costs are rarely reported, and we attempted to show drivers' insurance types and injury costs per driver's license. We calculated insurance types and estimated costs by race according to a NHTSA-recommended method (Blincoe et al., 2002). We tabulated insurance types by race because the previous study by Svenson and Spurlock (2001) could only examine injury disparity by insurance types, as opposed to race. We calculated nonfatal injury cost because of the importance of conveying racial disparity in MVC injuries to policy makers and the general public. Our simple calculation shows that Blacks had the highest hospital charges per injury. For the severe injury category, Black drivers had medical care cost of about $116,059, while White drivers had an average of $60,717 in costs. The costs suggest that Blacks had more severe injury in the severe categories, as reported by others. For the nonsevere injury group, the cost for Black driv-ers was less than that for White drivers ($3332 vs. $3962).

18.3.3 Discussion

In phase II of MVC surveillance, we further linked the race-enhanced MVC injury database to the hospital discharge data to construct the MAIS injury score and derive related costs. We found that Black drivers had the highest clinically defined severe and nonsevere injuries per 100,000 licensed drivers. Compared to White drivers, Black drivers were 11% and 168% more likely to be severely and nonseverely injured, respectively. These findings persist when controlling for age, sex, and place of residence.

In addition to providing more accurate injury severity scores, crash and driver's license data, in combination with hospital discharge data, provided many opportunities to further advance our knowledge base and obtain a closer estimate of MVC injury burden, in terms of injury severity and costs, among different subgroups of the population. From crash data alone, police-reported injury severity is the only source from which to estimate injury burden; however, police officers' judgments are most often based on subjective observation and could be influenced by a variety of factors, such as occupants' gender, lighting condition, and nature of injury (Popkin et al., 1991). By using MAIS, we were able to have not only a more reliable measure of injury severity from ICD-9 codes, but also other information associated with hospital discharge data. With the linked database, we were also able to quantify disparities existing in insurance availability, where we found Black drivers were about twice as likely to be uninsured as drivers in other racial groups. A previous study showed that the uninsured were treated less than ideally when injured (Svenson and Spurlock, 2001), and our linked database allows further analysis by both insurance and race.

A major limitation for the phase II study is the sample size. This sample represents only 10.5% of reported MVC injuries, and it can only be used to infer hospital-based MVC injury disparity for drivers. As a result, the statistic power may not be sufficient to detect disparity, especially for the severely injured category. Since the KABCO-based injury surveillance is much quicker, as it does not require data linkage to hospital discharge data, one way to increase the statistical power is to borrow the strength from the results of the KABCO-based injury analysis A limitation shared by both phase I and II studies is the lack of SES variables. Since residence information can be as detailed as the census tract through geocoding, the next phase is to include census tract SES information.

18.4 PHASE III PROJECT: GEOREFERENCING MAIS-BASED INJURY EVENT TO CENSUS TRACT FOR SES ANALYSIS

18.4.1 METHOD

In the above assessments, we could not include any SES variables. Neither police-reported data nor hospital discharge data have income-related variables. However, since hospital discharge data have been geocoded and used for poverty-related analyses (see Chapters 8 through 11), we could simply reversely add the census tract poverty variable in the internally indexed hospital discharge data back to the MAIS-based injury data. Normally, people link hospital data to obtain hospital discharge–related information, such as length of stay, E-code, and total charge, and the MAIS. All of the information could be directly derived from the hospital discharge data. Geocoded information was not on the traditional hospital discharge data, but on the enhanced version, and we can retrieve it in a similar manner. After adding the poverty variable, we have enhanced MVC injury data (1) from no race to including race (via driver's license linkage), (2) from no clinical-based injury score to including MAIS (via hospital discharge linkage), and (3) from no census tract variable to including poverty data from linking the indexed and geocoded hospital discharge data back to the MAIS-based injury data.

When using licensed drivers as the at-risk population, it was a perfect match to the injured drivers of in-state residents. However, we could not use them anymore, as each driver's license does not have a census tract code to which we could attach poverty or other SES information. For this reason, we used those in the 15–84 age group from the 2010 census of population for Nebraska as the at-risk population. We extracted the decennial census by race at the census tract level and attached the poverty levels of ACS 2006–2010 to each census tract. The three race categories were White alone, Black alone, and all others. We adopted this scheme because almost all the surveillance practices by race within the state of Nebraska use the single races of White, Black, American Indian, and Asian.

However, not all individuals within each census tract can drive, even though we restrict the age to 15–84. In addition, residents in poor neighborhoods were slightly less likely to own a car and drive to work. Both car ownership and driving to work status could be used as adjustment factors. Since household may not be compatible to the individual approach used for injured drivers, we used the percentage of age 16 and over who drive or carpool to work as an adjustment factor. In particular, we used the percentage to weight the selected census tract population; if 100% of people within a census tract drive to work, then the total selected population was used. If 50% of people drive to work, then the at-risk population would be 50% of the census tract population. In the preliminary analysis of crude injury rates (Table 18.A2), we found that using the adjusted at-risk population is more reasonable, although both adjusted and unadjusted population produced an SES gradient for MVC injury rates.

We ran a log-rate model similar to those in phases I and II by (1) dropping rural–urban status and (2) adding poverty level. If we kept rural–urban status in the multiway contingency table analysis, we would face the small-cell issue, causing less robust estimates.

18.4.2 RESULTS

For the severe injury model (Table 18.5, first two columns), there was no significant interaction term. Age, sex, and racial effects from this model were mostly consistent with the model estimates

TABLE 18.5
Race and SES Effects of Severe and Nonsevere MVC Injuries

	Severe Injury		Nonsevere Injury	
Age Group (reference = 65+)				
15–24	1.090	0.902–1.317	3.037	2.860–3.225
25–44	0.787	0.658–0.941	2.199	2.075–2.331
45–64	0.778	0.652–0.929	1.463	1.378–1.555
Sex (reference = Male)				
Female	0.488	0.431–0.552	1.195	1.161–1.231
Race (reference = White)				
Black	1.49	1.123–1.979	2.108	1.785–2.489
Other race	0.801	0.618–1.038	0.901	0.773–1.051
Poverty Level (reference = <5%)				
5% ≤ poverty < 10%	1.582	1.335–1.876	1.238	1.184–1.295
10% ≤ poverty < 15%	1.457	1.216–1.746	1.286	1.226–1.348
15% ≤ poverty < 20%	1.267	1.013–1.586	0.853	0.799–0.911
Poverty > 20%	1.48	1.194–1.834	1.569	1.484–1.657
Interaction Terms				
Black*pvt3			1.587	1.279–1.969
Black*pvt4			1.242	1.036–1.488
Other race*pvt4			1.435	1.162–1.774

Note: Injury severity was determined by the MAIS scheme. Poverty is based on the percentage of people below the federal poverty line in the ACS 2006–2010. Log-rate model estimates are based on 2010 census of the population.

when drivers were used as the at-risk population in Table 18.4. After controlling for these demographic factors, we found that poverty was a significant risk factor associated with severe MVC injury. Comparing with the least poor neighborhoods, drivers from the other four levels of neighborhoods were between 27% and 48% more likely to have an MVC injury. One can also look at poverty as a control factor when examining racial disparity. Without the poverty control, Black drivers were 57% more likely than White drivers to have a severe injury; with the poverty control, the percentage was reduced to 49, so it is a modest reduction.

The results from the nonsevere injury model (Table 18.5, last two columns) were also fairly consistent with those from the severe injury model. Drivers from poorer neighborhoods in most parts were associated with a 24%–57% increase in nonsevere injury. The only exception was the 15%–20% poverty neighborhood category, which had a 15% reduction compared to the neighborhood with <5% of the people under the poverty line. In addition, three race and poverty interaction terms were significant. Focusing on the 15%–20% poverty neighborhood category, we found that both Black and other race drivers were associated with increased risk of nonsevere injuries. This implies that White drivers from this neighborhood group contributed most to the reduced risk of nonsevere injuries.

Even though MVC is spatial in nature, the spatial aspect of disparity is rarely reported in the literature. We want to point out that there is a lot that can be done with MVC data. For instance, crash locations are routinely collected and geocoded by the Highway Safety Department, and there has been some use of such data. The use of accident location without connecting the driver's residential location makes it hard to study spatial aspects of disparities. Since some of the drivers' data can be linked to hospital data, and hospital data have been geocoded, we can then reversely provide injured drivers' census tract location without getting into geocoding.

Figure 18.1 displays all local highway traffic crash locations for 2006 overlaid with major roads (blue), other roads (brown), and 10 median family income categories in quantiles in Douglas County, where the city of Omaha is located. The map had all accidents, not just those that were matched with hospital records. It shows that most off-highway crashes in the county were concentrated in low-income neighborhoods. One could argue that most poor neighborhoods are located in the central city, near downtown, and people from all over the city tend to have an MVC injury there. While there is some truth in that, drivers from both ends of income categories tend to crash locally.

Table 18.6 cross-tabulates accidental locations and residential locations of injured drivers georeferenced at the census tract level. Note that the sample in this table is MAIS based, as we only had those linked to hospital records. The numbers are in row percent, so that we can read the distribution of MVC events from resident location in the row to accident location in the

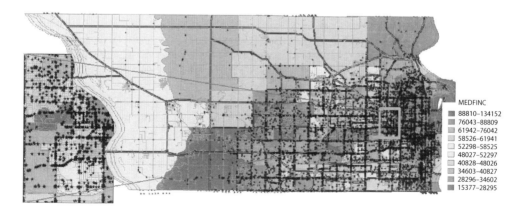

FIGURE 18.1 **(See color insert)** Nonhighway traffic crashes and median family incomes by census tracts in Douglas County, Nebraska.

TABLE 18.6
Poverty Levels by Residential and Crash Locations in Omaha, 2006–2011

Accident	a_pvt1	a_pvt2	a_pvt3	a_pvt4	a_pvt5	N
r_pvt1	39.65	25.72	14.53	4.78	15.31	4055
r_pvt2	24.17	31.71	16.00	6.49	21.63	3699
r_pvt3	22.51	23.93	23.07	6.95	23.54	2332
r_pvt4	14.50	20.08	11.81	14.37	39.24	1524
r_pvt5	8.21	15.11	9.53	6.92	60.23	4262

Note: Numbers are in row percent. a_pvt, driver's accident location poverty level; r_pvt: driver's residential location poverty level.

column. First, if all MVCs happened locally within each poverty neighborhood category, we would not see any off-diagonals being populated with numbers. In fact, only one diagonal has more than 50% of the events—from poverty group 5 (>20%) to poverty group 5 (60.23%), and the next closest to it is 39.65%, from poverty group 1 (<5%) to poverty group 1. Hence, not all MVCs happened locally in terms of staying in their own neighborhood categories. Interventions then should also focus on two ends of poverty categories by introducing age and other factors that will not be discussed here.

18.4.3 DISCUSSION

In this phase of the MVC surveillance project, we have gone to an uncharted SES and spatial surveillance for MVC injuries, where many perspectives could be taken that include explicit and nonexplicit spatial components. Following the traditional multivariate approach, we introduced both race and poverty variables. We found that both race and poverty followed some gradients, in which Black drivers and drivers from poor neighborhoods were between 25% and 60% more likely to have an MVC injury than the comparison groups. Although there are some interactions among nonsevere injury between race and poverty, they were limited. Certainly, the lack of significance in interaction for the severe injury category could be due to small sample size. It almost implies that if we used KABCO-based injury data, which does not require data linkage to hospital data, we would find some significant interaction due to its large sample size. However, it would require further work due to having to geocode the police-reported injury data, which would be very taxing.

Spatial disparity surveillance requires getting to neighborhood-level units. This whole area of study is wide open. In the current study, we just touched the tip of the iceberg, where we used drivers' residence locations and accident locations in poverty categories to reveal that the high concentration of people from the poorest neighborhoods had the highest concentration in accidents within their neighborhood categories. Internally, we could display data by census tract without Health Insurance Portability and Accountability (HIPAA) violation, but it would not be that useful since the findings cannot be revealed to the general public. Alternatively, we could do kernel density mapping, which would not reveal exact location. However, kernel density mapping essentially is a univariate mapping method, with age group and other variables involved in surveillance; kernel density mapping is not particularly revealing without further smoothing and multilayer presentations. It suggests that model-based disease mapping without a spatial unit might work, that is, in HIPAA compliance, and reveal sufficient location information (e.g., overlay spatial units on top of kernel or other smoothing techniques). As we said, there are many ways to conduct spatial disparity surveillance when data are spatially relevant. Unlike model-based estimation developed in log-linear, logistic, and other multivariate analyses, spatial-model-based estimation needs to be developed and refined for surveillance and intervention.

18.5   CHAPTER SUMMARY AND CONCLUDING REMARKS

This chapter used primarily data from the CODES program. The traditional CODES programs link MVC data to hospital E-code data for injury surveillance and hospital cost estimation. Since both CODES and hospital data do not have race variables, program data analysts simply ignored race in their previous reports before we conducted these analyses in phases. In phase I, we linked drivers to driver's licenses to obtain the drivers' race information, and reported racial disparities according to the police-reported injury score or KABCO. We found that Black–White disparities were greater for nonsevere injuries than severe injuries. In phase II, we linked enhanced race information in phase I study to hospital E-code data so that clinically derived MAIS can be used for assessing racial disparities. We found that results from MAIS were by and large consistent with the results from KABCO. In phase III study, we used drivers' data from phase II with MAIS to reversely add census tract information from the Nebraska hospital discharge data, with the latter already being geocoded. We found that both SES and racial gradients exist for MVC injuries with Black drivers, and those living in poor neighborhoods had independent effects on both severe and nonsevere injuries.

APPENDIX 18A: SENSITIVITY ANALYSIS

TABLE 18.A1
KABCO-Based Model Estimates from the Same Sample as in Table 18.4

Parameter	MAIS Model OR	95% CI		KABCO Model OR	95% CI	
Age group (reference = 65+)						
15–24	2.94	2.58	3.34	4.25	4.86	4.55
25–44	1.46	1.29	1.65	1.82	2.08	1.95
45–64	1.22	1.07	1.38	1.42	1.62	1.52
Sex (reference = Male)						
Female	0.76	0.70	0.81	1.28	1.38	1.33
Race (reference = White)	1.00	1.00	1.00			
Black	1.49	1.21	1.83	1.67	2.04	1.85
Other race	0.76	0.65	0.87	0.94	1.26	1.09
Place of Residence (reference = Urban)						
Rural	1.41	1.31	1.52	0.79	0.85	0.82
Interaction Terms						
Black age 25–44				1.19	1.47	1.32
Other race age 15–24				0.63	0.89	0.75
Other race age 25–44				0.72	0.99	0.84
Black female	1.35	1.00	1.82	1.09	1.35	1.21
Other race female				1.05	1.33	1.18
Rural Black	0.29	0.13	0.62	0.50	0.81	0.63

TABLE 18.A2
MVC Injury Rate per 100,100 by Poverty Level Adjusting
for Percent of Driving to Work vs. No Adjustment

	Pvt1	Pvt2	Pvt3	Pvt4	Pvt5
MAIS_inj = Severe, No Adjustment					
White	64	100	90	70	81
Black	81	52	88	76	112
Other	51	55	68	81	42
MAIS_inj = Severe, with Adjustment					
White	80	131	120	97	129
Black	99	67	120	111	184
Other	63	72	92	119	70
MAIS_inj = Nonsevere, No Adjustment					
White	1104	1256	1275	843	1464
Black	2552	2747	2574	3021	3714
Other	1148	1583	1322	1206	1408
MAIS_inj = Nonsevere, with Adjustment					
White	1366	1653	1712	1168	2339
Black	3116	3514	3501	4404	6132
Other	1408	2059	1790	1763	2378

Note: At-risk population without adjustment is based on the 2010 census, age 15–84. The adjusted at-risk population is the population without adjustment times the percentage of those driving to work in each census tract.

REFERENCES

Abdalla, I.M., R. Raeside, D. Barker, and D.R.D. McGuigan. 1997. An investigation into the relationships between area social characteristics and road accident casualties. *Accident Analysis and Prevention* 29(5): 583–593.

Baker, S.P., E.R. Braver, L. Chen, J.F. Pantula, and D. Massie. 1998. Motor vehicle occupant deaths among Hispanic and black children and teenagers. *Archives of Pediatrics and Adolescent Medicine* 152(12): 1209–1212.

Blatt, J. and S.M. Furman. 1998. Residence location of drivers involved in fatal crashes. *Accident Analysis and Prevention* 30(6): 705–711.

Blincoe, A., E. Seay, T. Zaloshnja, E. Miller, E. Romano, S. Luchter, and R. Spicer. 2002. The economic impact of motor vehicle crashes, 2000. DOT HS 809 446. Washington, DC: National Highway Traffic Safety Administration, U.S. Department of Transportation.

Braver, E.R. 2003. Race, Hispanic origin, and socioeconomic status in relation to motor vehicle occupant death rates and risk factors among adults. *Accident Analysis and Prevention* 35(3): 295–309.

Briggs, N.C., R.S. Levine, W.P. Haliburton, D.G. Schlundt, I. Goldzweig, and R.C. Warren. 2005. The Fatality Analysis Reporting System as a tool for investigating racial and ethnic determinants of motor vehicle crash fatalities. *Accident Analysis and Prevention* 37(4): 641–649.

Campos-Outcalt, D., C. Bay, A. Dellapena, and M.K. Cota. 2003. Motor vehicle crash fatalities by race/ethnicity in Arizona, 1990–96. *Injury Prevention* 9(3): 251–256.

CDC (Centers for Disease Control and Prevention). 2010. Web-Based Injury Statistics Query and Reporting System (WISQARS) [online]. Atlanta, GA: National Center for Injury Prevention and Control, CDC. http://www.cdc.gov/ncipc/wisqars (accessed December 20, 2012).

CDC (Centers for Disease Control and Prevention). 2012a. 10 leading causes of death by age group, United States—2010. Atlanta, GA: National Center for Injury Prevention and Control, CDC. http://www.cdc.gov/injury/wisqars/pdf/10LCID_All_Deaths_By_Age_Group_2010-a.pdf.

CDC (Centers for Disease Control and Prevention). 2012b. 10 leading causes of injury deaths by age group highlighting unintentional injury deaths, United States—2010. Atlanta, GA: National Center for Injury Prevention and Control, CDC. http://www.cdc.gov/injury/wisqars/pdf/10LCID_Unintentional_Deaths_2010-a.pdf (accessed April 7, 2012).

Chakravarthy, B., C.L. Anderson, J. Ludlow, S. Lotfipour, and F. E. Vaca. 2010. The relationship of pedestrian injuries to socioeconomic characteristics in a large southern California county. *Traffic Injury Prevention* 11(5): 508–513.

Compton, C.P. 2005. Injury severity codes: A comparison of police injury codes and medical outcomes as determined by NASS CDS investigators. *Journal of Safety Research* 36(5): 483–484.

Cooper, P.J., J. Osborn, and W. Meckle. 2010. Estimating the effect of the vehicle model year on crash and injury involvement. *Journal of Automobile Engineering* 224(12): 1527–1539.

Durkin, M.S., L.L. Davidson, L. Kuhn, P. O'Connor, and B. Barlow. 1994. Low-income neighborhoods and the risk of severe pediatric injury: A small-area analysis in northern Manhattan. *American Journal of Public Health* 84(4): 587–592.

Farmer, C.M. 2003. Reliability of police-reported information for determining crash and injury severity. *Traffic Injury Prevention* 4(1): 38–44.

GHSA (Governors Highway Safety Association). 2012. MMUCC guideline: Model minimum uniform crash criteria. 4th ed., DOT HS 811 631. Washington, DC: GHSA, U.S. Department of Transportation.

Grossman, D.C., A. Kim, S.C. MacDonald, P. Klein, M.K. Copass, and R.V. Maier. 1997. Urban-rural differences in prehospital care of major trauma. *Journal of Trauma: Injury, Infection & Critical Care* 42(4): 723–729.

Harper, J.S., W.M. Marine, C.J. Garrett, D. Lezotte, and S.R. Lowenstein. 2000. Motor vehicle crash fatalities: A comparison of Hispanic and non-Hispanic motorists in Colorado. *Annals of Emergency Medicine* 36(6): 589–596.

Hasselberg, M., M. Vaez, and L. Laflamme. 2005. Socioeconomic aspects of the circumstances and consequences of car crashes among young adults. *Social Science & Medicine* 60(2): 287–295.

Heeringsa, S.G., B.T. West, and P.A. Berglund. 2010. *Applied Survey Data Analysis*. Baca Raton, FL: Chapman & Hall/CRC.

Imai, S. and C.J. Mansfield. 2008. Disparities in motor vehicle crash fatalities of young drivers in North Carolina. *North Carolina Medical Journal* 69(3): 182–187.

MacKenzie, E.J. and W. Sacco. 1997. *ICDMAP-90: A Users Guide*. Baltimore, MD: Johns Hopkins University School of Public Health and Tri-Analytics.

McDonald, G., G. Davie, and J. Langley. 2009. Validity of police-reported information on injury severity for those hospitalized from motor vehicle traffic crashes. *Traffic Injury Prevention* 10: 184–190.

McGlincy, M.H. 2004. A Bayesian record linkage methodology for multiple imputation of missing links. In *Proceedings of the American Statistical Association*. Alexandria, VA: American Statistical Association, pp. 4001–4008.

Mueller, K.J. 1999. Health status and access to care among rural minorities. *Journal of Health Care for the Poor and Underserved* 10(2): 230–249.

Naumann, R.B., A.M. Dellinger, E. Zaloshnja, B.A. Lawrence, and T.R. Miller. 2010. Incidence and total lifetime costs of motor vehicle-related fatal and nonfatal injury by road user type, United States, 2005. *Traffic Injury Prevention* 11(4): 353–360.

NE-CODES (Nebraska Crash Outcome Data Evaluation System). 2006. Non-Nebraska drivers involved in motor vehicle crashes occurring in Nebraska. Lincoln, NE: Department of Health and Human Services. http://dhhs.ne.gov/publichealth/Documents/non-ne-drivers.pdf (accessed September 20, 2012).

NHTSA (National Highway Traffic Safety Administration). 2006. Race and ethnicity in fatal motor vehicle traffic crashes 1999–2004. DOT HS 809 956. Washington, DC: NHTSA, U.S. Department of Transportation.

NHTSA (National Highway Traffic Safety Administration). 2008. Traffic safety facts: Rural/urban comparison. DOT HS 810 812. Washington, DC: NHTSA, U.S. Department of Transportation.

NHTSA (National Highway Traffic Safety Administration). 2010. Traffic safety facts: Highlights of 2009 motor vehicle crashes. DOT HS 811 363. Washington, DC: NHTSA, U.S. Department of Transportation.

NHTSA (National Highway Traffic Safety Administration). 2012a. Traffic safety facts: Motor vehicle traffic crashes as a leading cause of death in the United States, 2008 and 2009. DOT HS 811 620. Washington, DC: NHTSA, U.S. Department of Transportation.

NHTSA (National Highway Traffic Safety Administration). 2012b. Traffic safety facts: Early estimate of motor vehicle traffic fatalities in 2011. DOT HS 811 604. Washington, DC: NHTSA, U.S. Department of Transportation.

NHTSA (National Highway Traffic Safety Administration). 2012c. Fatality Analysis Reporting System (FARS) encyclopedia. Washington, DC: NHTSA, U.S. Department of Transportation. http://www-fars.nhtsa.dot.gov/QueryTool/QuerySection/SelectCriteria.aspx (accessed April 27, 2012).

Popkin, C.L., B.J. Campbell, A.R. Hansen, and R.R. Stewart. 1991. Analysis of the accuracy of the existing KABCO injury scale. Chapel Hill: University of North Carolina Highway Safety Research Center.

Probst, J., S. Laditka, J. Wang, and A. Johnson. 2007. Effects of residence and race on burden of travel for care: Cross sectional analysis of the 2001 US National Household Travel Survey. *BMC Health Services Research* 7(1): 40.

Roudsari, B., S. Ramisetty-Mikler, and L.A. Rodriguez. 2009. Ethnicity, age, and trends in alcohol-related driver fatalities in the United States. *Traffic Injury Prevention* 10(5): 410–414.

Ryb, G.E., P.C. Dischinger, and S. Ho. 2009. Vehicle model year and crash outcomes: A CIREN study. *Traffic Injury Prevention* 10(6): 560–566.

Svenson, J.E. and C.W. Spurlock. 2001. Insurance status and admission to hospital for head injuries: Are we part of a two-tiered medical system? *American Journal of Emergency Medicine* 19(1): 19–24.

U.S. Census Bureau. 2011a. Nebraska's 2010 census population totals and demographic characteristics. Washington, DC: U.S. Census Bureau. http://www2.census.gov/census_2010/01-Redistricting_File-PL_94–171/Nebraska/ (accessed June 12, 2011).

U.S. Census Bureau. 2011b. Urban area criteria for the 2010 census; notice. *Federal Register* 76(164): 53030–53043.

U.S. Census Bureau. 2012a. Statistical abstract of the United States: 2012. Table 1103: Motor vehicle accidents—number and deaths: 1990 to 2009. Washington, DC: U.S. Census Bureau. http://www.census.gov/compendia/statab/2012/tables/12s1103.pdf (accessed January 2, 2013).

U.S. Census Bureau. 2012b. American FactFinder. Washington, DC: U.S. Census Bureau. http://factfinder2.census.gov/faces/nav/jsf/pages/index.xhtml (accessed July 6, 2012).

Vivoda, J.M., D.W. Eby, and L.P. Kostyniuk. 2004. Differences in safety belt use by race. *Accident Analysis and Prevention* 36(6): 1105–1109.

Whitlock, G., R. Norton, T. Clark, M. Pledger, R. Jackson, and S. MacMahon. 2003. Motor vehicle driver injury and socioeconomic status: A cohort study with prospective and retrospective driver injuries. *Journal of Epidemiology and Community Health* 57(7): 512–516.

Zhang, Y. and G. Lin. 2013. Disparity surveillance of nonfatal motor vehicle crash injuries. *Traffic Injury Prevention* 14(7): 697–702.

19 Linking Cancer Screening and Cancer Registry Data for Outcome Assessments

19.1 INTRODUCTION

The lack of socioeconomic status (SES) data in state cancer registries has been recognized for years. Occupation and smoking status are available in the data dictionary, but they tend to have a very high number of missing values. A challenge in cancer disparity studies is to find a proxy measure of SES. Recent studies generally use census county and census tracts as proxy measures. A few studies combine personal SES and area SES measures (Sprague et al., 2011). Generally, these studies found that low-SES neighborhoods tend to be associated with late-stage cancer (Koh et al., 2005; Lincourt et al., 2008; McCarthy et al., 2010; Yin et al., 2010; Lin et al., 2011) and elevated mortality (Morgan et al., 2011; Sprague et al., 2011). Another way to assess income effect is through special population. The Medicaid population can be considered poor (Schrag et al., 2009). Koroukian (2003) showed that the proportion of advanced-stage breast and cervical cancer in Ohio was more than double for Medicaid patients than other patients. Lantz and Soliman (2009) showed that Medicaid expansion for cancer care could shorten the duration between diagnosis and the initiation of treatment among White women in the NBCCEDP. Similar findings were also reported in California. Another, more restricted population is from a breast and cervical cancer screening program that not only had a low-income requirement, but also had an uninsured requirement, including Medicaid. We chose this population in the current study.

The National Breast and Cervical Cancer Early Detection Program (NBCCEDP) was introduced in 1990 by the CDC to screen low-income and uninsured women for breast and cervical cancers for free or at a low cost in cooperation with state health agencies. There was a special issue in *Cancer Causes and Control* that celebrated the 25-year anniversary of the program (White and Wong, 2015). Nebraska was one of the early states to receive funding under the NBCCEDP and started the Every Woman Matters (EWM) program in 1992 (Feresu et al., 2008). EWM covers the cost for screening and follow-up tests, including cervical and breast biopsies to diagnose breast and cervical cancer. In 2001, Nebraska passed a law that women diagnosed through EWM would be eligible to receive treatment through the state Medicaid program. In 2010, about 800 Nebraska providers, including physicians, clinics, hospitals, and family planning agencies, served the EWM program. To be eligible to participate in the EWM program, the participants must be age 40–74 years, meet income eligibility guidelines (<225% below the federal poverty level), have no insurance, including the Medicaid coverage, not belong to any health maintenance organization, and not be enrolled in Medicare Part B program. About 34,489 women were screened through the EWM program between 1995 and 2008.

Most previous studies evaluating the effectiveness of the NBCCEDP-funded programs have focused on screening and timeliness of treatment (Liu et al., 2005; Richardson et al., 2010; Wheeler et al., 2013). One of the earliest evaluation studies was from the Florida program (Tamer et al., 2003), which linked the Florida Breast and Cervical Cancer Early Detection Program (BCCEDP) to the Florida Cancer Registry. After case matching by age, sex, race, insurance, and census tract SES, the study found no difference in the stage at diagnosis between BCCEDP screening program cases and matched breast cancer cases from the registry. However, a recent study using a similar design in Texas found a higher proportion of late-stage diagnoses from the cases in the state Breast Cancer and Cervical Cancer Screening (BCCS) program than the matched cases from the Texas Cancer Registry (Rajan et al., 2014). The study by Lobb et al. (2010) attempted to disentangle area and individual SES effects. It assigned cases from the Massachusetts Breast Cancer Screening Program, together with those from the Massachusetts Cancer Registry, into different poverty neighborhood categories. Since all participants were on a low income, if a participant resided in a low-income neighborhood, then it is individual low and neighborhood SES low. If, on the other hand, a participant resided in a high-income neighborhood, then it is individual low and neighborhood SES high. The study found that program participants had stages of diagnosis to comparable to those of nonparticipants. It also found that the registry cases (nonparticipants) from poor neighborhoods tended to have an advanced stage of breast cancer, suggesting that the screening program could do a lot more to reach those from poor neighborhoods.

In Nebraska in particular, it was reported that although African American women were enrolled in equal or greater proportion to the program, they were less likely than White women to receive clinical breast cancer exams (Feresu et al., 2008). Since the study was based on the number of screened (numerator) versus the number of asymptomatic enrollees (denominator), it does not really reflect the program enrollment efforts, but rather how many enrollees were asymptotic or symptomatic. According to personal communication with the program director, the program enrolled and screened 37,900 women during 1997–2007, of which 2,523 reported themselves as Black, or about 6.7% of the total EWM population. Hence, the finding could be just a slightly different time frame or the difference between asymptomatic versus asymptomatic enrollees. In addition, the study did not compare the program enrollment population with the state population as a whole; the Nebraska program was deemed effective to capture the underserved and the minorities, given the high percentage of rural residences and the dominance of the White population in the state.

While not a head-to-head comparison, it was found that treatment patterns for women diagnosed with early-stage breast cancer through three state-based NBCCEDP programs from 1992 to 1995 appear to be comparable those of other women during the same period (Richardson et al., 2001). A recent update by the same lead author found that women screened by the NBCCEDP programs received diagnostic follow-up and initiated treatment within preestablished program guidelines (Richardson et al., 2010). The only direct head-to-head treatment comparison was from Massachusetts. Women diagnosed through a screening program in Massachusetts had stage and treatment patterns that were similar to those of other breast cancer patients in the state, except the use of radiation therapy (RT) (Liu et al., 2005).

At the time of this writing, there were three publications matching state breast cancer screening data with cancer registry data for survival assessments. Rajan et al. (2014) compared breast cancer treatment and survival differences between Texas BCCS participants and geographically matched nonparticipants age 40–64. The matched nonparticipants had to (1) come from the same census tracts as the participants and (2) reside in the two lowest-quintile census tracts indexed by a principal SES component scale. The authors referred to the American Joint Committee on Cancer (AJCC) breast cancer stages for treatment guidelines, and then mapped Surveillance, Epidemiology, and End Results (SEER) staging to the AJCC staging for treatment assessments that included surgery, chemotherapy (CT), and RT. It was found that participants had lower rates of breast surgery and higher rates of CT than nonparticipants, and there were no differences in survival rates between the two groups. Johnson et al.'s (2015) article is on survival surveillance in nature. It used cause-specific death among breast cancer patients age 30–64 to derive survival proportions by diagnosis stage. The study found that the Idaho breast cancer screening participants were more likely to have a late-stage diagnosis than nonparticipants, which, together with a low rate of surgery among participants, contributed to about 5% in survival proportions 5 years after the diagnosis. Finally, Bhuyan et al. (2014) used Nebraska EWM–Nebraska Cancer Registry (NCR) linked data to compare participants and nonparticipants in mortality and treatment outcomes among those age 40–74. The nonparticipants in the study were selected from participants' census tracts. This case study is built on Bhuyan et al.'s study and assesses short-term (3-year mortality) and long-term survival differences between participants and nonparticipants of Nebraska's EWM program.

19.2 METHOD

19.2.1 Data Source

The main data source is the NCR, which contains information on patient demographics, tumor site, tumor stage, first course of treatment, and treatment facility. Within the state, the NCR is used for analyzing long-term trends, planning, and evaluating cancer control programs. The NCR has received the North American Association of Central Cancer Registries' Gold Standard Award for data quality every year since 1995. Even though the EWM program started in 1992, in order to ensure data quality in terms of case completeness in race and ethnicity and vital statistics, we used NCR data starting from 1995. In addition, it normally takes 2 years for the NCR to have a complete update of the most recent data. We requested the NCR data in early 2014, which would have updated vital statistics at least up to the end of 2011. The EWM program is also required to provide information to the NCR on women diagnosed through it, so that staging and other diagnosis information can be reported to the Centers for Disease Control and Prevention (CDC) or the funding agency.

For the current project, all EWM participants from 1992 to 2009 were linked by name, diagnosis group, birth date, social security number, and zip code by the NCR program. The resulting matched records were used to identify women diagnosed with breast cancer through EWM. All other unmatched records in the NCR file were used to identify women diagnosed through other sources. After linkage, individual identifiers, such as single age, date of birth, and address, were deleted from the data analysis. We used both EWM participants and women diagnosed with breast cancer through EWM interchangeably, as some participants could be triggered by their participation in the program and by being screened outside of the EWM screening cycle. In addition, all the NCR data were geocoded at the census tract level. After attaching the 2000 census tract–level poverty status, census tract information was deleted.

19.2.2 Inclusion and Exclusion Criteria

NCR records of women newly diagnosed with breast cancer, using International Classification of Disease codes C50.0–C50.9 (except 9590–9989) from 1995 to 2008 were selected together with an indicator variable indicating whether the patient was a participant of the EWM program. We found that EWM participants were significantly dropped in 2009, suggesting the data from the EWM program were incomplete at the time of this data request. We therefore decided to use 2008 as the ending year, which would leave 3 years at the end of 2011 for mortality comparisons. Only women aged 40–74 years were included in the study, as most women screened through EWM program are within that age range. Women younger than 40 years are eligible for EWM breast cancer services only if they meet certain high-risk conditions, and women older than 74 years are only eligible if they do not have Medicare Part B coverage and have income below 225% of the federal poverty level. In fact, the overwhelming majority of the patients were between 40 and 65. For this reason, we also ran preliminary analyses for those age 40–65 and found that the results were almost identical to those of the full sample we selected here. In addition, women whose cancer was identified solely by a death certificate and women with carcinoma *in situ* were excluded. Based on the above criteria, we had 11,973 women in the sample; among them, 529 were EWM participants.

19.2.3 Variables

The primary outcomes of interest were 3-year mortality and long-term survival differences between EWM participants and nonparticipants. Demographic controls included age, race, and marital status.

We used 3-year mortality because the data we had would not allow us to estimate 5-year mortality for the 2008 sample year. In addition, the preliminary analysis showed that 1-year and 3-year models were similar in terms of mortality differences between EWM participants and nonparticipants, while the 5-year result was not significant. Age was categorized as (1) 40–50 years, (2) 51–60 years, and (3) 61–74 years. Marital status was considered an important predictor for cancer survival due to the availability of informal care by a spouse (McCarthy et al., 2010). We coded marital status as married and unmarried, which includes widowed, among others. Race was categorized as (1) Non-Hispanic White, (2) non-Hispanic Black, and (3) other races and ethnicity. Originally, we attempted to have a separate Hispanic group. However, a preliminary analysis showed that Hispanic was not significant in all the proposed models. We therefore grouped Hispanic with all others.

Ideally, we could use tumor stage based on the tumor, metastases, and nodes (TMN) system developed by the AJCC. However, the official TMN system was implemented in the NCR in 2004, which left those prior to 2004 relying on the SEER staging. To be consistent, we used SEER's staging as (1) local, (2) regional, (3) distant, and (4) upstaged.

Treatment by RT, resection surgery, hormone therapy, and CT was categorized as having been received or not. Previous studies have shown that surgery information from the cancer registry was fairly complete, while information on CT and RT was not (Mallin et al., 2013; Chapter 13). For both CT and RT, we provided updates by using outpatient data from 2005 to 2009. Although it was technically possible to update through 1995, we were told by the Nebraska Hospital Association that the data collection was not complete or population based until 2005. Hence, we updated treatment information for 2005–2008, which can be viewed as systematic data enhancement after 2005 from NCR.

Place of residence was based on the same rural–urban dichotomy throughout the book. Douglas, Sarpy, and Lancaster Counties each had a population of more than 100,000, and so were urban counties; the rest were rural. Census tract of residence was also used to categorize neighborhood poverty according to the percent of population under the 2000 poverty line: (1) <5% poverty rate, (2) 5%–15% poverty rate, and (3) >15% poverty rate. Lobb et al.'s study in Massachusetts used 20% as a cutoff point for poor and not poor census tracts, which does not seem fit in Nebraska, because not many census tracts had 20% or more of the population living under the poverty line.

19.2.4 STATISTICAL ANALYSES

Descriptive statistics were used to compare participants' and nonparticipants' age, race, marital status, stage at diagnosis, and all other outcomes of interest. Multivariate logistic regressions were used to calculate the 3-year mortality differences, while the Cox proportional hazard model was used for survival differences. Model covariates included age, race and ethnicity, location, treatment, and neighborhood poverty level. Logistic regression was also used to compare the treatment received by the women diagnosed through EWM with that received by women diagnosed through all other sources. These treatment models were adjusted only for age. Ninety-five percent confidence intervals (CIs) for all odds ratios (ORs) were estimated, and an OR was considered significant if it did not contain 1.0.

There are several ways to compare EWM participants versus nonparticipants. A straightforward comparison is to include all breast cancer patients during the study period, and compare their outcomes while controlling all other variables. This design implies that EWM participants had low income, but their participation in the program may ameliorate income effects in terms of stage of diagnosis and treatment options. If the participants' outcomes were comparable with those of the nonparticipants, then the EWM program would be achieving its goal literally. Alternatively, we can conceptualize potential life or cost saving by first comparing participants and nonparticipants in a more or less comparable income level. Since we do not have an income variable, we could select all those patients who live in the EWM participants' neighborhoods. This design implies that those residing in the same neighborhood tend to have similar incomes regardless of EWM participation status. If the outcomes of participants were comparable to those of nonparticipants within these neighborhoods, then participating in the screening program might simply have helped those non-insured and underinsured to achieve a level playing field with their neighbors who may or may not have insurance. If, on the other hand, the EWM participants had a lower mortality than nonpartici-pants living in the same neighborhoods, then this may suggest two things: (1) the program saved participant's lives, and (2) more could be done to reach nonparticipants in these neighborhoods so that more potential lives could be saved.

We follow both approaches described above. We first used all the breast cancer records that met the inclusion and exclusion criteria. The process of matching nonparticipants' neighborhoods is as follows:

1. Generate a list of census tract numbers that EWM participants came from.
2. Generate an indicator variable by matching EWM participants' census tract numbers to all records within the breast cancer sample for the study. If there is a match, the indicator variable is 1; else, it is 0.
3. Select all the records with an indicator of 1. This sample had 6628 breast cancer patients who met the inclusion and exclusion criteria.

After all the individual matching between program data, and neighborhood matching at the census tract level, the data were provided to us without any identifiers. The study protocol was sub-mitted to the institutional review board (IRB) at the University of Nebraska Medical Center. It was found that the IRB review was not necessary because the smallest identifiable information was the census tract poverty level and rural–urban status within the state.

19.3 RESULTS

Table 19.1 presents descriptive statistics for the study samples, with the first two columns being nonparticipants and the last column being participants. Comparing the first and last columns, we found that participants were younger, likely to be married, and non-White. In particular, the percent of non-Hispanic Blacks for EWM participants was 4.9%, whereas the corresponding number for all nonparticipants was 2.6%. Nonparticipants were more likely to be diagnosed at localized and distant stages, while participants were more likely to be diagnosed at the regional stage. Except CT, there were no significant differences in RT and curative surgery operations. It should also be mentioned that when the three treatment modalities in the table and other treatment options, such as hormone treatment, were jointly considered (not shown), there were almost no patients that had zero treatment. A higher percentage of participants came from neighborhoods with >15% people under the poverty level (24.4% vs. 13.3%), which is consistent with the EWM program objectives. A significantly higher percentage of women diagnosed through EWM were from rural counties than from urban metropolitan counties (71.9% vs. 29.9%).

After geographic matching, we found that while other characteristics remained broadly consis-tent with the above comparisons, those from neighborhoods with >15% people under the poverty level were similar between the participants and geographically matched nonparticipants. Also, the geographically matched nonparticipants had significantly higher mortality rates than participants.

Next, we examined potential treatment disparities (Table 19.2). Using resection surgery as the dependent variable, we found no significant differences between the participant and two nonpartici-pant groups in a logistic regression adjusting for age groups and tumor stage. When further restrict-ing the sample to stages 1 and 2 following Rajan et al. (2015), we found that EWM participants were significantly less likely to have a resection surgery, while adjusting for age and the two remaining stages. The odds ratios were 0.505 (95% CI 0.311–0.819) and 0.541 (95% CI 0.327–0.892), respec-tively, in contrast to all nonparticipants and the matched nonparticipants. However, if we compare the resection surgery rates for stages 1 and 2, the percentages were fairly close: 98.05%, 97.9%,

TABLE 19.1

Patient Characteristics by EWM Program Status: 1995–2008

	All Nonparticipants	Nonparticipants Matched to EWM Census Tract	EWM Program Participants
Total *N*	11264	6099	529
	86.8	88.7	
Age Group			
40–50	21.3	20.5	32.5
51–60	30.1	29.8	41.6
61–74	48.5	49.7	25.9
Married	68.8	67.9	51.8
Race			
NH White	94.7	94.6	86.6
NH Black	2.6	2.9	4.9
Other race	2.7	2.5	8.5
SEER Stage			
Localized	63.1	62.6	54.8
Regional	30.0	30.3	39.3
Distant	4.2	4.3	2.8
Upstaged	2.6	2.8	3.0
Treatment History			
Resection	94.5	93.9	93.8
RT	50.5	50.5	50.9
CT	50.6	50.7	59.9
NB Poverty Rate			
<5% poverty rate	28.3	18.2	10.4
5%–15% poverty rate	58.3	64.3	65.2
>15% poverty rate	13.3	17.5	24.4
Place of Residence			
Urban	48.4	38.3	29.9
Rural	51.6	61.7	70.1
1-year mortality	3.4	3.0	1.1
5-year mortality	9.7	10.8	6.8

Note: All in percent. NB, neighborhood; EWM, Every Woman Matters. Except RT and resection, all the *p*-values between participants and nonparticipants were significant at the 0.05 level for the chi-square test.

and 96.18%, respectively, for all nonparticipants, the matched nonparticipants, and the participants, even though they were statistically different according to the chi-squared test.

We found no difference in CT and RT in a set of logistic regressions. Using CT or RT as a dependent variable, we first ran a logistic regression and did not find a significant difference while controlling for age groups and stages. Second, we ran the regression model, restricting it to those who

TABLE 19.2

Treatment Difference: Matched and Unmatched Samples of EWM and Other Patients

Treatment	Model I: Unmatched		Model II: Geomatched	
	OR	95% CI	OR	95% CI
Full Sample				
Resection surgery	0.708	0.454–1.102	0.773	0.491–1.120
RT	0.916	0.767–1.095	0.902	0.752–1.083
CT	1.031	0.839–1.267	1.01	0.819–1.247
Stages 1 and 2 Sample Only				
Resection surgery	0.505	0.311–0.819	0.541	0.327–0.892
RT	0.936	0.779–1.124	0.919	0.762–1.109
CT	1.071	0.864–1.328	1.036	0.832–1.291

Note: Each treatment represents a separate age- and stage-adjusted model. Model I: All patients in Nebraska regardless of where they live. Model II: Patients who were geographically matched to EWM participant neighborhoods.

had resection surgeries, and still were unable to find any significant difference. Finally, we further restricted the sample to those who had breast conservative surgeries, which should be followed with RT. Again, we did not find any significant difference. We calculated RT rates by using those receiving breast conservative surgeries as the denominator. We found they were 84.92%, 84.29%, and 82.82%, respectively for all nonparticipants, the matched nonparticipants, and the participants, and they were not statistically different either. We therefore conclude there was no treatment difference between low-income EWM participants and nonparticipants in the current study.

Given that there were no differences in treatments, we would not expect much difference in mortality either. Table 19.3 shows otherwise. In both full and geographically matched samples, 3-year mortality differences between the participants and nonparticipants were significant. In the full sample, after controlling for age, race, cancer staging, treatment types, place of residence, and neighborhood poverty level, the EWM participants had lower odds of 3-year mortality than did non-EWM women (OR 0.633, 95% CI 0.437–0.917). In addition, effects of the control variables were consistent with the literature. Older and unmarried women were more likely to die within 3 years of the diagnosis. Tumor stage at the time of diagnosis was the most important predictor of 1-year mortality. Women diagnosed at a distant stage had a higher likelihood of 3-year mortality than women diagnosed at the local stage. Women receiving resection surgery and RT had significantly lowers odds of mortality than women receiving no corresponding treatment modalities. Place of residence was not significant. Finally, women residing in a census tract in the poverty-rate neighborhoods (>15%) were more likely to die within 3 years than women residing in areas with less than a 5% poverty rate. The above results were broadly consistent when the geographically matched sample was introduced.

Since the sample did not allow us to calculate 5- or 10-year mortality rates, we used the Cox survival model to estimate long-term survival effects between EWM participants and nonparticipants (Table 19.4). In both full and geographically matched models, there were no differences between participants and nonparticipants. In both models, all other control variables had effects similar to those of the logistic models. In addition, we also ran models by excluding noncancer causes of death, and the results were similar (not shown). We therefore concluded that there were no long-term survival benefits for EWM participants even though they had better odds of surviving in the short term, such as 1–3 years.

TABLE 19.3

Logistic Regression of Unmatched and Matched Samples of 3-Year Mortality Models

	Model I: Unmatched		Model II: Geomatched	
	OR	95% CI	OR	95% CI
Age (ref = 40–50 years)				
51–60	1.018	0.826–1.256	1.081	0.824–1.419
61–74	1.622	1.34–1.963	1.804	1.406–2.314
Marital Status (ref = Married)				
Unmarried	1.563	1.354–1.804	1.602	1.335–1.924
Race and Ethnicity (ref = NH White)				
NH Black	1.617	1.137–2.298	1.545	0.996–2.395
Other	1.054	0.727–1.529	1.386	0.882–2.181
Tumor Stage (ref = Local)				
Regional	3.033	2.565–3.587	2.969	2.399–3.674
Distant	19.201	14.877–24.781	17.745	12.685–24.825
Unstaged	3.064	2.205–4.257	2.764	1.805–4.233
Treatment (ref = no)				
Resection surgery	0.334	0.264–0.423	0.347	0.256–0.472
RT	0.661	0.573–0.764	0.663	0.552–0.797
CT	0.908	0.773–1.067	0.968	0.789–1.188
Place of Resident (ref = Rural)				
Urban	0.947	0.805–1.113	0.976	0.792–1.204
Neighborhood Poverty Rate (ref = <5%)				
5%–15% poverty rate	1.344	1.105–1.635	1.469	1.091–1.979
>15% poverty rate	1.723	1.363–2.179	1.978	1.43–2.737
EWM Program Status (ref = no)				
Participants	0.633	0.437–0.917	0.586	0.403–0.854

Note: Model I: All patients in Nebraska regardless of where they live. Model II: Patients who were geographically matched to EWM participant neighborhoods.

19.4   DISCUSSIONS AND CONCLUSIONS

The motivation of this chapter was twofold. On the one hand, lack of income measures is an important drawback for cancer registry data. Linking cancer screening program participants, who were all relatively low income and underinsured, provided an indirect way of measuring income effects. On the other hand, breast cancer screening programs funded by the CDC have been implemented for more than 20 years, but few studies have evaluated their mortality outcomes. Hence, the purpose of this study was to evaluate treatment and mortality under the premise that the same stage of cancer diagnosis should have the same treatment, and the same treatment should lead to similar mortality outcomes. Compared to all non-EWM participants, women diagnosed through EWM were more likely to survive within 3 years of diagnosis. However, this short-term survival advantage could be due to lead time bias of the screening, as it was a nonfactor in the survival models.

TABLE 19.4

Cox Proportional Hazard Models for Geographically Matched and Unmatched Samples

	Model I: Unmatched		Model II: Geomatched	
	HR*	**95% CI**	**OR**	**95% CI**
Age (ref = 40–50 years)				
51–60	1.164	1.043–1.301	1.203	1.042–1.388
61–74	1.866	1.689–2.062	1.946	1.709–2.216
Marital Status (ref = Married)				
Unmarried	1.449	1.351–1.554	1.489	1.36–1.63
Race and Ethnicity (ref = NH White)				
NH Black	1.471	1.226–1.765	1.438	1.147–1.804
Other	1.223	1.003–1.492	1.379	1.059–1.796
Tumor Stage (ref = Local)				
Regional	2.243	2.066–2.435	2.334	2.098–2.596
Distant	8.854	7.721–10.154	8.947	7.459–10.734
Unstaged	2.609	2.186–3.116	2.738	2.191–3.422
Treatment (ref = no)				
Resection surgery	0.532	0.468–0.604	0.55	0.466–0.649
RT	0.717	0.668–0.769	0.696	0.636–0.762
CT	0.928	0.857–1.004	0.934	0.845–1.034
Place of Resident (ref = Rural)				
Urban	1.074	0.992–1.163	1.103	0.993–1.226
Neighborhood Poverty Rate (ref = <5%)				
5%–15% poverty rate	1.249	1.136–1.374	1.175	1.021–1.353
>15% poverty rate	1.385	1.231–1.558	1.314	1.119–1.542
EWM Program Status (ref = no)				
Participants	1.031	0.87–1.222	0.983	0.827–1.168

Note: HR, Hazard ratio. Model I: All patients in Nebraska regardless of where they live. Model II: Patients who were geographically matched to EWM participant neighborhoods.

One may suspect that some of those protective effects were due to relatively younger age for EWM participants that cannot be fully accounted for by age group variables. However, the 40–65 age group generated identical results in terms of directional effects. Some short-term mortality effects could be explained by slightly higher proportions of distant stage for both nonparticipants' samples. However, if we consider both regional and distant as the late stage, then women diagnosed with breast cancer through EWM are significantly more likely than women diagnosed through other sources to be at the late stage, a result consistent with a similar study in Florida (Tamer et al., 2003). Several studies of disparities in breast cancer screening reported that women who are racial minorities, of low SES, and from rural areas, as well as lack transportation, are less likely to seek mammography and are more likely to be diagnosed at later stages of the disease and have higher mortality rates (Coughlin et al., 2008; Coughlin and King, 2010; Edwards et al., 2009). But EWM participants in our study were less likely to die in 3 years, especially in the geographically matched sample, suggesting that participants might have learned a few prevention strategies that otherwise might be missed by their pairs living in the same neighborhoods. Second, marital status is a strong predictor or benefit factor for surviving from cancer due to spousal support both emotionally and physically (Aizer et al., 2013), which is also consistent with most breast cancer studies on mortality and survival (Kroenke et al., 2006, 2012; Silliman et al., 1998). However, marital status would work against EWM participants, as only 52% of them married, compared to 68% for nonparticants. Hence, it must not be marital status that benefits EWM participants. Third, the non-Hispanic Black patients were more likely to die in both models, but EWM participants were more likely to be non-Hispanic Black. Hence, it cannot be the race factor benefiting the participants. Finally, the poorer the neighborhood in which the patient lives, the greater the likelihood of mortality in either the short or long term. The poverty factor would also work against the participants, as a disproportionate number of them live in poor neighborhoods, even for the geographically matched sample. Hence, it cannot be the neighborhood SES that benefits the participants. Thus, the protective effects of EWM in the short term are likely also contributed by some factors that are not observed from the existing control variables.

It is likely that the long-term survival models reflect effects that cannot be fully mitigated by the hospital-based health care system for breast cancer screening and treatment (Gerend and Pai, 2008; Aizer et al., 2013). Social determinants of health and psychosocial factors such as lack of financial resources, transportation, good nutrition, health literacy, problematic patient–provider interactions, and patient distrust of the system tend to have long-term effects. Factors that could not be measured from the cancer registry–based data are access to primary care and informal care (Roetzheim et al., 2012). Racial and ethnic minority women from minority groups and women residing in areas with a high poverty rate may not have access to primary care providers for follow-up care after breast cancer treatment. Recent literature showed that Medicare beneficiaries with breast cancer had better outcomes if they made greater use of a primary care physician's ambulatory services (Wirtz et al., 2014). The hidden cost of cancer care in terms of loss of income, other competing work and family demands, and transportation might be especially burdensome for low-income and racial and ethnic minority patients in long-term follow-up care for breast cancer (Shavers and Brown, 2002). Efforts should be taken to engage breast cancer survivors in managing long-term surveillance and follow-up care (Field et al., 2008).

Additionally, no difference was found between the participants and nonparticants in treatments with resection surgery, RT, and CT therapy. If we include only stages 1 and 2 patients, the participants would be less likely to have resection surgery. These results were consistent with previous studies that have shown disparities in treatment modalities based on the social class of patients (Boyer-Chammard et al., 1999; Schoen et al., 2011). All the NBCCEDP-funded state screening programs are required to have diagnostic and treatment networks in place before starting the screening program. In 2001, under the federal Medicaid mandate, a Nebraska law mandated that women diagnosed with breast cancer through EWM receive treatment through the state Medicaid program. The Patient Protection and Affordable Care Act will provide women with greater access to preventive cancer screening and treatment. However, a significant gap is likely to remain for women who are uninsured and underinsured.

From the perspective extending the EWM program's benefit, there are several implications. First, the weakening of the survival benefit of the participants suggests that more could be done for them. Clinicians could pass on prognosis information to the EWM program, so that the program can further communicate with the EWM participants in terms of long-term survival. If lack of informal support is an issue, the program could also potentially provide some informal care options based on some rating of variables other than clinical treatment. In this way, the program could provide more personalized survivor support beyond cancer screening.

The present study has a number of limitations that also provide opportunities for future studies. First, the participants and nonparticants received similar clinical treatments; we could not compare the quality and timeliness of care between the two groups because the date of diagnosis to treatment could not be estimated solely from the NCR data. Second, the geographic matching of participants and nonparticants was not based on the traditional case matching approach.

If we strictly followed the case matching approach, we would likely use the proportion of cases (participants) living in particular poverty categories and sampled proportionally to the nonparticipants. Alternatively, we could match participants by year in each neighborhood. In either case, it would reduce the statistical power. Future studies should attempt to disentangle the neighborhood disadvantages among geographically matched nonparticipants. If the survival benefits were also due to the clinically supported care network and case manager, then less expensive benefits can be extended to patients from poorer neighborhoods through health education and case management. Third, the small sample size in the current study not only limits the ability to detect significant mortality differences, but also prevents us from a more elaborate design due to the power concern. Additionally, some data may be missing due to problems in linkage of databases. Future studies should examine the long-term determinants of breast cancer survival for women diagnosed through an NBCCEDP screening program, including access to a social support system and the availability of primary care after initial diagnosis and treatment.

REFERENCES

Aizer, A.A., M.H. Chen, E.P. McCarthy, M.L. Mendu, S. Koo, T.J. Wilhite, P.L. Graham, et al. 2013. Marital status and survival in patients with cancer. *Journal of Clinical Oncology* 31(31): 3869–3876.

Bhuyan, S., J. Stimpson, S. Rajaram, and G. Lin. 2014. Mortality outcome among medically underserved women screened through a publicly funded breast cancer control program, 1997–2007. *Breast Cancer Research and Treatment* 146: 221–227.

Boyer-Chammard, A., T.H. Taylor, and H. Anton-Culver. 1999. Survival differences in breast cancer among racial/ethnic groups: A population-based study. *Cancer Detection and Prevention* 23(6): 463–473.

Coughlin, S.S. and J. King. 2010. Breast and cervical cancer screening among women in metropolitan areas of the United States by county-level commuting time to work and use of public transportation, 2004 and 2006. *BMC Public Health* 10: 146.

Coughlin, S.S., S. Leadbetter, T. Richards, and S.A. Sabatino. 2008. Contextual analysis of breast and cervical cancer screening and factors associated with health care access among United States women, 2002. *Social Science and Medicine* 66(2): 260–275.

Edwards, Q.T., A.X. Li, M.C. Pike, L.N. Kolonel, G. Ursin, B.E. Henderson, and R. McKean-Cowdin. 2009. Ethnic differences in the use of regular mammography: The multiethnic cohort. *Breast Cancer Research and Treatment* 115(1): 163–170.

Feresu, S.A., W. Zhang, S.E. Puumala, F. Ullrich, and J.R. Anderson. 2008. Breast and cervical cancer screening among low-income women in Nebraska: Findings from the Every Woman Matters program, 1993–2004. *Journal of Health Care for the Poor and Underserved* 19(3): 797–813.

Field, T.S., C. Doubeni, M.P. Fox, D.S. Buist, F. Wei, A.M. Geiger, V.P. Quinn, et al. 2008. Underutilization of surveillance mammography among older breast cancer survivors. *Journal of General Internal Medicine* 23(2): 158–163.

Gerend, M.A. and M. Pai. 2008. Social determinants of black-white disparities in breast cancer mortality: A review. *Cancer Epidemiology, Biomarkers and Prevention* 17: 2913–2923.

Johnson, C.J., R. Graff, P. Moran, C. Cariou, and S. Bordeaux. 2015. Breast cancer stage, surgery, and survival statistics for Idaho's National Breast and Cervical Cancer Early Detection Program Population, 2004–2012. *Preventing Chronic Disease* 12: E36.

Koh, H.K., C.M. Judge, B. Ferrer, and S.T. Gershman. 2005. Using public health data systems to understand and eliminate cancer disparities. *Cancer Causes and Control.* 16(1): 15–26.

Koroukian, S.M. 2003. Assessing the effectiveness of Medicaid in breast and cervical cancer prevention. *Journal of Public Health Management and Practice* 9(4): 306–314.

Kroenke, C.H., L.D. Kubzansky, E.S. Schernhammer, M.D. Holmes, and I. Kawachi. 2006. Social networks, social support, and survival after breast cancer diagnosis. *Journal of Clinical Oncology* 24(7): 1105–1111.

Kroenke, C.H., Y. Michael, H. Tindle, E. Gage, R. Chlebowski, L. Garcia, C. Messina, J.E. Manson, and B.J. Caan. 2012. Social networks, social support and burden in relationships, and mortality after breast cancer diagnosis. *Breast Cancer Research and Treatment* 133(1): 375–385.

Lantz, P.M. and S. Soliman. 2009. An evaluation of a Medicaid expansion for cancer care: The Breast and Cervical Cancer Prevention and Treatment Act of 2000. *Womens Health Issues* 19(4): 221–231.

Lin, S., S.L. Gomez, S.L. Glaser, L.A. McClure, S.J. Shema, M. Kealey, T.H. Keegan, and W.A. Satariano. 2011. The California Neighborhoods Data System: A new resource for examining the impact of neighborhood characteristics on cancer incidence and outcomes in populations. *Cancer Causes and Control* 22(4): 631–647. doi: 10.1007/s10552-011-9736-5.

Lincourt, A.E., R.F. Sing, K.W. Kercher, A. Stewart, B.L. Demeter, W.W. Hope, N.P. Lang, Greene, and B.T. Heniford. 2008. Association of demographic and treatment variables in long-term colon cancer survival. *Surgical Innovation* 15(1): 17–25.

Liu, M.J., H. Hawk, S.T. Gershman, S.M. Smith, R. Karacek, M.L. Woodford, and J.Z. Ayanian. 2005. The effects of a national breast and cervical cancer early detection program on social disparities in breast cancer diagnosis and treatment in Massachusetts. *Cancer Causes and Control* 16(1): 27–33.

Lobb, R., J.Z. Ayanian, J.D. Allen, and K.M. Emmons. 2010. Stage of breast cancer at diagnosis among low-income women with access to mammography. *Cancer* 116(23): 5487–5496.

Mallin, K., B.E. Palis, N. Watroba, A.K. Stewart, D. Walczak, J. Singer, J. Barron, W. Blumenthal, G. Haydu, and S.B. Edge. 2013. Completeness of American Cancer Registry treatment data: Implications for quality of care research. *Journal of the American College of Surgeons* 216(3): 428–437.

McCarthy, A.M., T. Dumanovsky, K. Visvanathan, A.R. Kahn, and M.J. Schymura. 2010. Racial/ethnic and socioeconomic disparities in mortality among women diagnosed with cervical cancer in New York City, 1995–2006. *Cancer Causes and Control* 21(10): 1645–1655.

Morgan, J.W., M.M. Cho, C.D. Guenzi, C. Jackson, A. Mathur, Z. Natto, K. Kazanjian, H. Tran, D. Shavlik, and S.S. Lum. 2011. Predictors of delayed-stage colorectal cancer: Are we neglecting critical demographic information? *Annals of Epidemiology* 21(12): 914–921.

Rajan, S.S., C.E. Begley, and B. Kim. 2014. Breast cancer stage at diagnosis among medically underserved women screened through the Texas Breast and Cervical Cancer Services. *Population Health Management* 17(4): 202–210.

Richardson, L.C., J. Royalty, W. Howe, W. Helsel, W. Kammerer, and V.B. Benard. 2010. Timeliness of breast cancer diagnosis and initiation of treatment in the national breast and cervical cancer early detection program, 1996–2005. *American Journal of Public Health* 100(9): 1769–1776.

Richardson, L.C., J. Schulman, L.E. Sever, N.C. Lee, and R.J. Coate. 2001. Early-stage breast cancer treatment among medically underserved women diagnosed in a national screening program, 1992–1995. *Breast Cancer Research and Treatment* 69(2): 133–142.

Roetzheim, R.G., J.M. Ferrante, J.H. Lee, R. Chen, K.M. Love-Jackson, E.C. Gonzalez, K.J. Fisher, and E.P. McCarthy. 2012. Influence of primary care on breast cancer outcomes among Medicare beneficiaries. *Annals of Family Medicine* 10(5): 401–411.

Schoen, C., M.M. Doty, R.H. Robertson, and S.R. Collins. 2011. Affordable care act reforms could reduce the number of underinsured US adults by 70 percent. *Health Affairs* (Millwood). 30(9): 1762–1771.

Schrag, D., B.A. Virnig, and J.L. Warren. 2009. Linking tumor registry and Medicaid claims to evaluate cancer care delivery health care. Financing Review 30(4): 61–73.

Shavers, V.L. and M.L. Brown. 2002. Racial and ethnic disparities in the receipt of cancer treatment. *Journal of the National Cancer Institute* 94(5): 334–357.

Silliman, R.A., K.A. Dukes, L.M. Sullivan, and S.H. Kaplan. 1998. Breast cancer care in older women: Sources of information, social support, and emotional health outcomes. *Cancer* 83: 706–711.

Sprague, B.L., A. Trentham-Dietz, R.E. Gangnon, R. Ramchandani, J.M. Hampton, S.A. Robert, P.L. Remington, and P.A. Newcomb. 2011. Socioeconomic status and survival after an invasive breast cancer diagnosis. *Cancer* 117(7): 1542–1551.

Tamer, R., L. Voti, L.E. Fleming, J. MacKinnon, D. Thompson, M. Blake, J.A. Bean, and L.C. Richardson. 2003. A feasibility study of the evaluation of the Florida breast cancer early detection program using the statewide cancer registry. *Breast Cancer Research and Treatment* 81(3): 187–194.

Wheeler, S.B., K.E. Reeder-Hayes, and L.A. Carey. 2013. Disparities in breast cancer treatment and outcomes: Biological, social, and health system determinants and opportunities for research. *Oncologist* 18(9): 986–993.

White, M.C. and F.L. Wong. 2015. Preventing premature deaths from breast and cervical cancer among underserved women in the United States: Insights gained from a national cancer screening program. *Cancer Causes and Control* 26(5): 805–809.

Wirtz, H.S., D.M. Boudreau, J.R. Gralow, W.E. Barlow, S. Gray, E.J. Bowles, and D.S. Buist. 2014. Factors associated with long-term adherence to annual surveillance mammography among breast cancer survivors. *Breast Cancer Research and Treatment* 143(3): 541–550.

Yin, D., C. Morris, M. Allen, R. Cress, J. Bates, and L. Liu. 2010. Does socioeconomic disparity in cancer incidence vary across racial/ethnic groups? *Cancer Causes and Control* 21(10): 1721–1730.

20 Linking Environmental Variables to Parkinson's Disease

20.1 INTRODUCTION

Funded by the Centers for Disease Control and Prevention (CDC), environmental tracking systems have been established in many states. Many excellent examples, case studies, and national reports have come out of the systems that track known pollutants and environment hazards, and their potential impacts on health outcomes. In many situations, however, we do not know the potential impacts of some environmental factors on the incidence of a disease. In such situations, it might be appropriate to use an exploratory spatial surveillance approach, the topic of this chapter.

Most spatial disease surveillance practices deal with disease cluster detection. In the traditional disease cluster detection approach, one first detects the existence of spatial clusters. Once a disease cluster is detected, one needs to trace etiological factors that contribute to the cluster. A common approach to identifying etiology is to compare a set of ecological or environmental factors within and outside the clustered area and determine which factors have strong correlations with the clustered areas. Alternatively, one might take a nonspatial approach by exploring correlations between disease incidence and environmental exposures over the entire study area. If an exposure is associated with an observed disease pattern, one could go a step further by investigating potential relationships between exposure and health effect. Both approaches assume that the detected health effect will reflect some dose–response relationship between environmental exposure and population health. In this chapter, we demonstrate both approaches by correlating Parkinson's disease (PD) data and pesticide and herbicide data in Nebraska.

PD is caused by the deterioration of nerve cells in a particular area of the brain, but its etiology remains largely unknown. Some evidence suggests a genetic component, but most PD patients do not have the identified gene (Ross and Smith, 2007). Studies also point to environmental factors that may generate toxins to the brain. However, specific and consistent environmental risk factors have not been identified. For instance, both pesticide and herbicide exposures have been associated with the onset (Butterfield et al., 1993) or higher rates of PD, but a metadata analysis found that those same studies also identified rural living and farming as significant exposures (Priyadarshi et al., 2001). In the meantime, many studies failed to link PD to pesticides and rural living (Kuopio et al., 1999). Despite inconclusive findings, many studies of environmental factors on PD have emerged recently. For example, more than a dozen studies since 2001 have found significant relationships between pesticides and PD (see a review study by Van der Mark et al., 2012), although some studies also failed to detect significant relationships (Firestone et al., 2005, 2010). Ascherio et al. (2006) showed a significant relationship between pesticide exposures and PD. In a more focused study sample, Kamel et al. (2007) showed that licensed private pesticide applicators and spouses were more likely to develop PD due to their overall exposures to pesticides. However, since both studies were survey-based, they all call for further studies to identify specific pesticides or chemicals contributing to PD risk. Besides pesticide studies, there have been more than a dozen studies showing a general relationship between herbicides and PD risk (Ryu et al., 2013; León-Verastegui, 2012; Fitzmaurice et al., 2013; Pan-Montojo et al., 2012) and identifying some specific herbicides, such as paraquat (Lee et al., 2012; Costello et al., 2009; Tanner et al., 2011).

Nebraska is the only state that has a statewide and population-based PD registry. Based on the 1996 Nebraska state statute, the Nebraska Department of Health and Human Services (DHHS) established the Nebraska Parkinson's Disease Registry (NPDR) in 1997. The statute requires that pharmacies and physicians report PD-related information directly to the NPDR coordinator. Pharmacies report information semiannually about patients who received a PD prescription. Physicians are required to report information about patients who are newly diagnosed with PD within 60 days of diagnosis. If cases from a pharmacy report cannot be found in the existing physician reports, the NPDR coordinator contacts the prescribing physicians to ascertain probable cases. Patients can also submit a PD information form, which includes physician and pharmacy contact information for verification. PD cases are also identified for inclusion in the registry using state death certificate information. If a person's cause of death indicates PD and the person was not in the registry, he or she could be included. About 8.7% of the cases are from death certificates, with a disproportionately greater number of cases in the beginning years of the registry development.

The data quality of NPDR is not ideal. Bertoni et al. (2006) reported that 78% of PD cases from 1997 to 2001 had clinically confirmed PD, and most false PD cases had drug-induced parkinsonism. In addition, there was no race information. Data after 2006 are likely better, with increased use of neurologists for ascertaining cases. Nevertheless, the NPDR is the only statewide registry, and it presents an opportunity to conduct descriptive epidemiology (Strickland and Bertoni, 2004) and exposure analysis. In the following, we first briefly describe how to derive pesticide exposure data. We then present a cluster detection approach that links a detected cluster with pesticide exposures, or vice versa, following an ecological study design. Next, we present a case-control study using PD as cases and multiple sclerosis (MS), Alzheimer's disease, ischemic stroke, and diabetes as separate control groups. Finally, we offer some concluding remarks.

20.2   ENVIRONMENTAL AND DISEASE DATA PROCESSING

Remote sensing–based land use data and pesticide usage data were used to develop pesticide and herbicide exposure maps. The 2005 Nebraska land use data were downloaded from the Center for Advanced Land Management Information Technologies (CALMIT) at the University of Nebraska–Lincoln. This dataset, originally developed to assist in the works of agricultural irrigation and soil protection, was based on Landsat 5 Thematic Mapper (TM) imageries.

Following the surveillance approach that attempts to assess a host of pesticides and herbicides (Elbaz et al., 2009), we obtained pesticide and herbicide usage data from Wan (2015), who developed exposures in several steps. First, information about county-level usage of active pesticide ingredients in Nebraska in 2005 was derived from the estimated annual agricultural pesticide use database of the U.S. Geological Survey. This dataset was estimated by integrating the pesticide usage survey data from proprietary crop reporting districts (CRDs), as well as county-level harvested-crop acreage information. The dataset provides two indicators of estimated pesticide usage (EPest): EPest-low and EPest-high, among which EPest-low assigned zero value to CRDs that were surveyed but did not provide any feedback of pesticide usage information, and EPest-high estimated values for those CRDs based on the value of their neighbors. We adopted EPest-high because it covers more counties than EPest-low. In total, exposures at 1×1 kilometer resolution were derived for 15 herbicides and 5 insecticides. The 15 herbicides were 2,4-dichlorophenoxyacetic acid (2,4-D), acetochlor, alachlor, atrazine, bentazone, bromoxynil, butylate, dicamba, S-ethyl-N,N-dipropylthiocarbamate (EPTC), ethalfluralin, glyphosate, metolachlor, metribuzin, paraquat, and pendimethalin. The five insecticides were carbaryl, carbofuran, chlorpyrifos, phorate, and terbufos. Exposures were aggregated at the census tract level from 1×1 kilometer grid maps.

Cluster detection is a critical component of spatial disease surveillance. We used the Poisson spatial association model or the residual Moran's I method (Lin and Zhang, 2007). The spatial association is determined by the first-order spatial adjacency. If we relax this constraint, the method is equivalent to SatScan, a popular method for cluster detection. PD data were processed by the Joint Data Center. The following steps were taken to generate the data request:

1. In-state records were extracted from the NPDR for 2007–2011. Only records with a valid diagnosis date were extracted.
2. All records were geocoded to the census tract level, and they were linked to environmental exposure data at the census tract level.
3. Case-specific data from 2007 to 2011 were extracted from the enhanced Nebraska hospital discharge data (HDD) data that had the census tract information. Nine diseases were included: encephalitis, MS, Alzheimer's, Huntington's, ischemic stroke, pneumonia, diabetes, influenza, and PD. PD was excluded as a comorbidity condition. The Nebraska HDD data were then linked to environmental exposure data at the census tract level.
4. A case-specific file was generated by removing census tract identifiers and other identifiable information. Only 10-year age groups, sex comorbidity indicators, and exposures were included.
5. A county frequency count file was generated with only age group and county identifiers included. The age groups included were <35, 35–44, and every 10 years until 85 years old or over. There were 2075 PD cases from 2007 to 2011; among them, only 9 cases were in the 35–44 age group. If we used all of the population age 35+ years, then the crude rate would be 230.18 per 100,000. However, we decided to use all age 45+ years as the at-risk population. The crude rate was 314.81 per 100,000.

20.3 CLUSTER DETECTION AND EXPOSURE COMPARISON

20.3.1 GENERAL CLUSTER DETECTION

The Poisson spatial association model uses a window of adjacency matrix to systematically search for a spatial cluster. Since each county is centered just once, there are 93 likelihood ratio tests in the first round, and the task is to find the largest and most significant one. Once the first cluster is detected, it likely reduces residual Moran's I, and the next round of iteration is to find the second cluster. The process goes on until there is no significant residual Moran's I for autocorrelation. For simplicity and for comparing with the SatScan test, we only detected the cluster with the largest log-likelihood ratio test in the first round. The data were divided into four age groups (35–59, 60–69, 70–79, and 80+). The corresponding county-level at-risk populations were based on the 2009 population estimates. Note that four age groups by 93 counties equal 372 cells for incidence data. Due to data sparseness, there were 124 cells that had 0 cases, and 159 cells had less than 5 cases. The two together accounted for 76% of the county–age group combinations. We added 0.01 to each cell so that there would be no logs of 0. The cluster detection result may not be as robust as when most cells are populated with at least five incidence counts.

Figure 20.1 displays the PD distribution based on the crude PD incidence rate. Preliminary analysis found a single-region outlier of Lancaster County, which is an urban county. After removing this outlier, we found that the most significant cluster was centered at Colfax County, of nine counties included (light blue in the figure). If one inspects the map, the clustered area is not the darkest area or the area with highest crude rate. Nevertheless, the likelihood ratio test showed 49 deviance G^2 with one degree of freedom, which was highly significant. The odds ratio was 1.559 (95% confidence interval [CI] 1.3778–1.764). The number of cases within the cluster was 304 out of the total 2075 cases. The corresponding at-risk population within the cluster was 66,451 out of the total 659,034 population, resulting in a relative risk of 1.453 = ([304/66,451]/[2,075/659,034]).

Since the spatial association test is based on a regular window, we also ran SatScan with irregular shape cluster detection (see the detailed results in Appendix 20A). The results were broadly

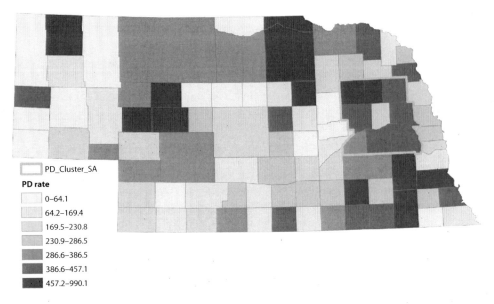

FIGURE 20.1 Crude PD rate per 100,000 by county in Nebraska: 2007–2011. PD_Cluster_SA is the most important cluster identified by the Poisson spatial association model. At-risk population: The 2009 population estimates age 35+ years from the Bureau of the Census. Cases: Nebraska Parkinson's Disease Registry, 2007–2011. Counties within the cluster: Butler, Cuming, Colfax, Dodge, Madison, Platte, Polk, Saunders, and Stanton, with the center being Colfax.

consistent. First, a Lincoln County outlier was detected by SatScan. Second, both SatScan and the spatial association test identified Cuming, Dodge, Madison, and Stanton Counties in the clustered region. Third, common pesticides and herbicides were also identified, which is discussed next.

Next, we compared 20 pesticides and herbicides within and outside of clustered areas using census tract as the unit for exposure. There were 43 census tracts within the clustered area and 489 outside the area. Three pesticides (carbaryl, carbofuran, and chlorpyrifos) and eight herbicides (including acetochlor, atrazine, butylate, paraquat, pendimethalin) were significantly different between the two groups (Table 20.1). All of the pesticides and herbicides had higher concentrations within the clustered area. Again, we have to be cautious not to interpret at this point, because environmental exposure data have some uncertainties too. For instance, atrazine was a significant factor. However, if one looks at the atrazine concentration map (Figure 20.2), the heavy use area is not in the clustered area. In fact, the high-usage areas are located in the middle and eastern regions of the state.

20.3.2 Focused Association Test

Most often, a focused test is aimed at a point source of pollution or exposure to detect a cluster surrounding it. However, one can conduct a more focused association test when the exposure area is known. For example, paraquat has been linked to PD in several studies. A focused association test would first identify paraquat clusters in the study area and then compare the PD cases within and outside of the clustered area. Figure 20.3 displays paraquat usage as exposure in 10 categories using the natural break classification from ArcGIS. The circled area in light blue is a paraquat cluster according to the local indicators of spatial association (LISA) test adjusted for multiple testings. Since the LISA test also uses spatial neighbors, we would have nine neighbors at most. We used the Bonferroni adjustment method, which multiplies the p-values by 9; if a p-value was <0.05, then it was significant. The resultant paraquat cluster includes 16 counties. The mean usage within the cluster was 1.78 (95% CI 1.49–2.07), in contrast to 0.46 (95% CI 0.38–0.54) outside of the cluster. The likelihood ratio test was 10 G^2 with one degree of

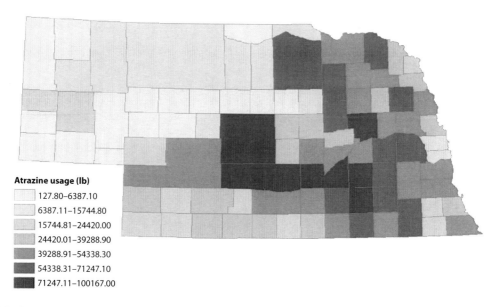

FIGURE 20.2 County-level atrazine usage in Nebraska: 2005.

TABLE 20.1
Pesticide and Herbicide Exposures within and outside of the PD Cluster

Exposure	Within Cluster Mean	Within Cluster 95% CI	Outside of Cluster Mean	Outside of Cluster 95% CI
		Pesticides		
Carbaryl*	0.88	0.63–1.14	0.17	0.13–0.21
Carbofuran*	0.37	0.21–0.52	0.17	0.14–0.20
Chlorpyrifos*	1.10	0.82–1.38	0.45	0.38–0.51
Phorate	0.51	0.35–0.67	0.29	0.16–0.43
Terbufos	0.06	0.05–0.08	0.12	0.08–0.15
		Herbicides		
24D	3.63	2.78–4.49	2.41	1.94–2.88
Acetochlor*	46.17	35.14–57.20	11.54	9.46–13.61
Alachlor	2.21	1.70–2.71	1.68	1.40–1.96
Atrazine*	52.54	40.09–65.00	21.54	18.24–24.84
Bentazone*	0.50	0.38–0.62	0.26	0.16–0.35
Bromoxynil*	0.29	0.21–0.37	0.14	0.12–0.17
Butylate*	3.53	2.69–4.37	0.69	0.53–0.85
Dicamba	0.56	0.43–0.70	0.48	0.38–0.58
EPTC	1.67	1.27–2.07	0.97	0.61–1.32
Ethalfluralin	0.03	0.01–0.04	0.03	0.01–0.06
Glyphosate*	70.44	54.48–86.40	29.38	25.28–33.47
Metolachlor	1.37	1.05–1.68	1.16	0.87–1.45
Metribuzin	0.49	0.34–0.64	0.34	0.27–0.42
Paraquat*	1.61	1.24–1.98	0.54	0.45–0.62
Pendimethalin*	5.18	3.39–6.97	1.68	1.32–2.03

Note: The exposures were based on census tract–level usage.

*Significant difference between the level inside and outside of the cluster.

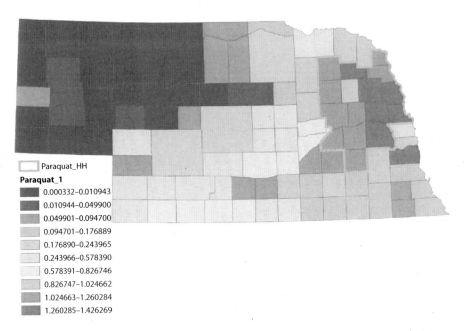

FIGURE 20.3   **(See color insert)** Paraquat usage in kilograms by census tract in Nebraska: 2005. Paraquat_HH is a high-value cluster based on the LISA adjusted for multiple testings. (Paraquat usage concentrations are from Wan, N., *Applied Geography* 56: 99–106, 2015.)

freedom, suggesting that the effect was much weaker than that detected earlier for the PD cluster, and the odds ratio showed a weak but significant association (odds ratio [OR] 1.203, 95% CI 1.078–1.341). Those residing inside of the cluster were 20% more likely to have PD than those living outside of the cluster.

20.4 USING CASE-CONTROL FOR EXPOSURE SURVEILLANCE

In this section, we use a case-control design to explore the relationship between the geographic distribution of PD and agriculture chemical exposures. Although long-term exposure should be used in the case-control study (Baldi et al., 2003; Dick et al., 2007), increased PD risk was also associated with a previous year of exposure to organic pesticides such as rotenone (Dhillon et al., 2008). Following this approach, we left a 2-year lag by using 2007–2011 PD cases from the NPDR. However, we did not have controls readily available. Ideally, in spatial matching, we can identify a control subject by looking for a PD patient's neighbor who does not have PD. However, we do not have access to such information from the general population. We decided to use HDD data because they are large enough and have corresponding years of data, and we can restrict controls to patients who had no PD. The next question is, which disease can be used as the control? The idea is to find a number of diseases similar to PD (so that PD and controls may come from the same patient population). Although we were unable to find controls pertinent to PD in studies using clinical databases, Schull et al. (2005) used psychiatric conditions (dementia, psychoses, schizophrenia, bipolar disorders, and paranoid disorders) as a control to examine emergency department (ED) visits due to major influenza-related conditions using ED claims data. In their study, the psychiatric group was not known to be related to influenza infection. We took a slightly different approach. We attempted to find diseases that are somewhat related to PDs, but are known to be not related to agricultural chemicals. In the absence of the empirical study, we relied on the work by Kukull and Bowen (2002) in the *Oxford Textbook of Public Health*. In that chapter, Alzheimer's disease and ischemic stroke are focal neurological diseases, whereas MS, PD, and diabetes are peripheral neuropathies. We selected four diseases as controls: MS, Alzheimer's disease, ischemic stroke, and diabetes. They were all from the 2007–2011 hospital discharge data. International Classification of Diseases, Ninth Revision (ICD-9) code 340 was used to select MS; code 3310 was used for Alzheimer's disease. For ischemic stroke, we used the codes of 43491 and 43820. For diabetes, we used a set of codes for diabetes mellitus with complications (24901, 24910, 24911, 24920, 24921, 24930, 24931, 24940, 24941, 24950, 24951, 24960, 24961, 24970, 24971, 24980, 24981, 24990, 24991, 25002, 25003, 25010, 25011, 25012, 25013, 25020, 25021, 25022, 25023, 25030, 25031, 25032, 25033, 25040, 25041, 25042, 25043, 25050, 25051, 25052, 25053, 25060, 25061, 25062, 25063, 25070, 25071, 25072, 25073, 25080, 25081, 25082, 25083, 25090, 25091, 25092, and 25093).

Controls were selected from unique patients from the HDD who had a selected comorbidity condition, but did not have PD as a diagnosis for up to 10 diagnosis codes. For example, if we want to have MS as the control, we would select all the MS patients who did not have PD as one of the comorbidities. Through the exclusion criteria, we found that the HDD data had 3089 PD cases based on the ICD-9 diagnosis code of 332, which could also be used as cases. For this reason, we ran separate models using these PD cases from the Nebraska HDD as cases while using the same control variables. Since NPDR does not have race and ethnicity variables, we could only use age and sex (Table 20.2) to adjust the odds ratio. The tabulations show that all controls were quite different in the age and sex distributions. PD cases were older and had more males in proportions.

We used a logistic regression in the case-control study. Due to a relatively high correlation among environmental exposures, we set correlation coefficients to be less than 0.6 in a backward selection algorithm in SAS 10.2. It turned out that there was no significant environmental exposure for MS, and we therefore only included Alzheimer's disease, ischemic stroke, and diabetes in the results (Table 20.3).

When the cases were from NPDR (Table 20.3, first two columns) and the control was PD–Alzheimer's disease, a unit increase in terbufos pesticide was associated with a nearly 52% increase in having PD versus having Alzheimer's disease. In contrast, dicamba was associated with a reduced effect on PD, which does not mean dicamba had a protective effect. Since the geographic distribution of dicamba depends on crops grown in the field, this result could only mean that areas with less use of dicamba had fewer PD cases. When the control was ischemic stroke, four herbicides were detected to have some associations.

When the cases were from the Nebraska HDD (last two columns of Table 20.3), acetochlor was associated with a 2.3 % increase in PD incidence, in contrast to Alzheimer's disease. When the control was diabetes, acetochlor and ethalfluralin were associated with increased odds of

TABLE 20.2

Age and Sex Distribution of Cases and Controls

	PD-NPDR	PD-HDD	MS	Alzheimer's	Ischemic Stroke	Diabetes
Total *N*	2,075	3,889	1,868	6,380	8,421	14,129
			Age Groups			
<45	0.48	0.36	26.23	0.03	4.69	19.59
45–54	4	1.9	23.55	0.44	7.75	16.13
55–64	14.41	6.97	26.98	1.83	13.85	20.87
65–74	25.06	17.43	15.2	9.01	19.58	18.33
75–84	41.98	41.91	6.75	39.15	28.77	17.06
85+	14.07	31.42	1.28	49.53	25.35	8.01
Males	55.28	53.07	23.93	34.08	47.42	52.18

Note: PD-NPDR, PD cases from the PD registry; PD-HDD, PD cases from the HDD.

TABLE 20.3

Pesticide and Herbicide Exposures Detected from the Case-Control Design for PD

	Cases from NPDR		Cases from Nebraska HDD
	Alzheimer's Disease		
Dicamba	0.831 (0.758, 0.911)	Acetochlor	1.023 (1.01, 1.036)
Terbufos	1.515 (1.175, 1.953)		
	Ischemic Stroke		
Acetochlor	1.018 (1.001, 1.035)	Pendimethalin	0.989 (0.98, 0.997)
Bentazone	1.238 (1.135, 1.35)		
Dicamba	0.808 (0.735, 0.888)		
Ethalfluralin	1.299 (1.075, 1.571)		
	Diabetes		
Acetochlor	1.02 (1.006, 1.035)	Acetochlor	1.027 (1.014, 1.039)
Ethalfluralin	1.291 (1.078, 1.547)	Ethalfluralin	1.192 (1.015, 1.399)

Note: Each disease represents each control category, and it is estimated separately. All results are age and sex adjusted. All exposures are based on 2005 data; all cases and controls were based on 2007–2011 data.

having PD. When the control was ischemic stroke, pendimethalin was associated with a minor reduction in PD.

When both NPDR and Nebraska HDD cases were considered, we found that acetochlor was associated about a 2% increase in PD in four out of six case-control combinations, while ethalfluralin was significant for three out of six case-control combinations. It should be pointed out that our surveillance used the backward selection method to find the most significant pesticides and herbicides. Another way to conduct surveillance is to go through each pesticide and herbicide one at a time to identify a significant single agent separately in the logistic model. For example, we found that paraquat was significantly associated with a small increase in PD incidence (OR 1.063, 95% CI 1.023–1.106) in the PD–diabetes case-control design, but not in any other case-control designs.

20.5   CONCLUSION AND DISCUSSION

In this chapter, we touched upon only the tip of the iceberg of environmental exposure and spatial surveillance. We used PD because of the unique dataset collected by the NPDR. Recognizing data quality issues, we attempted to methodologically associate PD with specific pesticides and herbicide exposures at the census tract level. It is possible to do so at a finer geographic scale, but that would require a more detailed analytical study after surveillance results.

Given the methodological emphasis, we provided both cluster detection and case-control approaches to linking environmental exposures. A general cluster detection approach is to detect spatial clusters first, and then compare exposures within and outside the cluster. Using this approach, we found a PD cluster in the northeast part of Nebraska, where a set of agricultural chemicals can be differentiated. A more focused cluster detection was to identify a cluster of the known risk factor, such as paraquat, and then associate it with the incidence. Using the focused test, we found a significant spatial cluster of paraquat in northeast Nebraska. We also learned that those residing inside of the cluster were 20% more likely to have PD than those living outside of the cluster.

The case-control design is nonspatial. It assesses if cases and controls could be differentiated by exposures that were observed in the past. We used the 2005 exposure data and 2007–2011 case-control data. Furthermore, due to the lack of standard controls in a population, we tapped HDD data for control cases in MS, Alzheimer's disease, ischemic stroke, and diabetes. Both acetochlor and ethalfluralin were associated with increased PD risk, and they were significant in most cast-control pairs. Interestingly, no significant exposures were detected for the PD-MS case-control pairs. In a single-agent exposure surveillance, we found that paraquat is weakly associated with PD incidence.

One feature of this current study was to borrow expertise from the outside. As part of public health data infrastructure building, we wanted to generate fine-scale agriculture exposures across the entire state. Since remote sensing and spatial data integration are expertise not typically presented within the public health agency, it was necessary to collaborate with outside experts. In total, 20 exposure maps were generated on 1×1 kilometer grids, but we could have done a lot more. Examples include arsenic in water, lead poisoning in the built environment, radon, and radioactive materials typically regulated by the environmental health division within public health. In addition, more environmental exposure maps, such as agricultural pharmaceuticals, could be generated from the infrastructural building approach.

A major limitation of the current study is that environmental exposures of PD patients are unlikely to be acute and short term. In other words, exposures to PD could be accumulated over a long period of time, and using 1-year pesticide and herbicide exposure data requires a lot of implicit assumptions: the 2005 exposures represent at least a few years before and a few years after 2005. The PD registry patients stayed in the current address during the exposure period. Data generated by the data analyst from the Nebraska DHHS had high quality in terms of diagnosis dates and location variables, but they still suffered the problem of false positives. Given these warnings, we made little attempt to interpret the detected geographic cluster and pesticides and herbicides. Methodologically, however, we could explore further. For instance, we found that acetochlor appeared in cluster detection and case-control analyses. According to the CDC biomonitoring program, acetochlor is a chloroacetanilide-type herbicide with restricted usage for preemergent control of grasses and broadleaf weeds on agricultural cropland. Acetochlor has low acute toxicity, but its human health effects at low environmental doses from low environmental exposures are unknown. Acetochlor has been associated with testicular atrophy, renal injury, and neurological movement abnormalities (USEPA, 2006). The U.S. Environmental Protection Agency (USEPA) considers acetochlor likely to be carcinogenic in humans, but no ratings have been assigned with regard to human carcinogenicity. It seems that it might be worthwhile to further look into the relationship between PD and acetochlor exposure.

APPENDIX 20A: RESULTS FROM SATSCAN TEST AND ASSOCIATED PESTICIDES AND HERBICIDES WITHIN AND OUTSIDE OF THE CLUSTER

The results from cluster detection found a single-region outlier of Lancaster County. In addition, it identified two connected clusters using the elliptical shape option in SatScan. We used county boundary rather than the elliptical option to depict the clustered area. Since the SatScan separately reported two adjacent clusters, we connected them into one (light blue in Figure 20.3), and then ran a log-rate model to evaluate it jointly using the likelihood ratio test. The result showed the adjusted odds ratio of 1.698 for the cluster effect; with one degree of freedom, the area accounted for 56 deviance, which was highly significant. If one looks at the shape of the cluster, it goes along the Elkhorn River basin, which is one of the most intensely farmed areas in the state. The six counties in the clustered area are Holt, Madison, Stanton, Antelope, Cuming, and Dodge. The numbers of cases within the cluster was 229, out of the total 2075 cases. The corresponding at-risk population within the cluster was 43,864 out of the 659,034 total population, resulting in a relative risk of 1.658 ([229/43,864]/[2,075/659,034]). Note that relative risk is most often used in spatial cluster detection,

TABLE 20.A1

Pesticide and Herbicide Exposures within and outside of the SatScan Cluster

Exposure	Within Cluster			Outside of Cluster		
	Mean	95% CI L	95% CI U	Mean	95% CI L	95% CI U
			Pesticides			
Carbaryl*	0.61	0.309	0.906	0.20	0.160	0.245
Carbofuran*	0.48	0.241	0.715	0.17	0.139	0.200
Chlorpyrifos	0.84	0.486	1.200	0.48	0.412	0.545
Phorate	0.37	0.182	0.554	0.31	0.174	0.444
Terbufos	0.08	0.023	0.129	0.11	0.080	0.146
			Herbicides			
24D	3.42	2.001	4.849	2.45	1.996	2.908
Acetochlor*	35.39	20.835	49.955	13.08	10.899	15.255
Alachlor	1.90	1.223	2.577	1.71	1.438	1.982
Atrazine*	44.07	26.135	62.014	22.85	19.566	26.131
Bentazone	0.51	0.254	0.758	0.26	0.174	0.353
Bromoxynil	0.17	0.079	0.268	0.15	0.126	0.181
Butylate*	2.51	1.419	3.593	0.83	0.658	0.995
Dicamba	0.61	0.315	0.914	0.48	0.382	0.578
EPTC	1.68	0.848	2.508	0.98	0.641	1.328
Ethalfluralin	0.26	0.005	0.522	0.02	0.005	0.033
Glyphosate*	56.97	36.181	77.765	31.25	27.122	35.370
Metolachlor	0.96	0.566	1.361	1.19	0.911	1.469
Metribuzin	0.32	0.180	0.455	0.36	0.286	0.432
Paraquat*	1.27	0.826	1.707	0.58	0.500	0.668
Pendimethalin*	6.72	3.710	9.726	1.68	1.347	2.009

Note: The exposures were based on census tract–level usage.

*Significant difference between the level inside and outside of the cluster.

but the rate ratio inside and outside of the cluster is also used in place of relative risk. The rate ratio for the current cluster is 1.740 ([229/43,864]/[1,846/615,171]), which should always be greater than the relative risk in the presence of a high-value cluster. Mean exposures within and outside of the SatScan-detected PD cluster are listed in Table 20.A1.

REFERENCES

Ascherio, A., H. Chen, M.G. Weisskopf, E. O'Reilly, M.L. McCullough, E.E. Calle, M.A. Schwarzschild, and M.J. Thun. 2006. Pesticide exposure and risk for Parkinson's disease. *Annals of Neurology* 60(2): 197–203.

Baldi, I., A. Cantagrel, P. Lebailly, F. Tison, B. Dubroca, V. Chrysostome, J.F. Dartigues, and P. Brochard. 2003. Association between Parkinson's disease and exposure to pesticides in southwestern France. *Neuroepidemiology* 22(5): 305–310.

Bertoni, J.M., P.M. Sprenkle, D. Strickland, and N. Noedel. 2006. Evaluation of Parkinson's disease in entrants on the Nebraska State Parkinson's Disease Registry. *Movement Disorders* 21(10): 1623–1626.

Butterfield, P.G., B.G. Valanis, and P.S. Spencer. 1993. Environmental antecedents of young onset Parkinson's disease. *Neurology* 43: 1150–1158.

Costello, S., M. Cockburn, J. Bronstein, X. Zhang, and B. Ritz. 2009. Parkinson's disease and residential exposure to maneb and paraquat from agricultural applications in the central valley of California. *American Journal of Epidemiology* 169(8): 919–926.

Dhillon, A.S., G.L. Tarbutton, J.L. Levin, G.M. Plotkin, L.K. Lowry, J.T. Nalbone, and S. Shepherd. 2008. Pesticide/environmental exposures and Parkinson's disease in East Texas. *Journal of Agromedicine* 13(1): 37–48.

Dick, F.D., P.G. De, A. Ahmadi, N.W. Scott, G.J. Prescott, J. Bennett, S. Semple, et al. 2007. Environmental risk factors for Parkinson's disease and parkinsonism: The Geoparkinson study. *Occupational and Environmental Medicine* 64(10): 666–672.

Elbaz, A., J. Clavel, P.J. Rathouz, F. Moisan, J.P. Galanaud, B. Delemotte, A. Alperovitch, and C. Tzourio. 2009. Professional exposure to pesticides and Parkinson disease. *Annals of Neurology* 66(4): 494–504.

Firestone, J.A., J.I. Lundin, K.M. Powers, T. Smith-Weller, G.M. Franklin, P.D. Swanson, W.T. Longstreth Jr., and H. Checkoway. 2010. Occupational factors and risk of Parkinson's disease: A population-based case-control study. *American Journal of Industrial Medicine* 53(3): 217–223.

Firestone, J.A., T. Smith-Weller, G. Franklin, P. Swanson, W.T. Longstreth Jr., and H. Checkoway. 2005. Pesticides and risk of Parkinson disease: A population-based case-control study. *Archives of Neurology* 62(1): 91–95.

Fitzmaurice, A.G., S.L. Rhodes, A. Lulla, N.P. Murphy, H.A. Lam, K.C. O'Donnell, L. Barnhill, et al. 2013. Aldehyde dehydrogenase inhibition as a pathogenic mechanism in Parkinson disease. *Proceedings of the National Academy of Science USA* 110(2): 636–641.

Kamel, F., C. Tanner, D. Umbach, J. Hoppin, M. Alavanja, A. Blair, K. Comyns, et al. 2007. Pesticide exposure and self-reported Parkinson's disease in the agricultural health study. *American Journal of Epidemiology* 165(4): 364–374.

Kukull, W.A. and J.D. Bowen. 2002. Public health, epidemiology and neurological diseases. In *Oxford Textbook of Public Health*, ed. R. Detels, J. McEwen, R. Beaglehole, and H. Tanaka. Oxford: Oxford University Press, pp. 1369–1395.

Kuopio, A.M., R.J. Marttila, H. Helenius, and U.K. Rinne. 1999. Environmental risk factors in Parkinson's disease. *Movement Disorders* 14: 928–939.

Lee, P.C., Y. Bordelon, J. Bronstein, and B. Ritz. 2012. Traumatic brain injury, paraquat exposure, and their relationship to Parkinson disease. *Neurology* 79(20): 2061–2066.

León-Verastegui, A.G. 2012. Parkinson's disease due to laboral exposition to paraquat. *Revista Médica del Instituto Mexicano del Seguro Social* 50(6): 665–672.

Lin, G. and T. Zhang. 2007. Loglinear residual tests of Moran's I autocorrelation: An application to Kentucky breast cancer data. *Geographical Analysis* 39: 293–310.

Pan-Montojo, F., M. Schwarz, C. Winkler, M. Arnhold, G.A. O'Sullivan, A. Pal, J. Said, et al. 2012. Environmental toxins trigger PD-like progression via increased alpha-synuclein release from enteric neurons in mice. *Scientific Reports* 2: 898.

Priyadarshi, A., S.A. Khuder, E.A. Schaub, and S.S. Priyadarshi. 2001. Environmental risk factors and Parkinson's disease: A metaanalysis. *Environmental Research* 86(2): 122–127.

Ross, C.A. and W.W. Smith. 2007. Gene-environment interactions in Parkinson's disease. *Parkinsonism and Related Disorders* 13(Suppl. 3): S309–S315.

Ryu, H.W., W.K. Oh, I.S. Jang, and J. Park. 2013. Amurensin G induces autophagy and attenuates cellular toxicities in a rotenone model of Parkinson's disease. *Biochemical and Biophysical Research Communications* 433(1): 121–126.

Schull, M.J., M.M. Mamdani, and J. Fang. 2005. Influenza and emergency department utilization by elders. *Academic Emergency Medicine* 12(4): 338–344.

Strickland, D. and J.M. Bertoni. 2004. Parkinson's prevalence estimated by a state registry. *Movement Disorders* 19(3): 318–323.

Tanner, C.M., F. Kamel, G.W, Ross, J.A. Hoppin, S.M. Goldman, M. Korell, C. Marras, et al. 2011. Rotenone, paraquat, and Parkinson's disease. *Environmental Health Perspectives* 119: 866–872.

U.S. Environmental Protection Agency (USEPA). 2006. Report of the Food Quality Protection Act (FQPA) Tolerance Reassessment Progress and Risk Management Decision (TRED) for acetochlor. EPA 738-R-00-009. Washington, DC: USEPA, March. http://www.epa.gov/pesticides/reregistration/REDs/acetochlor_tred.pdf (accessed June 2015).

Van der Mark, M., M. Brouwer, H. Kromhout, P. Nijssen, A. Huss, and R. Vermeulen. 2012. Is pesticide use related to Parkinson's disease? Some clues to heterogeneity in study results. *Environmental Health Perspectives* 120: 340–347.

Wan, N. 2015. Pesticide exposure modeling based on GIS and remote sensing data. *Applied Geography* 56: 99–106.

Index